Visible and Near Infrared Absorption Spectra of Human and Animal Haemoglobin

VISIBLE AND NEAR INFRARED ABSORPTION SPECTRA OF HUMAN AND ANIMAL HAEMOGLOBIN

DETERMINATION AND APPLICATION

Willem G. Zijlstra
University Hospital, Department of Paediatrics
University of Groningen
Groningen, the Netherlands

Anneke Buursma
Department of Medical Physiology
University of Groningen
Groningen, the Netherlands

Onno W. van Assendelft
Department of Health and Human Services
US Public Health Service
Centers for Disease Control and Prevention
National Center for Infectious Diseases
Atlanta, Georgia, USA

CRC Press
Taylor & Francis Group
Boca Raton London New York

CRC Press is an imprint of the
Taylor & Francis Group, an **informa** business

First published 2000 by VSP

Published 2019 by CRC Press
Taylor & Francis Group
6000 Broken Sound Parkway NW, Suite 300
Boca Raton, FL 33487-2742

First issued in paperback 2019

ISBN 13: 978-0-367-44740-3 (pbk)
ISBN 13: 978-90-6764-317-7 (hbk)

Visit the Taylor & Francis Web site at
http://www.taylorandfrancis.com

and the CRC Press Web site at
http://www.crcpress.com

To Esso J. van Kampen,
our companion in much of this work.

CONTENTS

FOREWORD

This book, viewed in scientific terms is a record of improvements in hemoglobin measurement. In human terms, it is much more. It tells the story of a lifelong passion; a passion that began, innocently enough as an attempt to improve the measurement of that oxygen carrying protein upon which all higher forms of life depend.

The measurement of hemoglobin in blood is one of the fundamental tests of the clinical laboratory and is a test of great antiquity. In the late nineteen fifties hemoglobin measurement was in disarray. A survey of laboratories in the Netherlands was performed in early 1960. On a sample that had been determined by reference laboratories to contain 16.3 g/dL of hemoglobin, less than one quarter of the laboratories measured results between 15.9 and 16.7 g/dL. The range of reported results ran from 12.3 to 18.3 g/dL!

The early 1960s proved to be a landmark period in hemoglobinometry for other reasons. The complete elucidation of the composition of human hemoglobin was accomplished almost simultaneously by Braunitzer *et al.* (1961) and Hill *et al.* (1962). Van Kampen and Zijlstra, using two different methods for the determination of hemoglobin iron, measured the millimolar absorptivity (extinction coefficient) of cyanmethemoglobin as $\varepsilon_{HiCN}(540) = 11.0\ L \cdot mmol^{-1} \cdot cm^{-1}$. With knowledge of the structure of the hemoglobin molecule and the publication of a reference method for spectrophotometric measurement of a stable hemoglobin derivative, a reference method existed against which clinical methods for hemoglobin measurement could be compared.

The choice of cyanmethemoglobin as the hemoglobin pigment on which to base hemoglobinometry proved to be prescient. Solutions of hemoglobin in which all of the pigment has been transformed into cyanmethemoglobin (and virtually all hemoglobins in the presence of $Fe(CN)_6^{3-}$ and CN^- become HiCN) showed no change during years of storage. This stability made feasible the job of preparing, calibrating, storing and finally distributing reference preparations.

Today, a blood sample analyzed for its hemoglobin content in a reasonably equipped laboratory anywhere in the world will yield results that are indistinguishable for most clinical purposes. Hemoglobinometry has come a long way from that trial in the Netherlands in 1960. This outcome is due in no small measure to a small group of scientists who 40 years ago set out to solve a practical problem that for them became a lifelong passion. Van Kampen and Zijlstra have persisted through the past four decades making improvements large and small in hemoglobinometry. Van Assendelft who co-authors this volume began working with Zijlstra in 1961 just as the original HiCN method was reduced to routine practice. The Dutch Institute of Public Health (RIV; Dr Spaander, Dr Holtz and Dr Coster) organized national and international trials, provided logistical support, and manufactured, calibrated and distributed hemoglobin standards through the years.

The hallmark of this group of dedicated hemoglobin scientists has been an unwillingness to compromise on quality and a passion for capturing that last possible bit of methodological improvement. This volume makes the passion clear.

Prof. Brian S. Bull M.D.
President
International Council for
Standardization in Hematology

Loma Linda University
School of Medicine
Loma Linda, California
USA, January 1999

PREFACE

This book is primarily a compilation of the numerous quantitative data concerning the spectral and functional properties of haemoglobin that have resulted from half a century of research in the Department of (Chemical) Physiology of the University of Groningen and many of the methods that have been used in their production. It is mainly about spectrophotometry in its different forms for different purposes. Of course, we have tried to place our results in the context of the developments in this field that started in the second half of the nineteenth century, and that, with the recent popularisation of pulse oximetry, with the rapid improvement of routine methods for multicomponent analysis of haemoglobin, and with the advent of near infrared spectroscopy *in vivo* have led to a wide range of applications. In the next decades these probably will grow in number and importance.

Much of the data have been published earlier, though not in such detail. A considerable part, such as the absorption spectra of pig, horse and sheep haemoglobin, and part of the data on the oxygen affinity of human blood, have not yet appeared in print. Moreover, we surmise that to present this material together in a single volume may add to its clarity and usefulness. Especially, the part on applications (Chapters 16–20) has been written with a view to foster knowledge and insight of those who use these methods in the care of patients and in research. In our experience the rapid progress in technology has not been associated with a similar growth in knowledge of the principles, purposes and limitations of the various methods that have come into daily use for monitoring quantities pertaining to oxygen transport by the blood. We hope that this book may also help to fill this gap.

For the realisation of this book I have to thank many people. In the first place my co-authors for their willingness to join me in this task, for their efforts and for the fruitful and pleasant collaboration. Also the dean of the faculty of Medical Sciences and the heads of the Departments of Medical Physiology and Paediatrics for giving Mrs Buursma and me the opportunity to continue our research in this field after my retirement from the Chair of (Chemical) Physiology.

I also thank Professor Berend Oeseburg (University of Nijmegen), Ing Gerard Kwant and Dr Reindert Graaff for their permission to make use of our joint unpublished data, and for their critical reading of several chapters. For fruitful discussions and critical reading thanks are also due to Professor Gerrit Mook, Dr Albert Zwart, Dr Louis Hoofd (University of Nijmegen) and Drs Bart Maas.

Finally, I have to thank Bert Tebbes and Ing Jaap van der Leest for their help in preparing the figures, and the Publishers for their effort and friendly collaboration.

Groningen, July 1999 WGZ

CHAPTER 1

INTRODUCTION

The light absorbing properties of haemoglobin have, from the very beginning, played a substantial part in the chemical investigation of this compound, as well as in the study of its role in oxygen transport by the blood. Even before 1862, when Hoppe-Seyler isolated the blood pigment and called it **haemoglobin**, he had concluded from changes in the colour of blood under various circumstances that when carbon monoxide is absorbed by the blood it prevents the blood pigment from binding oxygen, thus impeding oxygen transport [16].

This introductory chapter gives a short survey of the development of spectrophotometry of haemoglobin and its many biological and medical applications in the study of oxygen transport by the blood.

VISUAL SPECTROPHOTOMETRY

Soon after the introduction of spectroscopy in chemical analysis by Bunsen and Kirchhoff [348], Felix Hoppe-Seyler [173] in Germany and George Stokes [377] in England almost simultaneously observed the light absorption bands of haemoglobin in the visible part of the spectrum and the changes which occur when oxygen is added to or removed from the solution. Karl Vierordt was the first (1876) to study not only the spectral changes of haemoglobin in solution, but also in a transilluminated finger. When the circulation through the finger was stopped, the two oxyhaemoglobin bands disappeared and the band of deoxygenated haemoglobin appeared [403]. Although this experiment may be regarded as the beginning of oximetry, it was followed up only a half century later when more suitable equipment became available.

Of more immediate consequence was that Vierordt had also shown the suitability of spectral analysis for biochemical applications [402]. From that time the spectroscopic study of haemoglobin and its derivatives developed rapidly. In 1878 Soret [366, 367] described the strong absorption bands of haemoglobin in the near ultraviolet, and Gustav Hüfner embarked on the development of a spectrophotometer; this resulted in 1889 in an instrument with which quite accurate measurements could be made [178].

In Hüfner's instrument, the light of a bright kerosene lamp passed through a cuvette of which the lower part was filled with a glass body (*Schulzescher Körper*); the upper part of the cuvette was filled with the solution of which the absorbance was to be measured. The part of the light which had traversed the glass body, then passed through a nicol prism and thus became polarised. After the two beams had passed through a dispersion prism and slits to isolate small wavebands, they reached a lens system through which both could be seen side by side. In the polarised light beam — the one which had not passed the solution to be measured — was a second nicol prism, which could be rotated in order to attenuate it until its intensity matched that of the beam which had traversed the solution. From the rotation of the second nicol prism, which could be read from a scale, the absorbance of the solution was calculated.

In his well-known investigation of the oxygen binding capacity of haemoglobin [180], Hüfner not only used this spectrophotometer for the determination of the total haemoglobin concentration of the solutions of which he measured the carbon monoxide capacity, but also for checking the purity of these solutions. To this end he used the central part of the region between the α- and the β-peak of oxyhaemoglobin (554–565 nm) and a region around the β-peak (531.5–542.5 nm). Table 1.1 shows some ratios of absorptivities of oxyhaemoglobin, deoxyhaemoglobin and carboxyhaemoglobin in these regions in comparison with the same quantities calculated on the basis of recent measurements [467]. The fair agreement of Hüfner's results with modern data clearly demonstrates that, at the time, reasonably accurate spectrophotometric measurements already could be made.

Considerable progress in the spectrophotometric study of haemoglobin was made by Drabkin and Austin, who published an admirable series of *Spectrophotometric studies* between 1932 and 1946 [94–97]. Already the first paper of the series [95] gives data on several haemoglobin deriva-

Table 1.1.
Absorption ratios of HHb, O_2Hb and COHb in two spectral regions according to Hüfner and as calculated from recent measurements

Ratio	Hüfner (1894)	Recent (1991)
$\varepsilon_{O_2Hb}(2)/\varepsilon_{O_2Hb}(1)$	1.581	1.448
$\varepsilon_{HHb}(2)/\varepsilon_{HHb}(1)$	0.761	0.745
$\varepsilon_{O_2Hb}(1)/\varepsilon_{HHb}(1)$	0.655	0.705
$\varepsilon_{COHb}(2)/\varepsilon_{COHb}(1)$	1.095	1.122
$\varepsilon_{COHb}(2)/\varepsilon_{O_2Hb}(2)$	1.037	1.044

$\varepsilon(1)$ = absorptivity in region 1 (554–565 nm); $\varepsilon(2)$ = absorptivity in region 2 (531.5–542.5 nm); HHb = deoxyhaemoglobin; O_2Hb = oxyhaemoglobin; COHb = carboxyhaemoglobin. Hüfner's ratios are from ref. [180]; the recent ratios have been calculated from data given in Chapter 8.

tives (oxyhaemoglobin, carboxyhaemoglobin, haemiglobin, haemiglobincyanide), prepared from haemolysed human, dog and rabbit blood, and measured with a König–Martens type spectrophotometer [93]. The total haemoglobin concentration was determined on the basis of the oxygen binding capacity of the solutions. After the absorptivities of haemiglobincyanide at 551, 545 and 540 nm had been determined ($\varepsilon_{HiCN} = 11.0, 11.5, 11.5$ $L \cdot mmol^{-1} \cdot cm^{-1}$, respectively), the total haemoglobin concentration in subsequent studies was determined after converting all haemoglobin in the solution into haemiglobincyanide.

In the second study haemoglobin solutions were produced from washed erythrocytes instead of from haemolysed whole blood, and nitric oxide haemoglobin and sulfhaemoglobin were added to the haemoglobin derivatives studied [96]. An important innovation was the introduction of a flow-through cuvette with a lightpath length of only 0.007 cm [97]. This allowed measurements at a total haemoglobin concentration as is present in whole blood and thus enabled spectrophotometric measurements of the oxygen saturation to be made. Absorptivities of very concentrated solutions of horse haemoglobin were determined, and it was demonstrated that Lambert–Beer's law is valid for haemoglobin solutions over a concentration range from 0.0001 to 1, where 1 corresponded with a concentration of 25 mmol/L. In a later study, accurate measurements were even made at concentrations around 38 mmol/L. This paper [94] reports mainly crystallographic results, but the ensuing pure solutions of various haemoglobin derivatives of human, horse and dog blood were used for new determinations of absorptivities on the basis of $\varepsilon_{HiCN}(540) = 11.5$ $L \cdot mmol^{-1} \cdot cm^{-1}$. To demonstrate the accuracy which was attained in this advanced stage of visual spectrophotometry, the absorptivities found for human oxyhaemoglobin and deoxyhaemoglobin in solutions of crystallised preparations have been compared with data from Chapter 8, in Table 1.2.

Table 1.2.
Absorptivities of oxyhaemoglobin and deoxyhaemoglobin from Drabkin [94] in comparison with recent data

	λ (nm)	ε (Drabkin)	ε (Table 8.1)
O_2Hb	578	15.38	15.36
	562	8.47	8.77
	542	14.68	14.52
HHb	555	13.57	13.35

Absorptivities expressed in $L \cdot mmol^{-1} \cdot cm^{-1}$. Drabkin's values of oxyhaemoglobin have been measured at a total haemoglobin concentration of about 35 mmol/L with a lightpath length of 0.007 cm; the measurements of deoxyhaemoglobin have been made at $ct_{Hb} \approx$ 0.1 mmol/L with a lightpath length of 1.0 cm.

PHOTOELECTRIC SPECTROPHOTOMETRY

Horecker was one of the first to use a photoelectric spectrophotometer to record absorption spectra of haemoglobin derivatives and to extend the measurements to 1000 nm, the near infrared spectral region [174]. He found that the absorptivity of oxyhaemoglobin increased again beyond 700 nm, whereas that of carboxyhaemoglobin decreased to near zero. On the basis of these findings Horecker and Brackett [175] developed a spectrophotometric method for the determination of the carboxyhaemoglobin fraction in blood. Combining the method with the principle of Evelyn and Malloy [114] — measuring the absorption change on the addition of a trace of potassium cyanide to the blood sample — they even succeeded in devising a method for measuring the fractions of carboxyhaemoglobin and methaemoglobin in a single blood sample.

The photoelectric spectrophotometer described by Cary and Beckman in 1941 [60], which soon became commercially available, brought accurate and precise spectrophotometric measurements within the reach of the average biochemical laboratory. The instrument contained a quartz prism, a special RCA C7032 phototube most sensitive for measurements at wavelengths below 600 nm, in combination with an RCA 919 caesium oxide phototube for measurements up to 1000 nm. An (automobile headlight) incandescent lamp, and a hot cathode hydrogen discharge tube for measurements between 350–220 nm were used as light sources. The instrument was robust and easy to operate. After balancing the voltage drop across a resistor in the phototube circuit with a slide wire potentiometer, transmission and absorptivity at the chosen wavelength could be read from the potentiometer scale. In the visible range the spectral band width was about 1 nm.

The 'Beckman DU' greatly advanced the development of spectrophotometric methods for the determination of haemoglobin and haemoglobin derivatives in biomedical research and medical practice. It was the principal instrument in the development of a standard method for determination of the total haemoglobin concentration in blood [468], of numerous methods for the determination of oxygen saturation, and of the fractions of carboxyhaemoglobin, haemiglobin and other haemoglobin derivatives in human blood.

A 1965 review of the state of the art of haemoglobin spectrophotometry [211] in which the international standardisation of the measurement of total haemoglobin, as well as the determination of haemoglobin derivatives are discussed, deals almost exclusively with measurements made with Beckman DU spectrophotometers. In the review, absorptivity data are presented of several haemoglobin derivatives, for the visible spectral region and for the Soret bands in the near ultraviolet region.

The rapid growth of data on the haemoglobin spectra is shown by the monograph on haemoglobin spectrophotometry [7], which appeared only five years later. It shows new absorptivity values, determined with Optica CF4 grating spectrophotometers, which are more suited for haemoglobin spectrophotometry than instruments equipped with a quartz prism, because of the better spectral resolution in the red and near infrared. The use of a recording instrument in addition to a spectrophotometer on which the absorbance values have to be read one by one, made it possible to obtain complete spectra from 390 to 1000 nm.

DIODE ARRAY SPECTROPHOTOMETRY

It was again a technical improvement which caused the next leap forward in the spectrophotometry of haemoglobin: the advent of the reversed optics, diode array spectrophotometer with built-in computing power (Hewlett-Packard HP8450A). The second review of haemoglobin spectrophotometry [210], which appeared in 1983 as an update of the one from 1965, already shows some spectra recorded with this new instrument, but the new method described for the simultaneous determination of five haemoglobin derivatives in a single blood sample is still performed with a conventional spectrophotometer, and most of the absorptivity data presented are based on earlier measurements [7, 211]. The diode array spectrophotometer made multicomponent analysis simpler and allowed easy and rapid determination of absorbance spectra. It was a lucky coincidence that these new possibilities arrived almost simultaneously with the identification of spectral differences between foetal and adult human haemoglobin [177, 446].

Until about 1980, it had generally been assumed that the spectral properties of mammalian haemoglobin did not differ between species, because the colour of haemoglobin was assumed to be solely dependent on the iron-porphyrin moiety, which is the same in all animals. Finding a difference between foetal and adult human haemoglobin of course immediately raised the question of possible differences between the haemoglobin spectra of different animals. These were soon shown to exist, but the differences proved to be so slight that it was not surprising that they had escaped observation for a long time. On the other hand, they are not small enough to be neglected in multicomponent analysis, as is demonstrated by the fact that the very discovery of the differences between foetal and adult haemoglobin originated from the finding of spurious fractions of carboxyhaemoglobin in umbilical cord blood [177, 446].

Hence it became mandatory to accurately determine the absorptivity spectra of the common haemoglobin derivatives of animals, before multicomponent analysis of haemoglobin derivatives in their blood could be performed. Because the differences are so slight it proved to be necessary to identify any differences between human and animal haemoglobin spectra by determining these spectra in multiple specimens from the same species. The diode array spectrophotometer has made it possible to obtain the substantial number of absorption spectra required within a reasonable period. The data presented in Chapters 8–14 have mainly been obtained with the help of this technique.

BIOMEDICAL APPLICATIONS

The spectrophotometry of haemoglobin and its derivatives and the ensuing absorptivity spectra are the basis of several methods and techniques which have gained wide biomedical application: haemoglobinometry, multicomponent analysis (MCA) of haemoglobin derivatives, oximetry and near-infrared spectroscopy (NIRS). Some of these methods are in daily use in clinical practice, some are mainly used in pathophysiological research, and some have, in special applications, provided more accurate quantitative data concerning the oxygen transport capability of human blood (Chapters 19 and 20).

Haemoglobinometry is concerned with the determination of the total haemoglobin concentration of blood. It is one of the oldest and most used clinical chemical procedures, for which numerous methods have been devised, most utilising the bright colour of haemoglobin. Accurate determination of total haemoglobin is of great significance in diagnosis and treatment of patients with many different diseases, but it is also the first essential step in the determination of the absorptivity spectra of haemoglobin derivatives, as well as for the calculation of the oxygen transport capability of the blood in patients in intensive care, in experimental subjects in exercise physiology, and in animals during (patho)physiological experiments. The measurement of total haemoglobin and its relation to the determination of the absorption spectra of the haemoglobin derivatives is dealt with in Chapter 4, the international reference method and its relation to the oxygen binding capacity of haemoglobin are discussed in Chapters 16 and 19.

Multicomponent analysis of haemoglobin derivatives, which is now routinely performed in the clinical laboratory, was preceded by numerous spectrophotometric methods for the determination of the fraction of single clinically relevant haemoglobin derivatives, *e.g.* carboxyhaemoglobin and haemiglobin. The carboxyhaemoglobin fraction of the blood has been a clinically important quantity from the very moment the mechanism of carbon monoxide poisoning was understood, and many procedures for its estimation have been described, for example, several spectrophotometric two-wavelength methods for measuring the carboxyhaemoglobin fraction in the two-component system carboxyhaemoglobin/oxyhaemoglobin or carboxyhaemoglobin/deoxyhaemoglobin [210, 211, 479].

Similar two-wavelength methods have been devised for haemiglobin and other haemoglobin derivatives, but the analysis of mixtures of more than two components remained complicated and tedious [319, 357, 494]. Multicomponent analysis became practical only in the 1980s with the use of diode array spectrophotometers and, almost simultaneously, various dedicated multiwavelength photometers became available for easy and rapid analysis of the fractions of haemoglobin derivatives in blood. These methods are discussed in Chapter 17.

The term **oximetry** is used in a general and in a restricted sense. In general, it means methods for the determination of the oxygen saturation of haemoglobin. In the restricted sense it means measuring oxygen saturation *in vivo*, foreshadowed by the early transillumination experiments of Vierordt. After 1930, German investigators devised methods for continuously measuring oxygen saturation in exposed blood vessels and in the skin. On the basis of this experience, oximeters for use in patients were developed. The application of the oximeters was restricted to tissues such as the ear or a finger which can easily be transilluminated. Many variations were developed in the following half-century: reflection oximeters for measuring anywhere on the skin, although used mainly on the forehead, and fibre optic oximeters for measuring in blood vessels and in the heart. However, only after the introduction of pulse oximetry around 1975, did oximetry become simple enough for routine clinical use.

Another recent development is **NIRS**, the application *in vivo* of near infrared spectroscopy. Near infrared light may still be detected after traversing several layers of tissue. Thus, by transilluminating a part of the body, for instance the neonatal skull, intracerebral measurements of deoxygenated and oxygenated haemoglobin, and even of the oxygenation state of cytochrome oxidase can be made. Chapter 18 gives a short history of oximetry and includes a critical evaluation of the present state of the art.

The application of fibre optic oximetry and MCA of haemoglobin derivatives, in combination with selective electrodes for the continuous measurement of p_{O_2} and pH, has made it possible to study the oxygen binding properties of haemoglobin in fresh human whole blood by simultaneous measurement of all relevant quantities. In Chapter 20, such a method is described briefly and new quantitative data on the oxygen affinity of human whole blood are presented.

CHAPTER 2

DEFINITIONS AND TERMINOLOGY

In the long history of spectrophotometry of haemoglobin and its application in the study of oxygen transport by the blood, many technical terms have been coined and many symbols and abbreviations have been introduced. Some have come into general use, have kept their meaning throughout history, and have remained understood. Other terms vanished when the concepts which they expressed lost their relevance. Also, the advent of new theories and techniques was often accompanied by the emergence of new terms and symbols and several synonyms came into being. Thus, the consistency, for example, of blood gas terminology was sometimes far from perfect. When later generations lost sight of the original meaning of a term and of the underlying concept, serious confusion ensued. A case in point is the concept of oxygen saturation [489, 495], which after the introduction of multicomponent analysis was confused with oxyhaemoglobin fraction (Chapter 17); this, in turn, led to a misunderstanding of the limitations of pulse oximetry (Chapter 18).

In the following sections, concepts, quantities, terms and symbols pertaining to spectrophotometry of haemoglobin and its biomedical applications are defined as concise and clearly as possible. Internationally accepted conventions [54, 107, 194] are followed as far as possible. When there is no consensus, a choice best accommodating the subject under discussion is made.

SPECTROPHOTOMETRY

Spectrophotometry is concerned with the spectral properties of material specimens and is applied in the laboratory for both qualitative and quantitative analyses. In qualitative analyses the wavelength range of interest, for example, the *ultraviolet* (UV), 200–380 nm, the *visible* (VIS), 380–780 nm, or the (near) *infrared* (IR), >780 nm, is scanned to locate **absorption bands**, *i.e.* regions of the spectrum in which the absorbance passes through a maximum value, that are specific for a given light absorbing substance. The presence of, for example, oxyhaemoglobin can be seen by scanning the 500 to 600 nm range and finding the two absorption bands around 542 and 577 nm, respectively.

In quantitative analyses, **Lambert–Beer's law** is applied, which states that the amount of light absorbed by a homogeneous sample is directly proportional to the thickness and to the concentration of the absorbing substance. The law may be expressed mathematically as

$$A = \log(I_0/I) = a \cdot b \cdot c, \quad (2.1)$$

where A is the absorbance, I_0 is the incident light intensity, I is the intensity of the light transmitted through the sample, a is a constant for a given absorbing substance at a specific wavelength, b is the thickness of the sample, and c is the concentration of the absorbing substance. Lambert–Beer's law is discussed in detail in Chapter 3.

In 1946 the Society for Applied Spectroscopy recognised a pressing need for standard nomenclature and formed a committee to prepare a list of definitions in the applied emission, absorption and Raman fields. The American Society for Testing Materials, ASTM, appointed additional members and in 1948 the Joint Committee on Nomenclature in Applied Spectroscopy was formed. With a single exception, the terms and definitions used in the following chapters conform to the recommendations of this committee [4, 5, 183].

The ratio of transmitted light intensity over incident light intensity, I/I_0, is called **transmittance**, symbol T, and is dimensionless. The logarithm of $1/T$, $\log(I_0/I)$, is called **absorbance**, symbol A, as used in equation 2.1. This symbol may be expanded to also denote the absorbing species and the wavelength of measurement, *e.g.* $A_{HiCN}(540)$ denotes the absorbance of haemiglobincyanide at 540 nm.

The product $a \cdot b \cdot c$ (equation 2.1), being a logarithm, must also be a pure number. Hence, it is essential that the units of each of the quantities are specified. The cell width thickness b is usually expressed in cm. In the following chapters the symbol l will be used to designate the cell width thickness as lightpath length; l is preferred to b, because it has recently been generally used in papers on absorption spectra of haemoglobin. When the concentration c is expressed in gram per litre, the proportionality constant a is called **absorptivity**, unit $L \cdot g^{-1} \cdot cm^{-1}$. When c is in mol per litre the proportionality constant is called the **molar absorptivity**, symbol ε, unit $L \cdot mol^{-1} \cdot cm^{-1}$.

In spectrophotometry of haemoglobin it is customary to use the **millimolar absorptivity**, symbol ε, unit $L \cdot mmol^{-1} \cdot cm^{-1}$. The millimolar absorptivity thus is the value of A when $c = 1$ mmol/L and $l = 1$ cm. These quantities and units are exclusively used throughout the following chapters. When the unit is explicitly given, the addition 'millimolar' is often omitted. The symbol ε may be expanded to also denote the absorbing species and the wavelength of measurement; $\varepsilon_X(\lambda)$ then denotes the absorptivity of haemoglobin derivative X at wavelength λ.

A wavelength where the molar absorptivities of two light absorbing substances are equal is called an **isosbestic wavelength**. A cross-over point in the absorption spectra of two substances is analogously called an **isosbestic point**.

HAEMOGLOBIN AND HAEMOGLOBIN DERIVATIVES

The term **haemoglobin** is used as a general term for the chromoprotein consisting of **haem**, an iron-porphyrin, and **globin**. Haemoglobin is present in the blood of most vertebrates as the principal vehicle in the transport of oxygen. The term 'haemoglobin' is used irrespective of the functional state or the presence of normal or abnormal variations. The **total haemoglobin** concentration correspondingly denotes all haemoglobin in a volume of blood.

Mammalian haemoglobin contains two pairs of polypeptide chains: α chains and β chains. Each chain is connected to a haem group. A single haem group and its globin chain is called a haemoglobin **monomer**. The complete haemoglobin molecule is a **tetramer**, composed of two α chains, two β chains and four haem groups.

The forms in which haemoglobin actually exists in the blood are called **haemoglobin derivatives**, although the relation between the so-called derivatives and haemoglobin as defined above is not in full agreement with the relation between derivatives and their parent compound as used in organic chemistry.

The most common haemoglobin derivatives are **oxyhaemoglobin**, symbol O_2Hb, and **deoxyhaemoglobin**, symbol HHb. Normally, almost all haemoglobin is present in one of these forms; they continually change into one another during normal oxygen transport.

Haemoglobin derivatives which have temporarily or permanently lost the capability of reversibly binding oxygen at physiological oxygen tension are called **dyshaemoglobins** [101, 103], symbol dysHb. The common dyshaemoglobins are **carboxyhaemoglobin**, symbol COHb, and **haemiglobin**, symbol Hi, also called methaemoglobin.

As to the terms 'methaemoglobin' and 'haemiglobin', there is a difference between the conventions proposed by NCCLS [107] and by ICSH [194]. NCCLS proposes the term 'methaemoglobin' and the symbol 'MetHb', whereas ICSH uses the term 'haemiglobin' and the symbol 'Hi'. We have chosen to follow ICSH because it is more practical in the formation of compounds. Thus we use **haemiglobincyanide** and HiCN, not cyanmethaemoglobin and CNMetHb. Analogously, we have opted to use SHb instead of SulfHb as the symbol for **sulfhaemoglobin**. Consequently, one gets a manageable symbol for **sulfhaemiglobin**, SHi, instead of SulfMetHb, and even for **sulfhaemiglobincyanide**, SHiCN, instead of CNSulfMetHb.

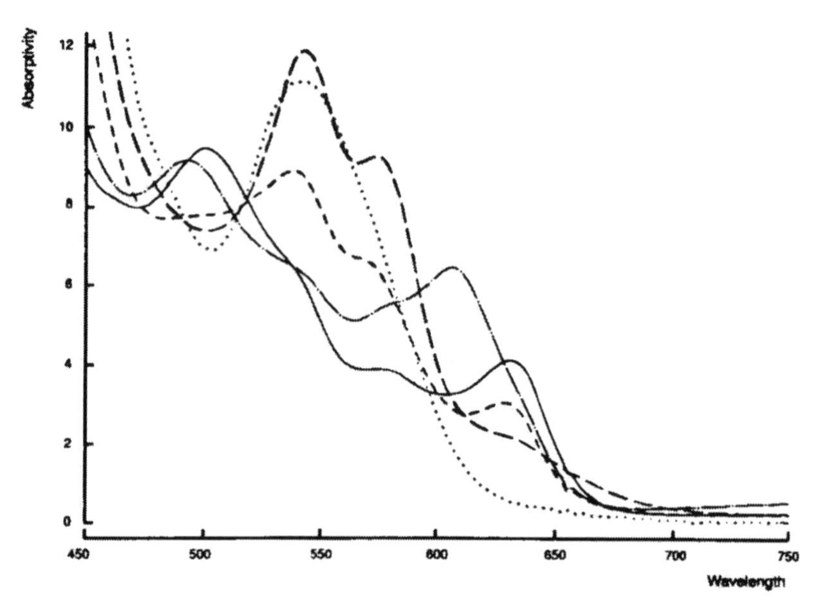

Figure 2.1. Absorption spectra of Hi (—), HiF (——·), $HiNO_2$ (- - - -), HiN_3 (— —), and HiCN (· · ·).

Absorptivity in $L \cdot mmol^{-1} \cdot cm^{-1}$, wavelength in nm. l = 0.010–0.016 cm. HP8450A diode array spectrophotometer. pH (Hi) ≈ 7.2.

COHb is haemoglobin which has bound carbon monoxide instead of oxygen. The binding of CO to haemoglobin is very tight: once bound, it hardly dissociates (p. 314). However, it can be reconverted to O_2Hb or HHb.

Hi is haemoglobin in which the iron has been oxidised to the ferric state. It can be reconverted to HHb or O_2Hb by reducing the iron to the ferrous state. Hi cannot bind oxygen, but becomes easily associated with negative ions. Thus it binds OH^-, depending on the pH of the solution, and several other anions: fluoride (F^-), nitrite (NO_2^-), azide (N_3^-), cyanide (CN^-) may be bound, although with differences in binding strength. A displacement series has been shown to exist:

$$OH^- < F^- < NO_2^- < N_3^- < CN^-$$

There are distinct differences in colour between the various Hi compounds. Hi (pH ≈ 7.2) is brown, HiF green, $HiNO_2$ reddish brown, HiN_3 and HiCN are brownish red. Figure 2.1 shows the absorption spectra of these five Hi compounds. The most stable of the Hi compounds is HiCN, which is used in the determination of the total Hb concentration (Chapters 4 and 16).

SHb is haemoglobin with an irreversibly bound sulphur atom in the porphyrin ring. The sulphur atom is bound to a pair of β-pyrrole carbon atoms at the periphery of a chlorine ring formed by saturation of that $\beta-\beta$ double bond of the corresponding protoporphyrin IX of haemoglobin [28]. This structure explains the considerable departure of the absorption spectrum

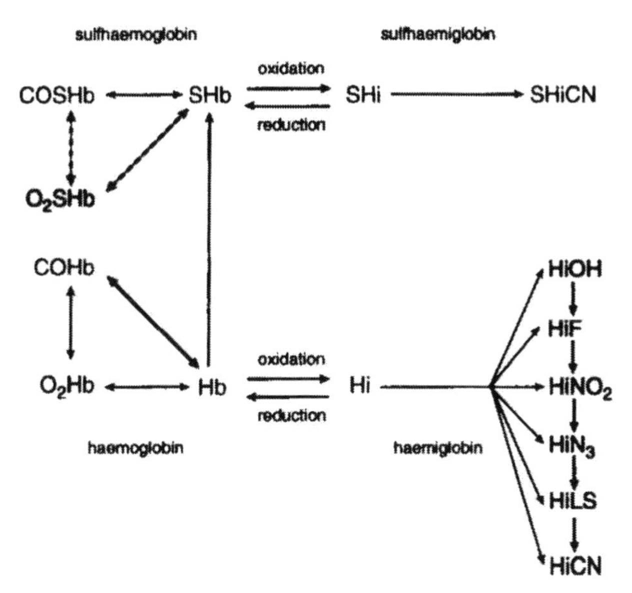

Figure 2.2. Interrelationship of haemoglobin derivatives.
For HiLS see Chapter 16.

of SHb from that of the other haemoglobin derivatives, as well as the very low oxygen affinity [59]. It binds O_2 only at very high pressure and its contribution to physiological O_2 transport is zero. Therefore, SHb is a dyshaemoglobin as defined above.

COSHb is formed more easily. This has been a confounding factor in the spectrophotometric determination of F_{COHb} [207, 210]. The iron in SHb can be oxidised to the ferric state, which gives rise to the formation of sulfhaemiglobin. Similar to Hi, SHi can bind a negative ion. Of the addition products, only SHiCN is of practical importance, because of the slight underestimation that it causes when ct_{Hb} (the total concentration of haemoglobin) is determined by the reference HiCN method [103, 210]. The interrelationship of the various haemoglobin derivatives is summarised in Fig. 2.2.

The term dyshaemoglobin has been coined to designate haemoglobin with changes in the iron-porphyrin moiety, which are of consequence for the reversible binding of oxygen. It serves to distinguish these alterations from changes in the globin moiety, which are specified by placing a letter after the symbol Hb. This is done in case of the **normal variants**, as HbA and HbF for adult and foetal haemoglobin, as well as for **abnormal haemoglobins**, such as sickle cell haemoglobin, HbS. HbF and most of the abnormal haemoglobins may of course exist in the form of the same derivatives as HbA. In case the distinction is relevant, one may, for instance, write O_2HbF instead of simply O_2Hb.

A disease like sickle cell anaemia is called a **haemoglobinopathy**, while a considerable amount of a dyshaemoglobin in the blood is called **dyshae-**

moglobinaemia. An abnormal haemoglobin may also be a dyshaemoglobin, for instance when HbS becomes oxidised to HiS or binds carbon monoxide forming COHbS.

Finally, it should be noted that variation in the globin moiety is present in the blood of most animals, although this has not been examined as thoroughly as in man. In the investigation of the spectral properties of animal blood the possible influence of such differences has not been taken into account. Consequently, with sheep haemoglobin is meant the mixture of haemoglobins as normally occurring in the blood of adult sheep, etc. Analogously, this also holds for human blood, since only the spectral difference between human HbA and HbF has been studied.

OXYGEN CARRYING PROPERTIES OF MAMMALIAN BLOOD

The **oxygen tension**, p_{O_2}, of the blood is defined as being equal to the partial pressure of oxygen in the gas in equilibrium with the blood. This partial pressure, in turn, is equal to the product of the substance fraction of oxygen and the total pressure of the gas: p_{O_2} is expressed in kPa or in mmHg (Torr), where 1 kPa = 7.5 mmHg.

Molecular oxygen in blood is present in two forms: oxygen bound to haemoglobin, O_2(Hb), and oxygen dissolved in blood but not bound to any other substance, O_2(free). The **substance concentration of total oxygen**, ct_{O_2}, is the sum of the substance concentrations of the two forms:

$$ct_{O_2} = c_{O_2}(\mathrm{Hb}) + c_{O_2}(\mathrm{free}), \tag{2.2}$$

ct_{O_2} is expressed in mmol/L. It is also called **oxygen content** [107] and expressed in mL(STPD)/L or mL(STPD)/dL, where STPD is Standard Temperature and Pressure Dry, *i.e.* 101.33 kPa (1 atm), 0°C, no water vapour.

The relation between c_{O_2}(free) and p_{O_2} is given by Henry's law:

$$c_{O_2}(\mathrm{free}) = \alpha_{O_2} \cdot p_{O_2}, \tag{2.3}$$

where α_{O_2} is the concentrational solubility coefficient in blood [54]; α_{O_2} = 0.010 mmol · L^{-1} · kPa^{-1} at 37°C.

The **oxygen capacity** of blood, B_{O_2}, is the maximum amount of haemoglobin-bound oxygen per unit volume of blood. B_{O_2} is expressed in mmol/L or in mL(STPD)/L or mL(STPD)/dL. The oxygen capacity is determined by the **concentration of active haemoglobin**, *i.e.* the sum of the concentrations of O_2Hb and HHb. Since the internationally accepted method for the determination of haemoglobin in blood measures the total haemoglobin concentration, ct_{Hb}, a correction for any dyshaemoglobin present in the blood must be made.

Hence, in the calculation of B_{O_2} from ct_{Hb} the dyshaemoglobin concentration, c_{dysHb}, is taken into account. When B and c are expressed in mmol/L and the molar concentration of haemoglobin reflects the monomer:

$$B_{O_2} = ct_{Hb} - c_{dysHb}. \tag{2.4}$$

When B is expressed in mL/L and c in g/L:

$$B_{O_2} = \beta_{O_2} \cdot (ct_{Hb} - c_{dysHb}), \tag{2.5}$$

where β_{O_2} is the volume of oxygen in mL(STPD) that can be bound by 1 g of haemoglobin. Theoretically, $\beta_{O_2} = 22394/16114.5 = 1.39$ mL/g, where 22394 is the molar volume of oxygen in mL(STPD) and 16114.5 is a quarter of the molar mass of human HbA in g [38]. The experimental value has been found near the theoretical one [101] (Table 19.2).

As the most frequently occurring dyshaemoglobins are COHb and Hi, the oxygen capacity may be approximated by

$$B_{O_2} = \beta_{O_2} \cdot (ct_{Hb} - c_{COHb} - c_{Hi}). \tag{2.6}$$

The **oxygen saturation** of the blood, S_{O_2}, is defined as the concentration of haemoglobin-bound oxygen divided by the oxygen capacity. This is equivalent to the concentration of O_2Hb divided by the sum of the concentrations of O_2Hb and HHb:

$$S_{O_2} = c_{O_2}(\text{Hb})/B_{O_2}, \tag{2.7}$$

$$S_{O_2} = [ct_{O_2} - c_{O_2}(\text{free})]/B_{O_2}, \tag{2.8}$$

$$S_{O_2} = c_{O_2Hb}/(c_{O_2Hb} + c_{HHb}). \tag{2.9}$$

It should be noted that there is a conceptual difference between $c_{O_2}(\text{Hb})$ and c_{O_2Hb}, although the two quantities are numerically equal when both are expressed in mmol/L.

In multicomponent analysis of haemoglobin, the various haemoglobin derivatives are usually expressed as a **fraction of total haemoglobin**, F:

$$F_{O_2Hb} = c_{O_2Hb}/ct_{Hb}, \tag{2.10}$$

$$F_{HHb} = c_{HHb}/ct_{Hb}, \tag{2.11}$$

$$F_{COHb} = c_{COHb}/ct_{Hb}, \tag{2.12}$$

$$F_{Hi} = c_{Hi}/ct_{Hb}. \tag{2.13}$$

Note the difference between oxyhaemoglobin fraction (or fractional oxyhaemoglobin [107]), F_{O_2Hb}, and oxygen saturation, S_{O_2}, which may be further elucidated by writing:

$$S_{O_2} = c_{O_2Hb}/(ct_{Hb} - c_{dysHb}). \tag{2.14}$$

S_{O_2} is determined by p_{O_2} and the **oxygen affinity** of haemoglobin, which is usually expressed by a graph of S_{O_2} as a function of p_{O_2}, the **oxygen dissociation curve** (ODC). This is an S-shaped curve (Fig. 20.2), of which the position as well as the exact shape are determined by several factors: pH, p_{CO_2}, temperature (T), 2,3-diphosphoglycerate (DPG), and dyshaemoglobins. Factors which increase the oxygen affinity of haemoglobin shift the ODC to the left, and a decrease in oxygen affinity results in a shift to the right. The position of the ODC is indicated with the help of the value of p_{O_2} at $S_{O_2} = 50\%$, denoted by p_{50}; p_{50} at $T = 37°C$, $p_{CO_2} = 5.33$ kPa and pH $= 7.40$ is called **standard**-p_{50} and denoted by p_{50}(std).

If it is desirable to indicate *where* a quantity is measured, a letter should be added to the symbol denoting the quantity, for example, aS_{O_2} may represent the arterial oxygen saturation. If it is desirable to indicate *the method used* to measure a quantity, a letter in parentheses should be added to the symbol denoting the quantity, for example, vS_{O_2}(f.o.) may be written for venous oxygen saturation measured by fibre optic oximetry. For a more extensive review of quantities and symbols pertaining to oxygen in blood see Wimberley *et al.* [417].

Oxygen saturation, S_{O_2}, and dyshaemoglobin fractions, F_{dysHb}, can be expressed as a number between 0 and 1, but are usually given in percent, *e.g.* $F = 0.3 = 30\%$. When the spread in this value is given as 2%, this means that F lies between 28 and 32%. When confusion with percent as relative measure may occur, the expression $\%S_{O_2}$ or $\%F$ is used.

CHAPTER 3

SPECTROPHOTOMETRY

One of the most commonly used methods of analysis in the clinical laboratory is absorption spectrophotometry. Especially in the study of haemoglobin, light absorption measurements have, from the very beginning, played an invaluable role. This chapter deals with the laws describing the relationship between the absorption of radiant energy as it passes through a medium and the properties of this medium, and with the application of these laws in the determination of the total haemoglobin concentration and of the analysis of mixtures of haemoglobin derivatives. In addition, a brief survey is given of the instrumentation and procedures used in these applications.

THE LAWS OF ABSORPTION OF RADIANT ENERGY

The use of absorption of radiant energy in quantitative analysis depends on the relationship between the amount of absorbing substance and the extent to which radiant energy is absorbed. Subject to certain limitations, this relationship is defined by two laws relating the degree of absorption to the characteristics of the absorbing substance. The first law, formulated by Bouguer in 1729, restated by Lambert in 1768, predicts the effect of sample thickness upon the fraction of radiation that is absorbed. Lambert concluded that each layer of equal thickness absorbs an equal fraction of the light that traverses it. When a monochromatic light beam of intensity I passes through an infinitesimal layer thickness db of absorbing matter, the reduction of intensity within the absorbing matter can be described mathematically as $dI = -k \cdot I \cdot db$, with $dI \ll I$, where dI represents the change in intensity and k is a proportionality constant, negative because I decreases as b becomes larger. This equation may be rearranged to state that the fraction of radiation absorbed is proportional to the thickness traversed: $dI/I = d\ln(I) = -k \cdot db$.

If I_0 represents the incident light intensity when $b = 0$, integration over I_0 to I and inverting the ratio I/I_0 gives $\ln(I_0/I) = k \cdot b$, where I is the intensity of the light transmitted through a sample of thickness b. By converting to log (base 10), one obtains the final form of **Lambert's law**:

$$\log_{10}(I_0/I) = k \cdot b \cdot \log_{10} e = k' \cdot b, \tag{3.1}$$

with the proportionality constant k' a function of the wavelength of the incident light and the temperature of the absorbing substance.

The second law, formulated by Beer in 1852, deals with the effect of concentration. Beer found that increasing the concentration of a light-absorbing substance in a solution had the same effect as a proportional increase in the light absorbing path. Hence, the proportionality constant k' (equation 3.1) is in turn proportional to the concentration c of the light absorbing solute: $k' = a \cdot c$, where a is a constant for a given light absorbing substance at a specific wavelength of incident radiation. By substitution in equation 3.1, one obtains

$$\log_{10}(I_0/I) = a \cdot b \cdot c, \tag{3.2}$$

where $\log_{10}(I_0/I)$ is defined as the absorbance A (Chapter 2). While **Lambert–Beer's law** is generally applied to the centre of the electromagnetic spectrum, from ultraviolet to infrared, it holds for absorption processes throughout the spectrum and forms the basis for quantitative analysis by the absorption of electromagnetic radiation.

There are three assumptions implicit in the derivation of Lambert–Beer's law: (1) the only interaction between the radiation and the solute is absorption; (2) the absorbing solutes act independently of each other, regardless of number and kind; (3) the absorption is limited to a volume of uniform cross-section.

Conformance to Lambert–Beer's law should never be assumed, but must be checked under the conditions of the procedure and for the concentration range to be met.

APPLICATIONS OF LAMBERT–BEER'S LAW

Spectrophotometry may be applied in both qualitative and quantitative analysis of one or more light absorbing substances, *e.g.* haemoglobin derivatives. In quantitative analyses, determination of the concentration of an absorbing substance can be performed if only one component is present in the solution, as well as when more components are present. If only a single haemoglobin derivative is present in the solution and the molar absorptivity is known, the concentration can be determined by directly applying Lambert–Beer's law. Using the symbols defined in Chapter 2:

$$A(\lambda) = \varepsilon(\lambda) \cdot c \cdot l, \tag{3.3}$$

and

$$c = A(\lambda)/[\varepsilon(\lambda) \cdot l]. \tag{3.4}$$

If the molar absorptivity is not known, a reference solution of known concentration against which the unknown solution is measured, is required:

$A_S(\lambda) = \varepsilon(\lambda) \cdot c_S \cdot l$ and $A_R(\lambda) = \varepsilon(\lambda) \cdot c_R \cdot l$, where the subscripts S and R denote the (unknown) sample and the (known) reference solution, respectively. Hence,

$$c_S = c_R \cdot A_S(\lambda)/A_R(\lambda). \tag{3.5}$$

If more than one component is present in the solution, as is generally the case in spectrophotometry of haemoglobin derivatives, a subdivision can be made into those instances where the light absorbing regions do not overlap at the wavelength(s) of measurement and those cases where the measurements are performed in mixtures of several light absorbing substances of which the absorption regions overlap at the wavelength(s) of measurement. In the analysis of such **multicomponent systems**, measurements should be made at wavelengths at which the molar absorptivities of the absorbing substances differ substantially.

All methods of quantitative analysis depend on the fact that the total absorbance at a given wavelength is equal to the sum of the absorbances of each of the individual components at that wavelength:

$$A(\lambda) = A_1(\lambda) + A_2(\lambda) + \cdots + A_n(\lambda),$$

or, according to equation 3.3:

$$A(\lambda) = \varepsilon_1(\lambda) \cdot c_1 \cdot l + \varepsilon_2(\lambda) \cdot c_2 \cdot l + \cdots + \varepsilon_n(\lambda) \cdot c_n \cdot l. \tag{3.6}$$

If the molar absorptivities are known and measuring is performed at n wavelengths, n equations of the above type are obtained, from which c_1, $c_2, \ldots, c_n$ can be solved. When there are exactly as many equations as there are unknowns, the system is called an **exactly determined system**; such a system has been described for the simultaneous determination of HHb, O_2Hb, COHb, Hi and SHb [443].

Before the advent of microcomputers and programmable calculators, solving n equations with n unknowns was a laborious process and the analysis of multicomponent systems was usually limited to analysis of 2-component systems. If the two components are convertible into each other, as, for example HHb and O_2Hb, the fraction of each component can be determined without knowledge of the four molar absorptivities involved [272].

The procedure can be simplified by choosing one of the two wavelengths involved at an isosbestic point in the absorption spectra of the two components (Chapter 2). If the concentration of one of the components is expressed as a fraction (F) of the total concentration (the sum of the concentrations of the two light absorbing substances, $ct = c_1 + c_2$), an equation of the following type can be derived [7, 210]:

$$F = c_2/(c_1 + c_2) = a \cdot [A(\lambda_1)/A(\lambda_2)] + b. \tag{3.7}$$

The constants a and b are determined by measuring the absorbance ratio of solutions containing component 1 and component 2 exclusively.

If neither of the two wavelengths is isosbestic for the two components, the analogous equation is:

$$F = \frac{\varepsilon_1(\lambda_1) - \varepsilon_1(\lambda_2) \cdot [A(\lambda_1)/A(\lambda_2)]}{\varepsilon_1(\lambda_1) - \varepsilon_2(\lambda_1) - [\varepsilon_1(\lambda_2) - \varepsilon_2(\lambda_2)] \cdot [A(\lambda_1)/A(\lambda_2)]}. \tag{3.8}$$

If care is taken that in converting component 1 into component 2 no change in total haemoglobin concentration occurs, *i.e.* that samples of 100% component 1 and 100% component 2 are obtained with the same ct, the absorbance of the two samples can be measured with the same concentration and light pathlength. It follows from equation 3.3 that in this case the set of four absorbance values, $A_1(\lambda_1)$, $A_2(\lambda_1)$, $A_1(\lambda_2)$, $A_2(\lambda_2)$, can be used instead of the corresponding absorptivities in equation 3.8.

In the early 1980s, **reversed optics, diode array spectrophotometers** were introduced. In these instruments the light transmitted by the sample is dispersed across an array of parallel detectors permitting the absorbances to be measured simultaneously over an entire spectral region, *e.g.* 200 to 800 nm. By measuring at more wavelengths than there are components in the system, more equations than there are unknowns may be obtained. Such an **overdetermined system** may yield more accurate results in photometric analysis than an exactly determined system does.

The Hewlett Packard HP8450A diode array spectrophotometer, which has been used for obtaining most of the data of Chapters 8–14, has a built-in microprocessor with software for **multicomponent analysis (MCA)**. The concentrations of up to 12 components in a mixture can be determined rapidly, provided the absorptivity spectra of the individual pure components have been introduced into the memory of the microprocessor. The spectra of the individual components are fitted on the measured spectrum of the mixture and, by means of a matrix calculation procedure, the concentration of each component corresponding to the best fit is displayed by the instrument [444].

INSTRUMENTATION

Light absorbance is measured with a spectrophotometer with, as basic components, a light source, a monochromator, a cuvette for holding the sample, and a light detector with read-out device.

Spectrophotometers may be single-beam or double-beam instruments. In single-beam spectrophotometers, the reference is inserted into the light beam, the instrument is balanced to zero absorbance, the sample is then inserted into the light beam and the absorbance is read. In double-beam spectrophotometers, a light beam passes through both the sample solution and the reference solution and a direct measurement of the difference in absorbance between the two solutions is made.

As **light source**, a tungsten filament lamp is generally used for the 325–1000 nm wavelength range, a hydrogen, deuterium, or xenon discharge lamp for the 180–350 nm wavelength range.

The simplest form of **monochromator** is a band pass filter transmitting one limited band of light only; filters may be wide-band (> 35 nm) or narrow-band. Glass prism monochromators are suitable only for the visible and near IR range; quartz glass is required for the UV range, sodium or potassium chloride has often been used for the IR range. Most present day monochromators consist of a diffraction grating, a large number of equally spaced, parallel lines etched on a glass or metal surface. Incident light is dispersed equally, resulting in a constant bandwidth for a given instrument slit width.

Most spectrophotometers are equipped with two **detectors**: a photomultiplier tube for the wavelength range of 180–700 nm, and a second photomultiplier tube or a photocell for measurements above 650 nm. An array of suitable photodiodes may also be used to cover the full range of 200–800 nm.

Any receptacle in which the sample is placed for photometric measurement is called a **cuvette**. Most cuvettes are made of glass, suitable for measuring in the 325–1000 nm range; measurements below 320 nm require the use of quartz or silica. Cuvettes with a rectangular or square cross-section are generally preferred for spectrophotometry. They usually come in pairs or sets, one cuvette for the sample, the other for the water or reagent blank. Matched sets have essentially zero cuvette correction between the individual cells to compensate for variations in multiple reflections from glass surfaces and for differences in the focal point on the radiant energy detector. Cuvettes should always be inserted into the instrument facing the same way with reference to the light beam. The optical surfaces should be cleaned with a soft dry cloth or lens paper and should not be touched during a series of measurements.

The **spectral resolution** of a spectrophotometer is its ability to resolve closely adjacent absorption or emission bands. The spectral resolution is expressed as the ratio $\lambda/\Delta\lambda$, where λ is the wavelength of the region being examined and $\Delta\lambda$ the separation of the two absorption or emission bands that can just be distinguished. Thus, when the sodium D-lines at 589.0 and 589.6 nm can be separated, the spectral resolution of the instrument is about 1000.

The **spectral bandwidth** is the total wavelength range of the light leaving the exit slit of the monochromator, measured between the points where the intensity is half the peak intensity. As the spectral bandwidth decreases, the spectral resolution increases.

The **natural bandwidth** of a light-absorbing substance is defined as the width of the wavelength range at half the height of the absorption peak. For HiCN, for example, the natural bandwidth of the absorption peak at 540 nm is 70 nm; for O_2Hb the natural band width is 26 and 38 nm for the absorption peaks at 577 nm and 542 nm, respectively.

As a rule of thumb in spectrophotometry, accurate absorbance measurements require that the spectral bandwidth of the measuring instrument used should be < 10% of the natural bandwidth of the substance being measured [157]. Hence, for accurate measurement of HiCN concentration, an instrument with a spectral bandwidth < 7 nm at the wavelength of measurement suffices; for accurate measurements of O_2Hb, an instrument with a spectral bandwidth < 2.5 nm is required. The following measurements of the absorbance of an O_2Hb solution at 414 nm (Soret band absorption maximum; natural bandwidth about 28 nm) with varying spectral bandwidth show the influence of the spectral resolution on the measured absorbance. With a spectral bandwidth of 0.5, 8.0, 10, and 20 nm, $A_{O_2Hb}(414)$ was 0.515, 0.511, 0.484 and 0.403, respectively.

The data presented in Chapters 8–14 have been obtained with Optica CF4 spectrophotometers and with an HP8450A spectrophotometer. The Optica CF4 is a conventional single-beam spectrophotometer, with two light sources: a tungsten incandescent lamp for the visible region and a hydrogen lamp for the UV. It contains as a monochromator a diffraction grating with coupled entrance and exit slits, opened and closed at the same values. The medium band width of the monochromatic light is determined by the slit aperture. At a slit of 0.2 mm aperture, the band width is 0.32 nm. The light transmitted through the sample is detected by one of two photomultiplier tubes; one of these is used for the spectral region of 185–700 nm, the second one, which is sensitive in the red and near IR spectral region extends the working range of the instrument to 1000 nm. In the determination of absorption spectra of haemoglobin derivatives it has proved to be practical to switch from the one phototube to the other at 630 nm. Of the two filters which can be inserted between lamp and monochromator, the red filter F2 is used in the measurements from 630 to 1000 nm to eliminate the second order spectrum of the grating.

The HP8450A spectrophotometer is called a reversed optics instrument, because the light passes through the cuvette before entering the analytical system. The two light sources, a deuterium and a tungsten halogen lamp, have a common axis and are turned on and off simultaneously. The light that has passed through the cuvette is focused on the entrance slit of the polychromator that disperses the light onto a diode array. This is accomplished by a servo-controlled beam director, a holographic grating and a pair of photodiode arrays. A full wavelength scan from 200 to 800 nm is completed in 1 s. After scanning a spectrum several times the mean values are displayed on the screen. The instrument operates in the single beam mode. First, a cuvette with a water blank is put in the light beam; the measured absorbance is taken up in the memory of the built-in microprocessor and used as zero point in the following measurements. The spectral bandwidth in the region of 400–800 nm is 2 nm. Sharp peaks at odd wavelengths, such as the α-peak of O_2Hb at 577 nm, cannot be accurately measured. Actually, only a quasi-continuous spectrum is obtained due to the spacing of the photodiodes.

SPECTROPHOTOMETER PERFORMANCE

Instrumental factors determining accuracy in spectrophotometry are wavelength accuracy, spectral bandwidth, stray light, photometric linearity and accuracy, and cuvette aberrations. These factors should be evaluated for each spectrophotometer in relation to the quantities to be measured, and checked regularly.

For the determination of **wavelength accuracy**, the best source remains the quartz mercury lamp providing pure spectral emission lines; especially the lines at 546.1 and 579.1 nm are suitable in the spectrophotometry of haemoglobin. Hydrogen and deuterium lamps can also be used, especially the lines at 486.1 and 656.3 nm. Filters made from didymium or holmium oxide glass have several distinct maxima which can be used for wavelength scale calibration. Solutions of holmium oxide in perchloric acid may also be used for this purpose [196, 411].

Radiation of wavelengths other than those for which the dispersing element of the monochromator is set may be present in the monochromator output to varying degree. This unwanted radiation, **stray light**, comes from the source and persists because some radiation in the beam is scattered by dust particles and is reflected by various surfaces. The problem is more severe when wide slit settings are required to obtain sufficient energy for adequate detector response. As the amount of stray light increases, the system becomes increasingly non-linear and may cause erroneous absorbance values to be measured, even causing false apparent absorption maxima to appear.

Sharp cut-off filters or solutions are used to test for or eliminate stray light. Examples are sodium nitrite 50 g/L, to assess stray light at wavelength < 380 nm, or potassium iodide 10 g/L to assess stray light at wavelength < 260 nm [53].

Photometric accuracy, *i.e.* the accuracy of the absorbance reading, can be verified with certified glass filters. NIST (National Institute for Standards and Technology, formerly National Bureau of Standards; USA), for example, has certified carbon yellow, cobalt blue, and copper green filters. Since these filters show considerable dependence of transmission on the accuracy of the wavelength setting, NIST also has certified Schott NG-type glass neutral density filters (SRM 930), supplied in sets of three, with absorbance values around 0.5, 0.7 and 1.0. These neutral grey filters and, preferably, the Corning HT yellow filter [77] have been used for the verification of the photometric accuracy of the instruments used to obtain the absorptivities presented in Chapters 8–14.

An easy way to check the photometric accuracy, as well as the linearity and repeatability of spectrophotometers in the visible spectral range, is by measuring the absorbance of certified HiCN reference solutions, of which the absorptivity and concentration are exactly known (Chapter 16).

CHAPTER 4

TOTAL HAEMOGLOBIN CONCENTRATION

The cornerstone of spectrophotometry of haemoglobin is an accurate method for measuring the total haemoglobin concentration. The determination of the absorptivity spectra of the common haemoglobin derivatives of man and several animals required numerous determinations of ct_{Hb} to be made rapidly and reliably. The availability of the internationally standardised HiCN method for measuring ct_{Hb} not only fulfilled this requirement, but also provided a solid basis for the determination of the absorptivity values, which thus became traceable to an internationally accepted standard. The obvious importance for clinical medicine of a reliable standardised method for measuring ct_{Hb} provided the impetus for the international effort to develop such a method, to supervise prescriptions for its use and to produce an international reference solution.

This chapter deals with the fundamentals and the history of the standardised method for measuring ct_{Hb}. A detailed description of the method, and of its use as a reference method in clinical medicine are discussed in Chapter 16. Its consequences for the determination of the oxygen binding capacity of haemoglobin are described in Chapter 19.

Around 1950 numerous methods, often yielding widely different results, were in use for the determination of ct_{Hb}. The need for improvement was clear and several attempts at standardisation were undertaken. In the United States a National Academy of Sciences — National Research Council (Division of Medical Sciences) Panel on the Establishment of a Hemoglobin Standard reviewed several photometric methods for the determination of ct_{Hb} [56]. In 1958 it was concluded that ct_{Hb} was best determined photometrically after conversion of all haemoglobin in the sample to haemiglobincyanide (HiCN). All forms of haemoglobin likely to occur in human blood, with the exception of sulfhaemoglobin, are determined by the method and the colour is suitable for measurement in filter-type photometers as well as in narrow band spectrophotometers, because its absorbance band around 540 nm is broad and relatively flat [57].

Further recommendations included [57]:

1. *That cyanmethemoglobin be adopted as a standard in clinical hemoglobinometry.*

2. *That the standard be characterized spectrophotometrically on the basis that the extinction coefficient of one milligram atom of iron (c = 1 mg atom of iron per liter, d = 1 cm) in the form of cyanmethemoglobin at a wave length of 540 mμ is 11.5.*
3. *That 0.338 percent (w/w) be accepted as the iron content of hemoglobin (molecular weight of 16,520 per gram atom of iron) in accordance with the recent recommendation of the Protein Commission of the International Union of Pure and Applied Chemistry, and that a factor of 1652 be used in calculating hemoglobin in mg per 100 ml from millimoles per liter.*
4. *That the standard be distributed as a single concentration of not less than 55 mg of cyanmethemoglobin per 100 ml.*
5. *That solutions be distributed in brown glass containers and in sterile condition.*

The value of $11.5\ \mathrm{L \cdot mmol^{-1} \cdot cm^{-1}}$ for the quarter millimolar absorptivity of HiCN at 540 nm, $\varepsilon_{\mathrm{HiCN}}(540)$, adopted by the panel was based on photometric data and oxygen binding capacity measurements on 11 human, 9 rabbit, and 14 canine blood samples [95, 96]. In Europe the value of $\varepsilon_{\mathrm{HiCN}}(540)$ was extensively reevaluated in the late nineteen fifties [263, 320, 484]. Based on the determination of haemoglobin iron by spectrophotometric methods with *o*-phenanthroline and α, α'-dipyridyl, and a titrimetric method with titanous chloride, a mean value of $11.0\ \mathrm{L \cdot mmol^{-1} \cdot cm^{-1}}$ was found for samples of horse haemoglobin, human whole blood, and human haemoglobin obtained through toluene erythrolysis.

In 1963, at the Ninth Congress of the European Society of Haematology in Lisbon, a symposium on erythrocytometric methods and their standardisation was held, and a Standardising Committee of the European Society of Haematology was established [34]. As one of its first recommendations, this committee proposed that the HiCN method [208] be the method of choice at the European level to measure ct_{Hb}, that the value of $11.0\ \mathrm{L \cdot mmol^{-1} \cdot cm^{-1}}$ be adopted for $\varepsilon_{\mathrm{HiCN}}(540)$, and that the value of 64458 [38] be accepted as representing the relative molecular mass of human haemoglobin A [35].

The standardising committee of the European Society of Haematology was enlarged to form the International Committee (now Council) for Standardisation in Haematology (ICSH) in 1964 during the tenth International Congress of Haematology in Stockholm; the HiCN method was adopted as the method of choice for the determination of ct_{Hb} at the European level and, provisionally, at the world level [58]. The committee also issued recommendations for the preparation and use of HiCN reference solutions [36, 189] and established an Expert Panel on Haemoglobinometry charged with drawing up recommendations for haemoglobinometry and for measuring and controlling HiCN reference solutions prepared by the Dutch Institute of Public Health (RIV), and offered to the committee for use as international standards.

Table 4.1.
Absorptivity of haemiglobin cyanide at 540 nm, as reported by various authors

Ref.	Material	$\varepsilon_{HiCN}(540)$	*SEM*	*n*	Method
263	Horse Hb	11.0	0.04	12	Fe: *o*-phenanthroline
263	Horse Hb	11.0	0.04	12	Fe: $TiCl_3$
320	Whole blood	11.09	0.03	11	Fe: $TiCl_3$
320	Whole blood	11.19	0.065	4	Fe: complexon
484	Hb solution (*a*)	10.99	0.01	123	Fe: α,α'-dipyridyl
484	Hb solution (*a*)	10.94	0.03	35	Fe: α,α'-dipyridyl
484	Hb solution (*a*)	11.05	0.02	101	Fe: $TiCl_3$
302	Whole blood	10.99	0.05	10	Fe: α,α'-dipyridyl
302	Whole blood	11.06	0.08	8	Fe: α,α'-dipyridyl
335	Hb solution (*b*)	10.95	0.03	46	Fe: α,α'-dipyridyl
282	Whole blood	11.02	0.03	10	Fe: X-ray emission spectrography
282	Washed cells	10.97	0.07	6	Fe: X-ray emission spectrography
375	Hb solution (*c*)	11.00	0.02	55	Fe: sulfosalicylic acid
384	Hb solution (*b*)	10.90	0.05	55	N analysis
195	Hb solution (*d*)	10.88	0.04	16	C analysis
163	Hb solution (*a*)	11.01	0.03	17	Titration of Hi with CN^-

$\varepsilon_{HiCN}(540)$ = millimolar absorptivity of haemiglobincyanide, in $L \cdot mmol^{-1} \cdot cm^{-1}$; *SEM* = standard error of the mean; ref. = reference; (*a*) toluene erythrolysis; (*b*) purified on CMC column; (*c*) purified on CMC or Sephadex column, or by dialysis against Na_2-EDTA; (*d*) purified by chromatography.

The recommendations for haemoglobinometry, with the HiCN method as method of choice, were finally adopted at the world level during the eleventh International Congress of Haematology in Sydney [190]. The World Health Organisation (WHO) established HiCN solutions as WHO International Reference Preparations (now International HiCN Standards) in 1968 [427].

The ICSH Expert Panel on Haemoglobinometry published its first recommendations for haemoglobinometry in 1967 [190] and has since reviewed and where necessary updated the recommendations [191, 192, 194]. Meanwhile various investigators verified the value of $11.0\ L \cdot mmol^{-1} \cdot cm^{-1}$ for $\varepsilon_{HiCN}(540)$. In most of these investigations iron analysis was used, but with different methods of determination. Van Oudheusden *et al.* [302] and Salvati *et al.* [335] used the α, α'-dipyridyl method, Stigbrand [375] a procedure involving sulfosalicylic acid, Morningstar *et al.* [282] X-ray emission spectrography. After the complete elucidation of the composition of the globin moiety of HbA [39], it became possible to determine ct_{Hb} and thus $\varepsilon_{HiCN}(540)$ on the basis of content of other atoms than iron. This was first accomplished by Tentori *et al.* [384] on the basis of nitrogen, then by Itano [195] on the basis of carbon. Finally, Hoek *et al.* [163] determined $\varepsilon_{\mathrm{HiCN}}(540)$ on the basis of the titration of haemiglobin with cyanide.

The results of most determinations of $\varepsilon_{\mathrm{HiCN}}(540)$ are presented in Table 4.1, in which the methods for preparing the solutions of human haemoglobin used in the measurements are also indicated. Only the results of Minkowski and Swierczewski [266] and of Wootton and Blevin [426] have not been included. The former have been omitted because the spread of the measurements has not been reported in the paper, the latter because these results [$\varepsilon_{\mathrm{HiCN}}(540) = 10.68 \pm 0.04$ (*SEM*), $n = 14$] are significantly different from all other results taken together. The data of Table 4.1 show that the value of $\varepsilon_{\mathrm{HiCN}}(540)$ has been determined to lie between 10.88 and 11.19 $\mathrm{L \cdot mmol^{-1} \cdot cm^{-1}}$.

Since there was no reason to prefer any one of the methods used to the others, it has been decided to take the mean of all results. This mean value and its standard error were calculated as follows:

$$x_g = (1/n_{\mathrm{tot}}) \cdot \Sigma_i (n_i \cdot x_i), \tag{4.1}$$

$$s^2/n_{\mathrm{tot}} = [1/n_{\mathrm{tot}} \cdot (n_{\mathrm{tot}} - 1)] \cdot \Sigma_i \left[(n_i - 1) \cdot s_i^2 + n_i \cdot (x_i - x_g)^2 \right], \tag{4.2}$$

where x_g is the mean value of all determinations; x_i is the mean value of the ith series of determinations; n_{tot} is the total number of determinations; n_i is the number of determinations in the ith series; s_i is the standard deviation of the ith series of determinations; and s is the standard deviation of all determinations [13, 486].

The mean value of $\varepsilon_{\mathrm{HiCN}}(540)$ was 10.99 $\mathrm{L \cdot mmol^{-1} \cdot cm^{-1}}$, standard error of the mean (*SEM*) 0.01 $\mathrm{L \cdot mmol^{-1} \cdot cm^{-1}}$, $n = 521$. This is the experimental justification of the ICSH decision to use $\varepsilon_{\mathrm{HiCN}}(540) = 11.0$ $\mathrm{L \cdot mmol^{-1} \cdot cm^{-1}}$ as the basis for haemoglobinometry.

CHAPTER 5

ABSORPTIVITY AT 540 nm OF HAEMIGLOBINCYANIDE

The first step in the determination of the haemoglobin absorption spectra of any animal species is to measure, with the highest attainable accuracy, the absorptivity of a single derivative at a single wavelength. On this anchor value all other absorptivity values of all haemoglobin derivatives are based. For this purpose the obvious choice is $\varepsilon_{HiCN}(540)$ as used in the reference method for measuring total haemoglobin in human blood. The determination of this value involves the measurement of ct_{Hb} by a non-spectrophotometric method. As described in Chapter 4, $\varepsilon_{HiCN}(540)$ of human blood has been determined on the basis of the Fe, N and C content of the haemoglobin molecule, with the Fe measurements made with various methods. As the basis for the measurement of $\varepsilon_{HiCN}(540)$ of animal haemoglobin, we selected the determination of Fe with the α,α'-dipyridyl method, which had also been used extensively in the determination of $\varepsilon_{HiCN}(540)$ of human haemoglobin [211, 484].

This chapter presents a general description of the methods used in the determination of $\varepsilon_{HiCN}(540)$, the results obtained in various animal species, and a comparison with the values previously obtained for human haemoglobin.

PREPARATION OF A HAEMOGLOBIN STOCK SOLUTION

For the preparation of a stroma-free haemoglobin solution with a concentration of about 8 mmol/L, the following procedure is used [211, 485]. About 90 mL of fresh blood is collected in a 200-mL iron-free Erlenmeyer flask containing 100 USP units of heparin per mL of blood. The erythrocytes are washed three times with 9 g/L NaCl solution, using iron-free tubes. One volume of erythrocytes is thoroughly mixed with one volume of distilled water and 0.4 volume of analytic grade toluene in a stoppered volumetric glass cylinder. After standing at 4 °C for at least 16 h, the mixture has separated into three clearly distinct layers: a top layer of toluene, a middle layer consisting of a turbid suspension of erythrocyte stromata, and a lower layer of a clear, stroma-free haemoglobin solution. The stroma-free haemoglobin solution is sucked from under the other layers with the help of a 50-mL pipette equipped with a pipetting balloon, and transferred to two 30-mL centrifuge tubes. The haemoglobin solution is then subjected to centrifugation for 20 min at 8000g, filtered through ash-free paper, and stored at 4°C.

All glassware used in this and the following procedure must be iron-free, all reagents of analytical grade, all water distilled or deionised. Glassware can be cleaned free from iron contamination by washing in a warm solution of non-ionic detergent, rinsing with deionised water, soaking in dilute nitric acid (0.75 mol/L), rinsing with deionised water and drying.

DETERMINATION OF THE IRON CONCENTRATION OF THE HAEMOGLOBIN STOCK SOLUTION

For the determination of the iron concentration of the haemoglobin stock solution with the α, α'-dipyridyl method, the haemoglobin must be decomposed in order to set the haemoglobin iron free in the ferrous state [211, 472, 484]. With a calibrated Ostwald pipette, 1.0 mL haemoglobin stock solution is transferred to at least four iron-free decomposition tubes. Add 1.0 mL nitric acid (HNO_3), 65%, 1 mL hydrogen peroxide (H_2O_2), 30%, and three drops of octylalcohol (octanol-1; $C_8H_{17}OH$) to each of the tubes. Slowly raise the temperature on an electric furnace to 220°C until a clear yellow solution is obtained; evaporate almost to dryness. To remove excess acid, add 1.5 mL deionised water and evaporate *in vacuo* at 95°C; repeat this once. Transfer the contents of the decomposition tubes to volumetric flasks and make up to 100 mL with water.

Mix, in test tubes, 5 mL of each of the solutions with 1.0 mL buffered α,α'-dipyridyl solution and 1.0 mL sodium sulphite solution. Heat the mixtures to 95–100°C in a water bath for 10 min. During heating the test tubes are covered with glass marbles and the upper end of the tubes is cooled with an air stream. A pink colour develops.

The α,α'-dipyridyl solution consists of 0.5 g α,α'-dipyridyl, 3.4 g sodium acetate trihydrate, 1.41 mL glacial acetic acid, and water up to 50 mL; the sodium sulphite solution contains 2.52 g sodium sulfite ($Na_2SO_3 \cdot 7H_2O$) and water up to 100 mL.

Determine the absorbance of the cooled pink solutions at 520 nm with a lightpath length of 1.000 cm. Correct for traces of iron in reagents and glassware by measuring the absorbance at 520 nm of a solution resulting from the same decomposition procedure, but with 1.0 mL of water in the decomposition tube, instead of haemoglobin stock solution (blank). Correct for the faint yellowish background colour due to the decomposition of haemoglobin by measuring the absorbance at 520 nm of a solution resulting from the same haemoglobin decomposition procedure but without addition of α,α'-dipyridyl and sodium sulfite solutions (background).

For calibration, measure a series of solutions of analytic grade ferric ammonium sulphate, $Fe_2(SO_4)_3{\cdot}(NH_4)_2SO_4{\cdot}24H_2O$, the iron content of which has been checked by gravimetric analysis. Prepare a stock solution containing 1.7272 g ferric ammonium sulphate per litre; this corresponds to an iron concentration of 3.582 mmol/L. Dilute the stock solution 20-fold. Add to 0.5, 1.0, 1.5, ..., 5.5 mL of the diluted stock solution 1 ml α,α'-dipyridyl solution and 1 ml sodium sulphite solution and make up to 7.5 mL with water. A series of solutions containing 11.94, 23.88, 35.82, ..., 131.34 μmol Fe per litre is thus obtained. After colour developement (p. 30), measure the absorbance of the solutions at 520 nm and use the ensuing calibration line to calculate the iron concentration of the haemoglobin stock solution.

The ferric ammonium sulphate salt used in the calibration is stored in an atmosphere corresponding with the water vapour pressure of the crystals (0.8 kPa).

The procedure is illustrated by the following example. The Ostwald pipette delivered 1.023 mL haemoglobin stock solution to six decomposition tubes and 1.023 mL deionised water to a seventh tube (blank). After decomposition, the resulting Fe^{2+}-containing solutions and the blank were made up to 100.0 mL in volumetric flasks. Of the resulting solutions, 5.00 mL was mixed with 1.00 mL dipyridyl solution and 1.00 mL sodium sulphite solution. After colour development, $A(520)$ was measured with a lightpath length of 1.000 cm. The absorbance measurements of the background (of the decomposed haemoglobin solutions as well as of the blank) were performed on 5 mL of the same solutions to which 2 mL of water was added instead of dipyridyl and sodium sulphite solution. The six corrected $A(520)$ values were then calculated according to:

$$A(520) = [A_{\mathrm{SFH}}(520) - A_{\mathrm{bgr\ SFH}}(520)] - [A_{\mathrm{bl}}(520) - A_{\mathrm{bgr\ bl}}(520)], \quad (5.1)$$

where SFH is decomposed stroma-free haemoglobin, bgr is background and bl is blank.

This yielded:

$$\begin{aligned}
&(0.3980 - 0.0020) - (0.0020 - 0.0010) = 0.3950\\
&(0.3980 - 0.0025) - (0.0020 - 0.0010) = 0.3945\\
&(0.3985 - 0.0020) - (0.0020 - 0.0010) = 0.3955\\
&(0.4005 - 0.0025) - (0.0020 - 0.0010) = 0.3970\\
&(0.3980 - 0.0025) - (0.0020 - 0.0010) = 0.3945\\
&(0.3990 - 0.0025) - (0.0020 - 0.0010) = 0.3955
\end{aligned}$$

and a mean value of $A(520) = 0.3953$.

The calibration had been carried out with a ferric ammonium sulphate preparation of 99.11% purity. The series of solutions (8 solutions instead of 11; the highest three concentrations omitted) thus contained 0.9911 times the concentration given previously: 11.833, 23.666, 35.499, 47.332, 59.165, 70.998, 82.831 and 94.664 μmol Fe per L. The corresponding absorbance values at 520 nm were 0.1028, 0.2053, 0.3035, 0.4098, 0.5115, 0.6151, 0.7163, 0.8164. This yields the regression equation: c_{Fe} =

115.75 × $A(520)$ − 0.0078. Substituting $A(520)$ = 0.3953 then gives an iron concentration, c_{Fe} = 45.756 μmol/L.

Of the stroma-free haemoglobin solution, 1.023 mL had been made up to 100.0 mL with water and 5 mL of the solution thus diluted had been mixed with 2 mL of reagent solutions. The iron concentration of the stroma-free haemoglobin solution thus was $(100/1.023) \times (7/5) \times 45.756 =$ 6.262 mmol/L. Consequently, the concentration of the stroma-free haemoglobin solution on the basis of the iron content, $ct_{Hb}(Fe)$, was 6.262 mmol/L.

DETERMINATION OF THE ABSORBANCE AT 540 nm OF THE HAEMOGLOBIN STOCK SOLUTION AFTER CONVERSION TO HiCN

For the determination of the absorbance of HiCN at λ = 540 nm, 1.0 mL of haemoglobin stock solution is transferred to a volumetric flask with the help of the same Ostwald pipette as is used in the determination of the iron concentration of the stock solution, and made up to 250 mL with reagent solution [211, 472]. The reagent solution contains 200 mg potassium hexacyanoferrate(III) (potassium ferricyanide; $K_3Fe(CN)_6$), 50 mg potassium cyanide (KCN) and 1.0 g sodium bicarbonate ($NaHCO_3$) per litre. After at least 30 min, when all haemoglobin has been converted to HiCN, the absorbance at 540 nm, $A(540)$, is determined with a lightpath length of 1.000 cm, with the same cuvettes as were used in the iron determination. Since $A(540)$ of the reagent solution at this wavelength is zero, water is used as a blank.

The procedure is illustrated by the example at the end of the following section.

CALCULATION OF THE ABSORPTIVITY AT 540 nm OF HiCN

Since a haemoglobin monomer contains only one atom of iron, the substance concentration of iron in the haemoglobin stock solution is equal to the substance concentration of total haemoglobin. The HiCN concentration in the solution of which $A(540)$ is measured differs only by a dilution factor from the haemoglobin concentration of the stock solution. Hence, $\varepsilon_{HiCN}(540)$ can now be calculated according to Lambert–Beer's law:

$$\varepsilon_{HiCN}(540) = A(540) \cdot V_1/[V_2 \cdot ct_{Hb}(Fe) \cdot l], \quad (5.2)$$

where V_1 is the volume of the volumetric flask, V_2 is the volume delivered by the Ostwald pipette, $ct_{Hb}(Fe)$ is the haemoglobin concentration calculated on the basis of the iron concentration of the stock solution, and l is the lightpath length [472]. As V_1 and V_2 are in mL, $ct_{Hb}(Fe)$ is in mmol/L and l is in cm, $\varepsilon_{HiCN}(540)$ is obtained in $L \cdot mmol^{-1} \cdot cm^{-1}$.

The calculation is illustrated by the following example. The absorbance at 540 nm as HiCN, of the stroma-free haemoglobin solution of which the haemoglobin concentration had been determined on the basis of the iron content, was measured as described in the preceding section. The Ostwald pipette delivered 1.023 mL stroma-free haemoglobin solution to each of three 250-mL volumetric flasks, which were then made up to volume with reagent solution and allowed to stand for more than 30 min. $A(540)$ was measured and found to be 0.2815 for all three HiCN solutions. $\varepsilon_{HiCN}(540)$ was then calculated with equation 5.2:

$$\begin{aligned}\varepsilon_{HiCN}(540) &= (0.2815 \times 250)/(1.023 \times 6.262 \times 1.000) \\ &= 10.986\ \mathrm{L \cdot mmol^{-1} \cdot cm^{-1}}. \qquad (5.3)\end{aligned}$$

$\varepsilon_{HiCN}(540)$ AS FOUND FOR HUMAN AND ANIMAL HAEMOGLOBIN

In addition to the determination of $\varepsilon_{HiCN}(540)$ of haemoglobin of various animals, a new series of measurements was made for human haemoglobin. The same procedures and the same instruments were used for the determination of $\varepsilon_{HiCN}(540)$ of haemoglobin in the animal species. For the preparation of the haemoglobin stock solutions, however, differences in procedure were unavoidable, mainly because of differences in the solubility of haemoglobin. These differences are described in Chapters 9–14 for each of the animals.

Four to six determinations of $\varepsilon_{HiCN}(540)$ were made of each of the haemoglobin stock solutions. In the tables and calculations, the number of stock solutions, which equals the number of animals from which blood was obtained, is denoted by N; the number of actual determinations is denoted by n. Table 5.1 presents the mean values obtained for the stock solutions of human and animal haemoglobin. Table 5.2 shows mean values with standard deviation and standard error, calculated for both the stock solutions and for the individual measurements.

$\varepsilon_{HiCN}(540)$ of horse haemoglobin had previously also been measured by Meyer-Wilmes. and Remmer [263], who performed two series of measurements on the basis of iron, one by $TiCl_3$ titration and one by the *o*-phenanthroline method. Both series provided a mean value of 11.0 $\mathrm{L \cdot mmol^{-1} \cdot cm^{-1}}$, with a standard error of 0.04. This is in excellent agreement with the values shown in Tables 5.1 and 5.2.

Among the measurements underlying the internationally accepted value of 11.0 $\mathrm{L \cdot mmol^{-1} \cdot cm^{-1}}$ for $\varepsilon_{HiCN}(540)$ (Table 4.1), there are three large series of determinations for human haemoglobin on the basis of iron as determined with the α,α'-dipyridyl method: two series from Zijlstra and van Kampen [484] and one from Salvati *et al.* [335]. When these 204 measurements are added to the 64 values of Table 5.2, one obtains a group of 268 determinations with a mean value of 10.962 and a standard error of 0.008 $\mathrm{L \cdot mmol^{-1} \cdot cm^{-1}}$.

Table 5.1.
$\varepsilon_{HiCN}(540)$ as determined in N stock solutions of human and animal haemoglobin

Human $N = 13$	Bovine [476] $N = 15$	Pig $N = 11$	Horse $N = 12$	Sheep $N = 9$	Dog [472] $N = 9$	Rat [477] $N = 12$
10.896	10.911	11.014	10.948	11.240	10.984	10.932
10.910	10.927	10.855	11.035	11.207	11.092	10.844
10.936	10.993	10.792	11.078	10.878	10.848	10.984
10.935	10.935	10.851	10.855	10.856	11.101	11.030
10.963	10.970	10.907	11.141	10.921	10.851	11.136
10.885	10.911	10.863	10.970	10.965	10.940	11.074
10.915	10.966	10.855	10.924	10.855	10.917	10.871
10.947	10.999	10.951	10.882	11.007	11.005	10.820
10.898	11.036	10.929	11.056	10.857	10.944	10.927
10.888	10.998	11.070	10.938			10.922
11.009	10.891	11.058	11.069			11.086
10.957	10.942		10.865			11.082
10.949	11.036					
	10.937					
	10.953					

$\varepsilon_{HiCN}(540)$ is expressed in $L \cdot mmol^{-1} \cdot cm^{-1}$.

Table 5.2.
Mean values of $\varepsilon_{HiCN}(540)$ for human and animal haemoglobin

	Human $N = 13$	Bovine $N = 15$	Pig $N = 11$	Horse $N = 12$	Sheep $N = 9$	Dog $N = 9$	Rat $N = 12$
Mean	10.930	10.960	10.922	10.980	10.976	10.965	10.976
SD	0.034	0.043	0.088	0.090	0.141	0.086	0.101
SEM	0.010	0.011	0.026	0.026	0.047	0.029	0.029
	$n = 64$	$n = 82$	$n = 59$	$n = 64$	$n = 57$	$n = 43$	$n = 55$
Mean	10.930	10.956	10.920	10.979	10.984	10.958	10.973
SD	0.049	0.074	0.103	0.102	0.183	0.100	0.118
SEM	0.006	0.008	0.013	0.013	0.024	0.015	0.016

$\varepsilon_{HiCN}(540)$ is expressed in $L \cdot mmol^{-1} \cdot cm^{-1}$. *SD* is standard deviation; *SEM* is standard error of the mean.

A comparison of the mean values of $\varepsilon_{HiCN}(540)$ as shown in Table 5.2 for the various animals with this overall value for human haemoglobin shows that the greatest differences are found in sheep, +0.20%, and in pigs, −0.38%. Although the latter difference proved to be statistically significant according to Student's t-test for unpaired samples, it has been decided to base the absorptivities of all haemoglobin derivatives of all animals examined up to now on the internationally accepted value for human haemoglobin of 11.0 $L \cdot mmol^{-1} \cdot cm^{-1}$. The main reason for this decision is that, as explained in Chapter 4, the internationally accepted value is based on several series of measurements by different methods. Between the results of some of these methods, there were slight but statistically significant differences. However, there was not, and still is not, a valid reason to prefer any of the methods used to any one of the others. Hence, all technically flawless results were accepted. Consequently, it seems justified to adhere to $\varepsilon_{HiCN}(540) = 11.0\,L \cdot mmol^{-1} \cdot cm^{-1}$ for all animals for which one of the accepted methods (Table 4.1) provides a value within the range that underlies the internationally accepted value.

$\varepsilon_{HiCN}(540)$ of haemoglobin of the frog (*Rana esculenta*) and the Mexican axolotl (*Ambystoma tigrinum*) has been determined by Salvati *et al.* [336]. They used the same methods as in their study of human haemoglobin [335] and found for the frog $\varepsilon_{HiCN}(540) = 10.79\,L \cdot mmol^{-1} \cdot cm^{-1}$ with a standard deviation of $0.09\,L \cdot mmol^{-1} \cdot cm^{-1}$ ($N = 4$; $n = 12$) and for the axolotl $\varepsilon_{HiCN}(540) = 10.64\,L \cdot mmol^{-1} \cdot cm^{-1}$ with a standard deviation of $0.19\,L \cdot mmol^{-1} \cdot cm^{-1}$ ($N = 5$; $n = 15$). The authors concluded that these values were not statistically different from their results for human haemoglobin [335].

CHAPTER 6

PREPARATION OF HAEMOGLOBIN DERIVATIVES

The crucial step in the procedure to obtain accurate absorption spectra of haemoglobin derivatives is the preparation of the samples. Ideally, these should be clear solutions of a single haemoglobin derivative. Hence, any contamination with other haemoglobin derivatives or other light-absorbing substances should be precluded and any turbidity should be removed. To this end, various methods have been used, which, in the course of time, have undergone considerable improvement and a procedure has been developed with which reasonably pure and clear solutions of several haemoglobin derivatives can be prepared. Most of the data presented in the following chapters have been obtained with this procedure.

In this chapter, the latest procedure is described in detail to enable future investigators to check the results presented in Chapters 8–14, and to determine haemoglobin spectra in other spectral regions and of other animals than described in these chapters. Earlier procedures are mentioned only in so far as results are presented in which these have been used.

SPECIMEN PROCUREMENT AND HANDLING

Fresh whole blood is collected in a glass vessel containing 100 USP units of sodium heparin per mL of blood. Adult human blood is obtained from healthy donors, foetal human blood by puncture of the umbilical cord immediately after delivery. Animal blood is obtained in different ways depending on the species in question (Chapters 9–14).

Plasma is removed by centrifugation and the erythrocytes are resuspended in isotonic saline solution (NaCl, 154 mmol/L). The total haemoglobin concentration is kept between 100 and 150 g/L. The ensuing erythrocyte suspension, after filtration through cotton wool, is the starting preparation for all haemoglobin derivatives.

HHb, O_2Hb, and COHb are prepared by tonometry with different gas mixtures. After tonometry for 2 h, 2 mL of a 100 mL/L non-ionic detergent solution, equilibrated with the gas mixture which flows through the tonometer, is introduced into the revolving tonometer. After erythrolysis, which is almost immediately effected by the detergent, tonometry is continued for about

20 min. The erythrolysate is then transferred directly from the tonometer to the spectrophotometer cuvettes through cotton wool filters.

The non-ionic detergent added to the erythrocyte suspension in the tonometer serves primarily to lyse the red cells, but also helps to clarify the ensuing haemoglobin solution. In the preparation of most of the haemoglobin solutions used in the determination of the absorption spectra of Chapters 8–14, Sterox SE, an alkylphenol(thiol)polyethylene oxide detergent (Hartmann–Leddon) has been used. It may be replaced by similar non-ionic detergents, *e.g.* Nonidet P40 (Shell). For other non-ionic detergents see Chapter 16.

Erythrolysis thus is effected only near the end of tonometry. This has been shown to give better results than when a stroma-free haemoglobin solution is made first, for instance through toluene erythrolysis as described in Chapter 5, and tonometry with the various gas mixtures is carried out with this haemoglobin solution. When haemoglobin has been removed from the protective environment of the red cells, it may more easily deteriorate during tonometry. HbF proved to be especially vulnerable, as shown by the formation of Hi during tonometry. By itself, this is not a serious side-reaction, because in determining the absorption spectra a correction for the presence of Hi can easily be made, but it may herald the occurrence of further deterioration, and lead to the presence of unknown products with different light absorbing properties.

Hi is prepared from O_2Hb erythrolysate by oxidation of the haemoglobin iron with $K_3Fe(CN)_6$. HiCN is prepared from an oxygenated erythrocyte suspension, which, after erythrolysis, is converted to HiCN through the addition of $K_3Fe(CN)_6$ and KCN. SHb is prepared from an oxygenated erythrocyte suspension by addition of Na_2S and HCl; erythrolysis is effected after the formation of SHb. Hi, HiCN and SHb are transferred to the cuvettes, through a cotton wool filter, with the aid of a glass syringe. In the following sections the preparation of clear solutions of practically important haemoglobin derivatives is described in greater detail.

TONOMETRY, ERYTHROLYSIS AND FILTRATION

Tonometry is carried out in cylindrical glass tonometers with a volume of about 300 mL [322]. The tonometers are clamped between inlet and outlet sockets fixed in a metal frame (Fig. 6.1). The use of greased ground glass ball and socket joints allows rotation of the tonometer along its horizontal axis by means of a motor-driven rubber band. A thin layer of erythrocyte suspension or haemoglobin solution is thus formed on the inner wall of the tonometer. The outlet socket is equipped with a hinge and blade spring insuring leak proof, gas tight seals, and allowing easy insertion and removal of the tonometer.

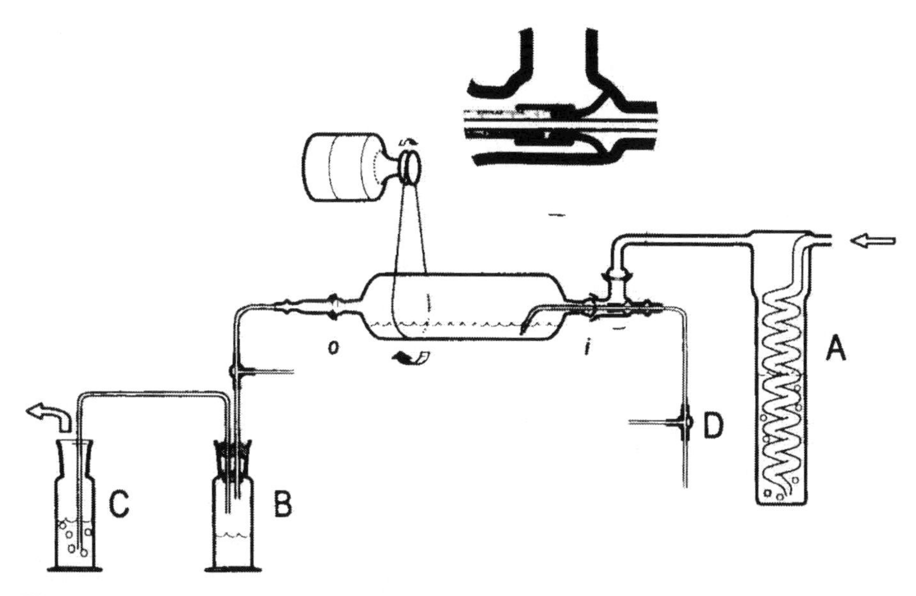

Figure 6.1. Diagram of a cylindrical glass tonometer for equilibration of erythrocyte suspensions with different gas mixtures.

The tonometer rotates along its horizontal axis between gas tight, greased glass bearings. The metal frame supporting the in- and outlet sockets is not shown. The incoming gas is humified in vessel A. In vessel B the detergent solution to be used for erythrolysis is equilibrated with gas that has passed through the tonometer. The water seal C prevents back-diffusion of air from the surrounding atmosphere. During tonometry fluid samples can be taken from the tonometer through the stopcock system D.

The gas mixtures are introduced into the tonometer through a glass spiral within a 50-mL flask half filled with water (A in Fig. 6.1). The bubbles formed at the bottom of the spiral become saturated with water vapour as they pass upwards along the outside of the spiral. The outlet of the flask is connected to the inlet of the tonometer by means of a ground glass ball and socket joint [322]. This socket is at the end of a side arm of the tonometer inlet socket piece, as shown in Fig. 6.1. The gas leaving the tonometer passes through a vessel (B) containing the non-ionic detergent solution subsequently used for erythrolysis, then escapes through a water seal (C). The latter is necessary to prevent any back diffusion of air from the surrounding atmosphere; it also causes a slight positive pressure in the tonometer helpful for easy retrieval of fluid samples from the tonometer. The gas flow through the tonometer usually is 40–50 mL/min.

In order to keep the erythrocyte suspensions cool during tonometry, the tonometers are covered with wet cotton gauzes, which are sprinkled at intervals with water.

As shown in Figs 6.1 and 6.2, a thin, stiff, bent nylon tube has been mounted in the gas inlet, through which a small PVC sampling tube is inserted into the

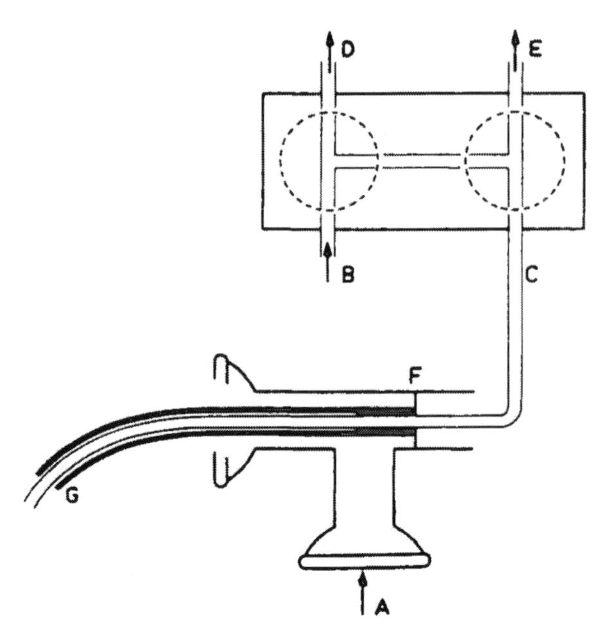

Figure 6.2. Diagram showing tonometer inlet socket piece and double 3-way stopcock.
Humidified gas enters the system through socket A. A thin PVC tubing C runs from the stopcock through the inlet into the tonometer to allow filling of, and sampling from, the tonometer. Tube C is held in place by means of a piece of curved, stiff nylon tubing G. This nylon tubing is attached to a glass septum F in the tonometer inlet socket piece. The double stopcock is used to fill a spectrophotometer cuvette directly from the tonometer or to fill a syringe for transferring a sample to a cuvette in a separate step, as described in the text (from ref. [322]).

tonometer. This tube serves for injecting the non-ionic detergent solution into the tonometer as well as for taking samples from the tonometer.

After tonometry for 2 h, 2.5 mL of the 100 mL/L non-ionic detergent solution is taken from the equilibration vessel (B in Fig. 6.1) with a 5-mL syringe; the syringe has previously been slowly rinsed several times with gas flowing from the tonometer. The syringe is then connected to the double 3-way stopcock at E (Fig. 6.2). Gas in the stopcock is removed by pressing some solution through the side branch; after turning the stopcock, the sampling tube C is made gas-free by allowing some of the erythrocyte suspension to escape through the side branch. The stopcock is turned again and 2 mL detergent solution is injected into the tonometer, causing rapid erythrolysis. Erythrolysate is drawn into the sampling tube and syringe in order to mix it with the remaining detergent solution, the mixture is pressed back into the tonometer, and tonometry is continued for 20 min. In the case of SHb, erythrolysis is not effected in the tonometer; the erythrocyte suspension is drawn into a glass syringe containing non-ionic detergent solution, as explained on p. 45.

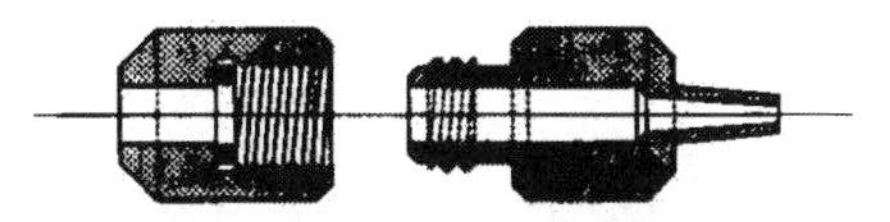

Figure 6.3. Diagram of filter cartridge for cotton wool filtration of erythrolysate and haemolysed blood.

The cotton wool is inserted into the right side section of the filter unit. Figure 6.4 shows the use of the filter in connection with a syringe and a needle for filling a cuvette. When cuvettes are filled directly from the tonometer the filter cartridge is connected to outlet E of the stopcock shown in Fig. 6.2.

Before the sample is introduced into the spectrophotometer cuvette it is passed through a cotton wool filter to clear the erythrolysate from scattering particles. During filtration any contact between haemoglobin solution and air should be prevented. This is accomplished by using the filter unit shown in Figs 6.3 and 6.4. The cartridge is made of stainless steel and can easily be cleaned and filled with a new piece of cotton wool. It has been demonstrated that even when haemolysed blood is introduced into spectrophotometer cuvettes through this filter, the solution is sufficiently clear of scattering material [443] that multicomponent analysis (Chapter 17) can be performed using the absorptivity values of partly purified haemoglobin solutions as presented in Chapters 8–14.

Two different procedures are used for transferring an erythrolysate from the tonometer to a spectrophotometer cuvette. If the erythrolysate requires further manipulation, it is first collected into a syringe and transferred to the cuvette in a separate step; if not, the sample is transferred directly to the cuvette.

To fill a syringe with equilibrated erythrolysate from the tonometer the syringe is attached to the double 3-way stopcock at E (Fig. 6.2). The syringe is filled by the following steps between which the stopcocks are turned. (1) The sampling tube C is rinsed with erythrolysate from the tonometer, using the side branch and outlet D; the fluid is propelled by the excess pressure in the tonometer. (2) About 1 mL erythrolysate is drawn into the syringe and discarded via outlet D. (3) The desired volume of erythrolysate is slowly drawn into the syringe.

For filling a spectrophotometer cuvette from the syringe a filter unit and a needle are placed on the syringe and the spectrophotometer cuvette is filled with the needle tip at the bottom of the cuvette (Fig. 6.4). In filling the cuvette, care should be taken that the glass walls do not become scratched by the oblique tip of the needle. Filling as shown in Fig. 6.4 is therefore mandatory.

To fill a spectrophotometer cuvette directly from the tonometer, a filter unit and a needle are connected to the 3-way stopcock at E (Fig. 6.2). The excess pressure in the tonometer is sufficient to drive a fluid sample through the cotton wool filter and the needle into the spectrophotometer cuvette. The

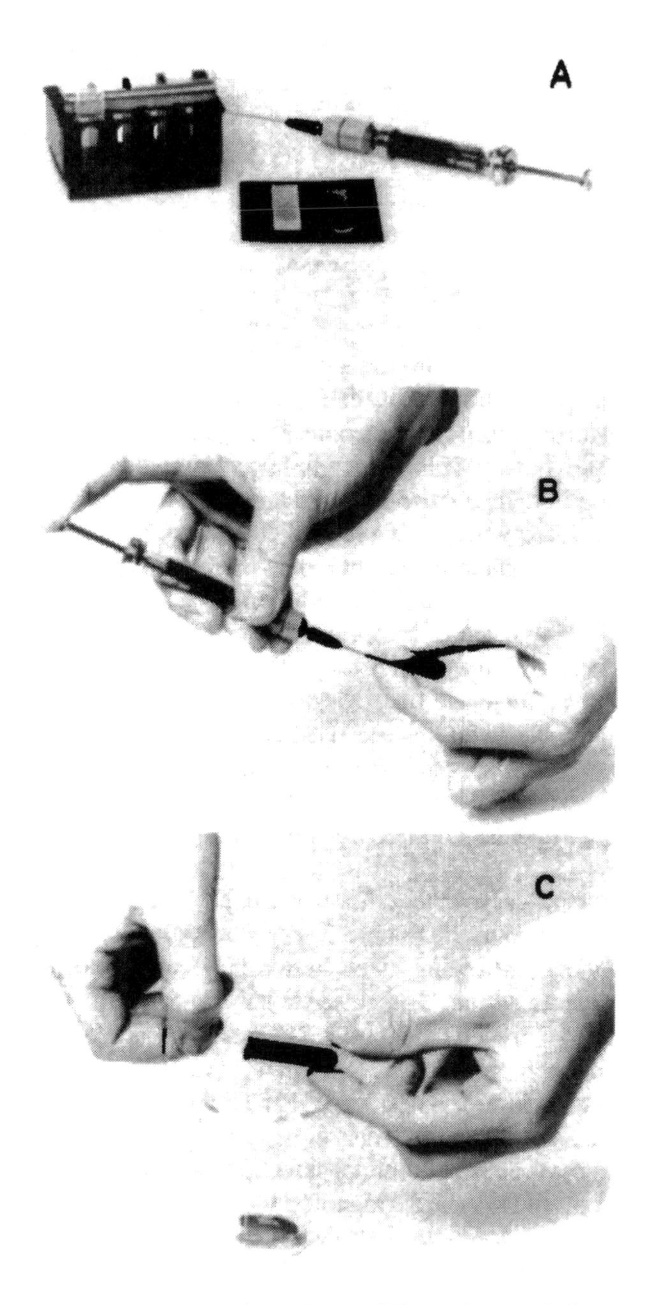

Figure 6.4. Filling a 0.100-cm cuvette and inserting a 0.090-cm plan parallel glass plate.

Cuvettes and glass inserts are stored in chromic acid solution, rinsed before use with deionised water followed by ethanol 96% and dried with an air jet. The filter cartridge is cleaned, dried and filled with new cotton wool before each measurement.

A. Cuvette holder with reference cuvette, cuvette, glass insert and 2-mL syringe with filter unit and needle, containing a sample. B. After discarding the first 10 drops, the sample is gently pressed through the filter into the cuvette. C. The plan parallel glass plate is inserted into the cuvette, leaving a lightpath of 0.010 cm (from ref. [443]).

right hand stopcock outlet E is connected to the sampling tube C and about 1 mL of erythrolysate is allowed to flow through filter and needle; this volume is discarded. The stopcock is closed and the needle is inserted into the cuvette with the tip at the bottom, as shown in Fig. 6.4B. After opening the stopcock, the pressure in the tonometer fills the cuvette.

PREPARATION OF HHb, O_2Hb AND COHb

Deoxyhaemoglobin, HHb, is prepared by tonometry of 25 mL of erythrocyte suspension with a gas mixture containing 95% N_2 and 5% CO_2. Tonometry and erythrolysis are carried out as described in the previous section. The HHb solution is transferred directly to a spectrophotometer cuvette.

In the preparation of HHb solutions, a minute amount of sodium dithionite ($Na_2S_2O_4$) may be added after all O_2 has been removed from the solution by tonometry with N_2/CO_2. This serves only to prevent any reoxygenation during further handling of the sample. Formerly, $Na_2S_2O_4$, which reacts with molecular O_2, was often used for the deoxygenation of O_2Hb solutions by adding an amount sufficient for binding all O_2. This, however, is accompanied by side reactions [82] which lead to a slight upward displacement of the HHb spectrum in the red/near infrared region. This shifts the isosbestic point of O_2Hb and HHb to a longer wavelength, 815 nm [7] (even 829 nm has been found [343]). After full deoxygenation by tonometry the isosbestic point is found at exactly 800 nm [272]. Even the very small amounts of dithionite used only to prevent reoxygenation have been seen to occasionally cause a slight shift in the HHb spectrum. Therefore, in the measurements underlying the results of Chapters 8–14, the two procedures, with and without dithionite, have been used side by side, but, generally, only the absorptivity values obtained with a trace of $Na_2S_2O_4$ have been reported.

When a trace of sodium dithionite is used to prevent reoxygenation, this is added to the sample as follows. Five mg of solid sodium dithionite is placed into the cylinder of a dry 10-mL glass syringe containing a mixing ring. The (metal) plunger is then replaced. The syringe is rinsed several times with gas from the tonometer, then connected to the sampling tube through outlet E (Fig. 6.2) and filled with 10 mL of erythrolysate as described in the previous section. Erythrolysate and dithionite are mixed by slowly turning the syringe several times. A cotton wool filter unit and a needle are placed on the syringe and a spectrophotometer cuvette is filled as described and shown in Fig. 6.4.

Oxyhaemoglobin, O_2Hb, is prepared by tonometry with a gas mixture containing 95% O_2 and 5% CO_2. Tonometry and erythrolysis are carried out as described in the previous section. The O_2Hb solution is transferred directly into a spectrophotometer cuvette.

Carboxyhaemoglobin, COHb, is prepared by tonometry in a chemical fume hood with a gas mixture containing 5% CO, 90% N_2 and 5% CO_2.

In addition to the wet cotton gauze mentioned above, the tonometer is also covered with aluminium film to exclude light. Tonometry, erythrolysis and filling of the cuvettes is identical to the procedure used for O_2Hb.

PREPARATION OF Hi, HiCN AND SHb

Haemiglobin, Hi, is prepared from O_2Hb. In a stoppered 25-mL Erlenmeyer flask, 10 mL of a O_2Hb solution ($ct_{Hb} \approx 80$ g/L) is mixed with 30 mg $K_3Fe(CN)_6$. After at least 15 min the pH is measured at room temperature and the solution is transferred to a 10-mL glass syringe with a metal plunger. The cuvette is filled from this syringe through a cotton wool filter as shown in Fig. 6.4. In measuring the absorptivities of Hi, it has been found that at high ct_{Hb} occasionally some precipitate is formed, even with generally well-soluble human haemoglobin. Therefore, ct_{Hb} was kept at about 80 g/L in the Hi solutions.

Haemiglobincyanide, HiCN, is prepared from O_2Hb. In a stoppered 25-mL Erlenmeyer flask, 10 mL of a O_2Hb solution ($ct_{Hb} \approx 120$ g/L) is mixed with 30 mg $K_3Fe(CN)_6$ and 6 mg KCN. After at least 30 min, the solution is transferred to a 10-mL glass syringe with a metal plunger. The cuvette is filled from the syringe as shown in Fig. 6.4.

Although it has been shown that the absorptivity of HiCN is independent of the concentration [481], a procedure resulting in a similar ct_{Hb} as used in the other haemoglobin derivatives has been employed for the determination of the absorptivity values given in Chapter 8.

Other addition products of Hi (Figs 2.1 and 2.2) can easily be prepared, although the differences in affinity between various ions should be taken into account. As there is little use for Hi derivatives other than HiCN, the spectra of these compounds have not been recorded in animal blood and no technical details as to the preparation of these compounds are given. However, a haemiglobin compound formed after addition of sodium laurylsulphate is described in Chapter 16.

The preparation of **Sulfhaemoglobin**, SHb, is hampered by the fact that 100% SHb solutions cannot be prepared. Moreover, ct_{Hb} of a haemoglobin solution containing SHb cannot be determined with the reference HiCN method, because SHiCN, formed on the addition of $K_3Fe(CN)_6$/KCN, has an absorption spectrum different from that of HiCN [100, 103, 210]. The following procedure is used. An erythrocyte suspension is fully oxygenated by tonometry with 95% O_2 and 5% CO_2. After centrifugation, the packed cells are divided into two equal parts. One part is incubated for 30 min with an equal volume of a freshly prepared solution of Na_2S and HCl (50 and 75 mmol/L, respectively; pH about 7.5) [357]; excess H_2S is removed by repeating the tonometry for 30 min. The erythrocyte suspension then contains O_2Hb and 15–25% SHb (suspension A).

The other part of the packed cells is mixed with an equal volume of O_2-saturated isotonic saline solution. This is suspension B, which contains only O_2Hb. Care is taken that ct_{Hb}, and thus the erythrocyte concentration, in the two suspensions, is exactly equal. Hence, measuring ct_{Hb} in suspension B yields ct_{Hb} for the two suspensions. Part of the two suspensions is used for determination of the SHb fraction of suspension A through measuring the oxygen binding capacity (β_{O_2}) of both suspensions by a titrimetric method [101, 104] (Chapter 19). The difference in β_{O_2} between the two suspensions yields the SHb fraction of suspension A [103]:

$$F_{SHb} = [\beta_{O_2}(\text{susp B}) - \beta_{O_2}(\text{susp A})]/\beta_{O_2}(\text{susp B}). \qquad (6.1)$$

Equal volumes of the two suspensions are then mixed with equal volumes of non-ionic detergent solution in glass syringes containing a mixing ball. The erythrolysate resulting from suspension B is used for determination of ct_{Hb}. The erythrolysate resulting from suspension A is transferred to a 0.1-cm cuvette through a cotton wool filter as shown in Fig. 6.4.

CHAPTER 7

DETERMINATION OF ABSORPTION SPECTRA

Absorptivity spectra of the various haemoglobin derivatives are determined by measuring the absorbance of clear solutions of single derivatives at various wavelengths, either one wavelength at a time with a conventional spectrophotometer, or a series of wavelengths simultaneously, with a diode array spectrophotometer. The measured absorbance is reduced to a lightpath length of 1.000 cm and to a concentration of 1.00 mmol/L. This requires knowledge of the lightpath length of the cuvette used in the measurements and of the total haemoglobin concentration, ct_{Hb}, of the solution. The latter is measured, after conversion of the haemoglobin derivative into HiCN, on the basis of the absorptivity of HiCN at 540 nm: $\varepsilon_{HiCN}(540) = 11.00$ $L \cdot mmol^{-1} \cdot cm^{-1}$. Consequently, $\varepsilon_{HiCN}(540)$ is the anchor value of all absorptivity values of all haemoglobin derivatives.

LIGHTPATH LENGTH

When measuring the absorbance of the solutions of single haemoglobin derivatives, acceptable accuracy can only be obtained when the absorbance is neither too high nor too low. It has been calculated that the random error in the absorbance measured with a single beam spectrophotometer is smallest when $A = 0.434$, and that above 0.700 and below 0.090 the imprecision increases steeply [7, 211]. With first generation conventional spectrophotometers, *e.g.* the Beckman DU (p. 4), it was mandatory to keep the absorbance measurements within these limits. The newer types of spectrophotometer are, however, capable of measuring at a considerably higher absorbance level, but to obtain accurate results it remains necessary to avoid absorbance values below 0.050.

Since the absorptivity of several haemoglobin derivatives varies over the visual and near-infrared range with a factor > 100, and since measurements are preferably made with $ct_{Hb} = 80-120$ g/L, it becomes necessary to adjust the length of the lightpath to keep the absorbance measurements within an acceptable range. Therefore, plan parallel glass cuvettes with $l = 0.10-1.00$ cm are used. To obtain $l < 0.10$ cm, plan parallel glass plates are inserted into 0.100-cm cuvettes, leaving an effective lightpath length of 0.007–0.016 cm (Fig. 6.4).

ABSORBANCE MEASUREMENT

Ideally, absorbance spectra should be measured continuously over the entire visual and near-infrared spectral range by means of a single instrument with a high spectral resolution, *i.e.* a spectral bandwidth of < 0.25 nm. In practice, this can hardly be realised. While the same incandescent lamp can be used over the spectral range above 350 nm, a different detector is necessary for the red/infrared region than for the shorter wavelengths. This requires at minimum a detector switch near 600 nm within the same instrument.

Of more importance is the fact that with conventional spectrophotometers the most accurate results are obtained when the absorbance at each wavelength is read from the potentiometer scale. This can be done with high spectral resolution — steps of only 0.1 nm are possible — but it is so laborious that this procedure is usually applied only to the most interesting parts of the spectrum in question. Diode array spectrophotometers provide an adequate solution of this problem in that they are able to measure complete spectra within a few seconds. A printed list of absorbances is easily obtained in addition to the plotted spectral absorbance curve. The spectral resolution, however, is limited. In the HP8450A used for most of the results shown in Chapters 8–14, the spacing of the diodes is such that the absorbance measurements are made in steps of 2 nm. Hence, the position of the α-peak of O_2Hb near 577 nm cannot be accurately determined, nor the absorbance at that wavelength accurately measured with this instrument.

Consequently, for the reliable localisation of absorption maxima and minima, a conventional spectrophotometer such as the Optica CF4 is the instrument of choice. The wavelength scale can be checked with the help of mercury or hydrogen emission lines, *e.g.* the mercury line at 546.1 nm, the hydrogen lines at 656.3 and 486.1 nm. To check the absorbance scale, a stable glass filter can be used, such as the Corning HT yellow filter (p. 23). This filter was used to check the spectrophotometers with which the measurements reported in Chapters 8–14 were made, an Optica CF4 grating spectrophotometer and an HP8450A diode array instrument.

MEASUREMENT OF TOTAL HAEMOGLOBIN

It is necessary to accurately determine ct_{Hb} in all samples of haemoglobin derivatives of which the absorption spectrum is measured. Therefore, part of the erythrolysate to be measured is transferred, through a cotton wool filter, from the tonometer or the syringe containing the erythrolysate into an

25-mL Erlenmeyer flask. With a calibrated Ostwald pipette, 0.500 mL of the haemoglobin solution is then transferred slowly to a 100 mL volumetric flask and made up to volume with the HiCN reagent solution as described on p. 32. After at least 30 min, in the case of COHb after at least 2 h, the absorbance is measured at 540 nm with a lightpath length of 1.000 cm against water as blank and ct_{Hb} calculated on the basis of $\varepsilon_{HiCN}(540) = 11.00\ L \cdot mmol^{-1} \cdot cm^{-1}$.

MEASUREMENT OF FOETAL HAEMOGLOBIN

The method of Jonxis and Visser [203] proved to be easy and sufficiently accurate for measuring the fraction of foetal haemoglobin in a specimen of human blood. The method is based on the difference in resistance to denaturation by alkali between foetal and adult haemoglobin. The spectral change that occurs when haemoglobin from neonatal erythrocytes is brought to pH > 11 is observed during 15 min. This change is solely the result of the denaturation of foetal haemoglobin, since the denaturation of the adult haemoglobin has already reached completion. The spectral change is measured at 576 nm.

Mix 0.1 mL heparin-anticoagulated blood, 10 mL of deionised water and 2 drops of 10% ammonia in a 20-mL cylinder glass (solution 1). $A(576)$ of solution 1 is measured with a lightpath length of 1.000 cm. This yields A_1. In a similar cylinder, mix 0.1 mL blood, 10 mL 0.06 mmol/L NaOH and two

Table 7.1.
Measurement results of a determination of foetal haemoglobin by alkali denaturation

Time (min)	A_2	R	log R
2	1.188	0.802	−0.096
3	1.161	0.771	−0.113
4	1.118	0.721	−0.142
5	1.090	0.689	−0.162
6	1.057	0.650	−0.187
7	1.027	0.616	−0.210
8	0.9985	0.583	−0.234
9	0.9724	0.552	−0.258
10	0.9496	0.526	−0.279
11	0.9261	0.499	−0.302
12	0.9048	0.474	−0.324
13	0.8826	0.448	−0.349
14	0.8643	0.427	−0.370
15	0.8467	0.407	−0.390
16	0.8296	0.387	−0.412

drops of 10% ammonia (solution 2). Time measurement starts at the moment of mixing and $A(576)$ of solution 2 is measured at 1-min intervals (A_2) for 15 min. The remaining part of solution 2 is then heated in a stoppered tube to 37°C for at least 15 min. After cooling $A(576)$ is measured. This yields A_3.

For each moment, the ratio $R = (A_2 - A_3)/(A_1 - A_3)$ is calculated and the log of the ratio plotted against time and extrapolated to time zero. R at time zero gives the fraction of foetal haemoglobin in the blood.

The procedure is illustrated by the following example. The absorbances at 576 nm were measured with a lightpath length of 1.000 cm by means of an HP8450A. The values obtained for A_2 at 1-min intervals are given in Table 7.1; for A_1 and A_3, 1.359 and 0.496, respectively, were found. The table also shows the corresponding values of R and $\log R$.

Linear regression of $\log R$ on time gives a y-intercept of -0.0494. Hence, $R = 10^{-0.0494} = 0.892$, which is the fraction of foetal haemoglobin in the blood.

ABSORPTIVITY CALCULATION

The absorptivity is calculated by dividing the absorbance measured at each wavelength by the total haemoglobin concentration in the sample and the lightpath length:

$$\varepsilon_X(\lambda) = A(\lambda)/(ct_{Hb} \cdot l), \tag{7.1}$$

where X may designate any haemoglobin derivative; ct_{Hb} is expressed in mmol/L, l in cm, thus $\varepsilon_{HiCN}(540)$ is obtained in $L \cdot mmol^{-1} \cdot cm^{-1}$.

For Hi, this calculation yields correct values of $\varepsilon_{Hi}(\lambda)$, because the Hi solutions are regarded to contain nc other haemoglobin derivative. For HHb, O_2Hb and COHb, however, the values obtained for $\varepsilon_X(\lambda)$ must be corrected for any contaminating Hi. Therefore, the fractions (F) of Hi in the solutions of the other haemoglobin derivatives must be measured. To this end, the cyanide addition method of Evelyn and Malloy [114] can be used. In the measurements underlying the data of Chapters 8–14, either a modification of this method [103] has been used or a multiwavelength haemoglobin photometer (Radiometer OSM3), which has been shown to give reliable results [482]. The absorptivity of haemoglobin derivative X is then calculated with the following equation:

$$\varepsilon_X(\lambda) = [\varepsilon(\lambda) - F_{Hi} \cdot \varepsilon_{Hi}(\lambda)]/(1 - F_{Hi}), \tag{7.2}$$

where $\varepsilon_X(\lambda)$ is the corrected absorptivity of HHb, O_2Hb or COHb at wavelength λ, $\varepsilon_{Hi}(\lambda)$ is the absorptivity of Hi at this wavelength, and $\varepsilon(\lambda)$ is the absorptivity calculated from the absorbance $A(\lambda)$ actually measured for HHb, O_2Hb or COHb.

Absorptivities determined in this manner for haemoglobin solutions from adult blood specimens were assumed to be those of pure HbA. For haemoglobin solutions prepared from umbilical cord blood of healthy human neonates, the measured absorptivities were assumed to be those of a mixture of HbA and HbF. A procedure similar to the one used for eliminating the effect of the possible presence of Hi in the samples of HHb, O_2Hb or COHb, was used to calculate the absorptivity values of pure HbF. The fraction (F) of HbF in the samples was determined by the alkali denaturation method as described. The absorptivities of the various derivatives of HbF were then calculated with the aid of the following equation:

$$\varepsilon_{HbF}(\lambda) = [\varepsilon_{HbA/HbF}(\lambda) - (1 - F_{HbF}) \cdot \varepsilon_{HbA}(\lambda)]/F_{HbF}, \qquad (7.3)$$

where $\varepsilon_{HbF}(\lambda)$ is the absorptivity of a derivative of HbF at wavelength λ, $\varepsilon_{HbA/HbF}(\lambda)$ is the absorptivity of the corresponding derivative as measured for the HbA/HbF mixture, and $\varepsilon_{HbA}(\lambda)$ is the absorptivity of the corresponding derivative of HbA.

The determination of the absorptivity of SHb is complicated by the fact that no pure SHb solutions can be prepared. When a mixture of O_2Hb and SHb is prepared, and ct_{Hb} and F_{SHb} have been measured as described in Chapter 6, $\varepsilon_{SHb}(\lambda)$ can be calculated from the absorbance of the erythrolysate of suspension A (p. 44) with an equation analogous to equation 7.3:

$$\varepsilon_{SHb}(\lambda) = [\varepsilon_{SHb/O_2Hb}(\lambda) - (1 - F_{SHb}) \cdot \varepsilon_{O_2Hb}(\lambda)]/F_{SHb}, \qquad (7.4)$$

where $\varepsilon_{SHb/O_2Hb}(\lambda)$ is the absorptivity measured at wavelength λ, for the erythrolysate of suspension A. For $\varepsilon_{O_2Hb}(\lambda)$ values from Tables 8.1 and 8.2 can be used or the values measured for the erythrolysate of suspension B.

INFLUENCE OF TEMPERATURE AND pH

Some influence of the temperature on the spectrophotometric determination of the oxygen saturation has been reported by Refsum [317]. Consequently, some investigators have made all measurements of haemoglobin spectra at a constant temperature, using thermostatted cuvettes [319, 357].

A thorough investigation of the influence of temperature on the absorption spectra of O_2Hb and HHb and of the underlying physico-chemical changes in the haemoglobin molecule has been published by Cordone *et al.* [78], who studied the absorption spectra of these haemoglobin derivatives in the temperature range of 300–20 K and the wavelength range of 350–1350 nm. The absorption peaks of O_2Hb in the visible range shifted upon cooling to a shorter wavelength and the absorbance increased; for HHb the observed change was mainly an increase in absorbance with decreasing temperature.

Useful quantitative data for O_2Hb, HHb and COHb in the wavelength range of 478 to 651 nm have been published by Steinke and Shepherd [372]. It may be concluded from these data that the absorption spectra of haemoglobin derivatives should preferably be determined, and measurements of oxygen saturation and dyshaemoglobin fractions preferably be made at constant temperature.

However, the absorptivity values presented in Chapters 8–14 have, for practical purposes, been measured at room temperature with instruments not equipped with a thermostatted cuvette compartment. The temperature in the laboratory was, on days that measurements were made, between 20 and 24°C. As all presented absorptivity values are the mean of the results obtained in several specimens, processed in the course of more than a year, these values can rightly be considered to be valid for 22°C.

This approach was deemed justified because the influence of temperature on the absorption spectra of most haemoglobin derivatives is but slight. As shown by the data of Siggaard-Andersen *et al.* [357] and of Steinke and Shepherd [372], the HHb spectrum is hardly dependent on temperature in the temperature range concerned. The temperature effect is most clearly present in the α and β peaks of O_2Hb and COHb. Table 1 of Steinke and Shepherd [372] describes, as the greatest difference, that of $\varepsilon_{O_2Hb}(540.5)$ of human haemoglobin: 14.485 and 14.168 $L \cdot mmol^{-1} \cdot cm^{-1}$ for 20 and 30°C, respectively. For 22°C linear interpolation gives 14.42 $L \cdot mmol^{-1} \cdot cm^{-1}$, with a range of 14.49 to 14.36 $L \cdot mmol^{-1} \cdot cm^{-1}$ when the temperature range is 20–24°C.

The influence of pH on the absorption spectra of haemoglobin derivatives is most striking in the case of Hi, because of the binding of OH^- to the ferric iron (Chapter 2). Therefore, the influence of pH on the Hi spectra of haemoglobin of the different species has been studied. Differences between the Hi spectra of different animals may occur through species-specific differences in the composition and/or structure of the haemoglobin molecule, and through differences in affinity of Hi for OH^-. If the only difference is a difference in OH^- affinity, a pair of pH values can be found at which the two Hi spectra coincide.

For accurate multicomponent analysis of haemoglobin derivatives, it is necessary that for Hi absorptivity values are used which are valid at the pH prevailing in the measured solution. Therefore, all data for Hi in the tables of Chapters 8–14 are related to the pH range which can be expected to ensue from the lysis of a fresh blood sample of the animal in question.

The influence of pH on the absorption spectra of the other haemoglobin derivatives is very small. In HHb there is no discernible effect; in O_2Hb a slight increase in absorbance has been shown to occur at the α- and β-peak [415]. Helledie and Rolfe [155] who studied the influence of pH at some wavelengths in the near-infrared found no effect on the absorbance of haemoglobin, but an appreciable effect on light absorption by erythrocyte suspensions. Wimberley *et al.* have reported a minute effect of pH on the measurement of F_{COHb} [416]. In the data of Chapters 8–14 no pH effects have been taken into account for haemoglobin derivatives other than Hi.

CHAPTER 8

ABSORPTION SPECTRA OF HUMAN HbA AND HbF

The absorption spectra of human HbA issued earlier from our laboratory [7, 210] have been determined over the spectral range of 390–1000 nm, thus including the Soret region where, between 390 and 440 nm, most haemoglobin derivatives have characteristic absorption peaks with absorptivity values an order of magnitude greater than those of the peaks in the visible range. Since the use for biomedical application of this part of the haemoglobin absorption spectra proved to be limited, it was decided, for practical reasons, to confine the following more extensive investigations to the spectral range of 450–1000 nm. This obviated an additional series of measurements of each haemoglobin derivative, for which dilution of the samples would have been necessary.

Two different procedures have been used for the determination of the absorption spectra of human haemoglobin. For most of the data the methods described in Chapters 6 and 7 have been used. This approach is called **standard procedure**. The absorption spectra of adult human haemoglobin have also been measured after first preparing a stroma free haemoglobin (SFH) solution as described in Chapter 5. From the SFH solution the various haemoglobin derivatives were prepared as detailed in Chapter 6, with the exception that no detergent solution was introduced into the tonometer. This approach is called **SFH procedure**. The major differences between the two procedures thus are that in the SFH procedure the haemoglobin solution is cleared at an early stage of erythrocyte remnants, and that tonometry is done with a haemoglobin solution instead of an erythrocyte suspension. Hence, more purified haemoglobin solutions are obtained, but there is an increased risk of some deterioration of some haemoglobin (Chapter 6). The SFH procedure proved to be unsuited to be used in the preparation of samples of haemoglobin derivatives of HbF.

ABSORPTIVITY OF HbA: STANDARD PROCEDURE

In applying the standard procedure for the determination of the absorptivities of HHb of human HbA, two measurements were made of each specimen, one with and one without the addition of a trace of $Na_2S_2O_4$, as described on p. 43. The results presented in Tables 8.1–8.4 and 8.9–8.11, however, are all from measurements in which $Na_2S_2O_4$ was used.

Table 8.1 shows the absorptivities of HHb, O_2Hb, COHb, Hi, HiCN and SHb of human HbA for the spectral range of 450–630 nm measured with an HP8450A in 0.100-cm cuvettes with 0.084 to 0.090-cm plan parallel glass inserts. To obtain reliable results, the measured absorbance should not be less than 0.050 (p. 47). No absorptivity value has been given when the measured absorbance was < 0.050. Table 8.2 shows the absorptivities of the same haemoglobin derivatives for the spectral range of 600–800 nm measured with the same spectrophotometer. The lightpath was adjusted so that the measured absorbance was greater than 0.050. No absorptivity value has been given when the spectrophotometer rejected the measurement because the absorbance of the sample was too high. The latter occasionally happened in measurements with $l = 0.200$ cm between 600 and 650 nm. The spectra represented by the absorptivities of Tables 8.1 and 8.2 are designated as the **standard spectra** of adult human haemoglobin.

Tables 8.1 and 8.2 show the mean values of the absorptivity of the numbers of specimens indicated at the head of the tables. The precision of the absorptivity measurements is shown for selected wavelengths in Table 8.3. It appears that by means of the standard procedure the absorptivities of the haemoglobin derivatives can be determined with a sufficiently high reproducibility to detect quite small differences in the absorption spectra of haemoglobin of different animal species in comparison with those of human haemoglobin. In fact, at the β-peak of O_2Hb and at the minimum of O_2Hb at 508.5 nm the coefficient of variation is only 0.4%. However, O_2Hb is easy to prepare; for the other haemoglobin derivatives the variation in the results is usually somewhat larger, even when the standard procedure is used for preparation of the samples. The coefficient of variation at the β-peak of COHb, for instance, is 0.8%. When the measured absorbance is prevented from becoming too low by adequately adjusting the lightpath (p. 47), the absolute error is approximately constant but the relative error increases considerably with decreasing absorptivity. At the maximum of Hi at 630 nm, where $\varepsilon_{Hi} = 3.895\ L \cdot mmol^{-1} \cdot cm^{-1}$, the coefficient of variation is 2.1%.

Figure 8.1 shows a composite picture of the absorption spectra of HHb, O_2Hb, COHb, Hi, HiCN and SHb of human HbA from 450 to 750 nm corresponding with the absorptivities of Table 8.1. The spectra have been recorded in a continuous run over the indicated spectral range with a constant small lightpath. For some haemoglobin derivatives the measured absorbance between 600 and 750 nm thus was < 0.050. These values are not given in Tables 8.1 and 8.2.

Haemiglobin is the only haemoglobin derivative of which the absorption is appreciably dependent on the pH of the solution (p. 52). The effect is shown

Table 8.1.
Absorptivities ($L \cdot mmol^{-1} \cdot cm^{-1}$) of various derivatives of adult human haemoglobin

λ (nm)	ε_{HHb} $N = 6$	ε_{O_2Hb} $N = 6$	ε_{COHb} $N = 8$	ε_{Hi} $N = 15$	ε_{HiCN} $N = 11$	ε_{SHb} $N = 6$
450	13.48	16.20	10.07	9.799	20.78	13.23
452	10.58	14.90	9.379	9.422	19.41	11.96
454	8.520	13.80	8.809	9.127	18.09	10.90
456	7.022	12.81	8.311	8.874	16.79	10.08
458	5.932	11.96	7.897	8.676	15.58	9.37
460	5.131	11.17	7.529	8.495	14.45	8.80
462	4.552	10.50	7.217	8.351	13.47	8.27
464	4.120	9.875	6.933	8.211	12.60	7.87
466	3.813	9.333	6.690	8.108	11.89	7.51
468	3.602	8.834	6.469	8.008	11.28	7.22
470	3.457	8.399	6.275	7.954	10.79	6.93
472	3.366	7.993	6.097	7.907	10.38	6.72
474	3.318	7.637	5.952	7.912	10.06	6.54
476	3.297	7.303	5.816	7.934	9.760	6.41
478	3.302	7.002	5.710	7.999	9.511	6.28
480	3.322	6.722	5.615	8.074	9.256	6.20
482	3.353	6.482	5.542	8.193	9.015	6.15
484	3.405	6.256	5.466	8.304	8.746	6.12
486	3.465	6.073	5.403	8.447	8.481	6.12
488	3.538	5.888	5.339	8.569	8.197	6.15
490	3.633	5.741	5.292	8.716	7.929	6.17
492	3.736	5.595	5.252	8.828	7.669	6.22
494	3.861	5.473	5.235	8.947	7.451	6.28
496	3.998	5.353	5.225	9.018	7.253	6.35
498	4.158	5.254	5.242	9.088	7.113	6.43
500	4.330	5.154	5.279	9.106	7.008	6.50
502	4.525	5.070	5.345	9.098	6.962	6.59
504	4.727	4.984	5.429	9.031	6.954	6.66
506	4.948	4.927	5.562	8.949	7.011	6.74
508	5.160	4.882	5.723	8.813	7.104	6.78
510	5.383	4.878	5.940	8.674	7.269	6.82
512	5.593	4.921	6.225	8.492	7.467	6.83
514	5.816	5.042	6.596	8.317	7.730	6.81
516	6.025	5.235	7.056	8.119	8.025	6.76
518	6.249	5.548	7.634	7.929	8.373	6.68
520	6.478	5.981	8.297	7.732	8.733	6.56
522	6.737	6.579	9.050	7.549	9.125	6.42
524	7.011	7.320	9.850	7.366	9.505	6.28
526	7.317	8.230	10.69	7.214	9.880	6.11

Table 8.1.
(Continued)

λ (nm)	ε_{HHb} $N = 6$	ε_{O_2Hb} $N = 6$	ε_{COHb} $N = 8$	ε_{Hi} $N = 15$	ε_{HiCN} $N = 11$	ε_{SHb} $N = 6$
528	7.644	9.250	11.50	7.067	10.20	5.95
530	8.010	10.35	12.29	6.944	10.48	5.74
532	8.411	11.42	12.99	6.826	10.69	5.65
534	8.866	12.40	13.61	6.728	10.84	5.56
536	9.355	13.22	14.05	6.611	10.94	5.54
538	9.901	13.90	14.30	6.496	11.02	5.54
540	10.47	14.32	14.27	6.344	11.05	5.62
542	11.06	14.52	14.04	6.170	11.08	5.69
544	11.62	14.34	13.59	5.948	11.06	5.90
546	12.15	13.83	13.08	5.709	11.03	6.09
548	12.59	12.99	12.55	5.437	10.94	6.31
550	12.96	12.01	12.11	5.161	10.82	6.57
552	13.21	11.02	11.78	4.895	10.63	6.74
554	13.35	10.17	11.63	4.659	10.41	6.90
556	13.36	9.482	11.65	4.457	10.14	6.98
558	13.29	9.015	11.87	4.297	9.854	7.02
560	13.11	8.767	12.25	4.176	9.544	7.03
562	12.88	8.769	12.80	4.094	9.243	7.00
564	12.59	9.042	13.43	4.043	8.931	6.95
566	12.25	9.614	14.04	4.019	8.629	6.84
568	11.85	10.50	14.43	4.009	8.323	6.69
570	11.42	11.68	14.46	4.010	8.013	6.43
572	10.96	13.05	13.99	4.010	7.692	6.10
574	10.49	14.37	13.07	4.011	7.355	5.80
576	10.02	15.26	11.76	4.004	7.005	5.62
578	9.564	15.36	10.24	3.986	6.638	5.68
580	9.141	14.42	8.638	3.954	6.266	5.95
582	8.734	12.60	7.089	3.900	5.889	6.41
584	8.333	10.28	5.710	3.829	5.517	6.93
586	7.903	7.938	4.550	3.744	5.156	7.33
588	7.422	5.887	3.618	3.652	4.803	7.59
590	6.875	4.262	2.891	3.557	4.456	7.94
592	6.271	3.062	2.330	3.468	4.116	8.12
594	5.629	2.216	1.899	3.386	3.780	8.37
596	4.976	1.634	1.567	3.319	3.451	8.65
598	4.342	1.234	1.307	3.264	3.132	8.97
600	3.759	0.961	1.104	3.226	2.828	9.37
602	3.242	0.767	0.939	3.200	2.539	9.88
604	2.806	0.625	0.806	3.187	2.271	10.54
606	2.452	0.519	0.697	3.185	2.027	11.41
608	2.169		0.607	3.194	1.807	12.49
610	1.946		0.530	3.215	1.611	13.81

Table 8.1.
(Continued)

λ (nm)	ε_{HHb} $N=6$	ε_{O_2Hb} $N=6$	ε_{COHb} $N=8$	ε_{Hi} $N=15$	ε_{HiCN} $N=11$	ε_{SHb} $N=6$
612	1.768		0.466	3.250	1.439	15.29
614	1.627			3.302	1.289	16.92
616	1.514			3.375	1.164	18.52
618	1.418			3.458	1.053	19.94
620	1.336			3.549	0.953	20.91
622	1.270			3.643	0.860	21.21
624	1.212			3.730	0.783	20.73
626	1.163			3.806	0.717	19.47
628	1.120			3.862	0.658	17.54
630	1.082			3.895	0.606	15.28

λ = 450–630 nm. Standard procedure; HP8450A.

Absorbance measured with l = 0.010–0.016 cm. N is number of specimens. $Na_2S_2O_4$ used to prevent reoxygenation of HHb. pH range (Hi) = 7.14–7.30.

Table 8.2.
Absorptivities ($L \cdot mmol^{-1} \cdot cm^{-1}$) of various derivatives of adult human haemoglobin

λ (nm)	ε_{HHb} $N=6$	ε_{O_2Hb} $N=6$	ε_{COHb} $N=4$	ε_{Hi} $N=7$	ε_{HiCN} $N=11$	ε_{SHb} $N=6$
600		0.957		3.227	2.828	9.37
602	2.63	0.775		3.204	2.539	9.88
604	2.374	0.641		3.191	2.271	10.54
606	2.123	0.539		3.191	2.027	11.41
608	1.921	0.460	0.550	3.199	1.807	12.49
610	1.753	0.397	0.507	3.223	1.611	13.81
612	1.618	0.346	0.463	3.257	1.439	15.29
614	1.508	0.305	0.422	3.313	1.289	16.92
616	1.416	0.271	0.384	3.379	1.164	18.52
618	1.338	0.242	0.348	3.466	1.053	19.94
620	1.272	0.218	0.316	3.557	0.953	20.91
622	1.215	0.197	0.287	3.650	0.860	21.21
624	1.166	0.180	0.261	3.746	0.783	20.73
626	1.123	0.164	0.238	3.815	0.717	19.47
628	1.085	0.150	0.216	3.875	0.658	17.54
630	1.051	0.139	0.196	3.905	0.606	15.28
632	1.022	0.129	0.179	3.900	0.561	12.94
634	0.996	0.120	0.163	3.855	0.520	10.73
636	0.974	0.114	0.148	3.767	0.485	8.80
638	0.953	0.108	0.136	3.620	0.454	7.26
640	0.936	0.104	0.125	3.416		6.02

Table 8.2.
(Continued)

λ (nm)	ε_{HHb} $N = 6$	ε_{O_2Hb} $N = 6$	ε_{COHb} $N = 4$	ε_{Hi} $N = 7$	ε_{HiCN} $N = 11$	ε_{SHb} $N = 6$
642	0.920	0.101	0.116	3.157		5.06
644	0.907	0.099	0.108	2.861		4.29
646	0.895	0.099	0.102	2.541		3.71
648	0.884	0.098	0.096	2.239		3.25
650	0.873	0.098	0.091	1.971		2.89
652	0.864	0.099	0.086	1.704		2.59
654	0.852	0.099	0.082	1.468		2.36
656	0.836	0.098	0.077	1.227		2.17
658	0.828	0.098	0.071	1.083		2.02
660	0.815	0.100	0.069	0.945		1.88
662	0.799	0.101	0.066	0.808		1.76
664	0.782	0.101	0.062	0.690		1.66
666	0.763	0.101	0.059	0.596		1.57
668	0.743	0.102	0.056	0.523		1.50
670	0.721	0.102	0.053	0.459		1.43
672	0.699	0.102	0.050	0.409		1.37
674	0.677	0.103	0.047	0.370		1.32
676	0.655	0.103	0.045	0.339		1.28
678	0.633	0.103	0.043	0.313		1.23
680	0.612	0.103	0.041	0.292		1.19
682	0.591	0.103	0.039	0.277		1.16
684	0.571	0.103	0.037	0.264		1.12
686	0.551	0.103	0.036	0.253		1.09
688	0.533	0.103	0.034	0.245		1.07
690	0.516	0.103	0.033	0.238		1.04
692	0.500	0.103	0.032	0.233		1.02
694	0.485	0.103	0.031	0.229		1.00
696	0.471	0.104	0.030	0.226		0.99
698	0.459	0.104	0.029	0.223		0.97
700	0.447	0.105	0.028	0.222		0.96
702	0.436	0.106	0.027	0.221		0.95
704	0.426	0.106	0.027	0.219		0.94
706	0.416	0.108	0.026	0.219		0.93
708	0.406	0.109	0.025	0.219		0.93
710	0.307	0.110	0.025	0.218		0.92
712	0.387	0.112	0.025	0.219		0.91
714	0.379	0.113	0.024	0.220		0.90
716	0.370	0.115	0.024	0.221		0.90
718	0.361	0.116	0.024	0.221		0.89
720	0.353	0.118	0.023	0.222		0.88
722	0.346	0.120	0.023	0.224		0.88
724	0.340	0.122	0.023	0.225		0.87
726	0.334	0.123	0.022	0.227		0.86

Table 8.2.
(Continued)

λ (nm)	ε_{HHb} $N=6$	ε_{O_2Hb} $N=6$	ε_{COHb} $N=4$	ε_{Hi} $N=7$	ε_{HiCN} $N=11$	ε_{SHb} $N=6$
728	0.330	0.125	0.022	0.229		0.85
730	0.327	0.127	0.022	0.231		0.84
732	0.326	0.129	0.022	0.233		0.83
734	0.327	0.131	0.021	0.235		0.82
736	0.329	0.133	0.021	0.238		0.81
738	0.333	0.135	0.021	0.241		0.81
740	0.339	0.137	0.021	0.245		0.80
742	0.346	0.139	0.020	0.248		0.79
744	0.355	0.141	0.020	0.251		0.79
746	0.366	0.143	0.020	0.254		0.78
748	0.377	0.145	0.020	0.258		0.77
750	0.388	0.148	0.019	0.261		0.76
752	0.399	0.150	0.019	0.265		0.75
754	0.409	0.152	0.019	0.269		0.74
756	0.415	0.154	0.019	0.274		0.73
758	0.417	0.157	0.019	0.278		0.72
760	0.414	0.159	0.018	0.283		0.71
762	0.407	0.161	0.018	0.287		0.69
764	0.396	0.163	0.018	0.292		0.68
766	0.381	0.166	0.018	0.298		0.67
768	0.364	0.168	0.018	0.302		0.66
770	0.347	0.171	0.018	0.309		0.65
772	0.330	0.173	0.018	0.314		0.64
774	0.314	0.175	0.017	0.319		0.62
776	0.299	0.178	0.017	0.325		0.60
778	0.285	0.180	0.017	0.331		0.59
780	0.273	0.183	0.017	0.338		0.57
782	0.262	0.185	0.017	0.345		0.56
784	0.253	0.188	0.017	0.351		0.54
786	0.245	0.190	0.017	0.357		0.52
788	0.238	0.193	0.016	0.364		0.50
790	0.233	0.196	0.016	0.371		0.49
792	0.228	0.198	0.016	0.377		0.47
794	0.223	0.201	0.016	0.385		0.45
796	0.219	0.203	0.016	0.392		0.43
798	0.216	0.206	0.016	0.398		0.41
800	0.215	0.208	0.015	0.406		0.40

λ = 600–800 nm. Standard procedure; HP8450A.

Absorbance measured with l = 0.010–0.016 cm (HiCN and SHb), l = 0.100 cm (Hi), l = 0.200 cm (HHb and O_2Hb), l = 1.000 cm (COHb). N is number of specimens. $Na_2S_2O_4$ used to prevent reoxygenation of HHb. pH range (Hi) = 7.14–7.30.

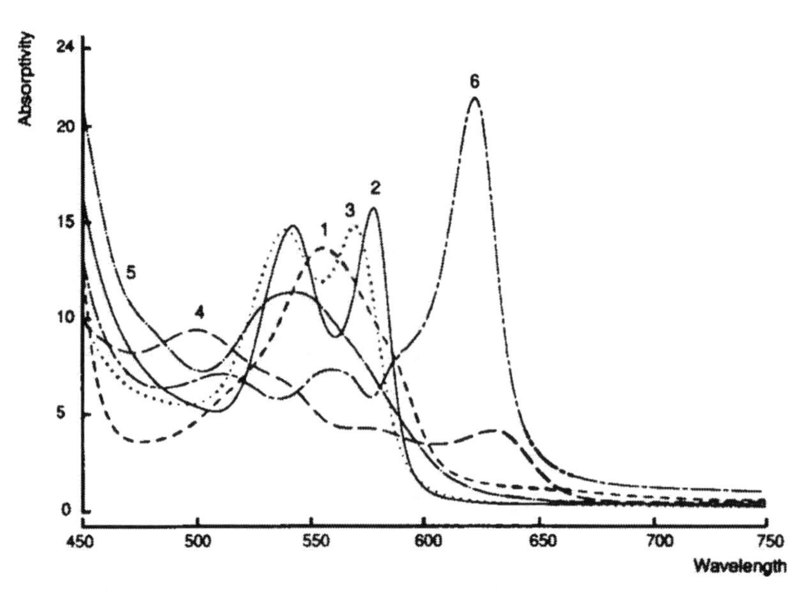

Figure 8.1. Absorption spectra of HHb (1), O_2Hb (2), COHb (3), Hi (4), HiCN (5) and SHb (6) of adult human haemoglobin.

Absorptivity in $L \cdot mmol^{-1} \cdot cm^{-1}$, wavelength in nm. l = 0.010–0.016 cm. HP8450A. pH (Hi) ≈ 7.21.

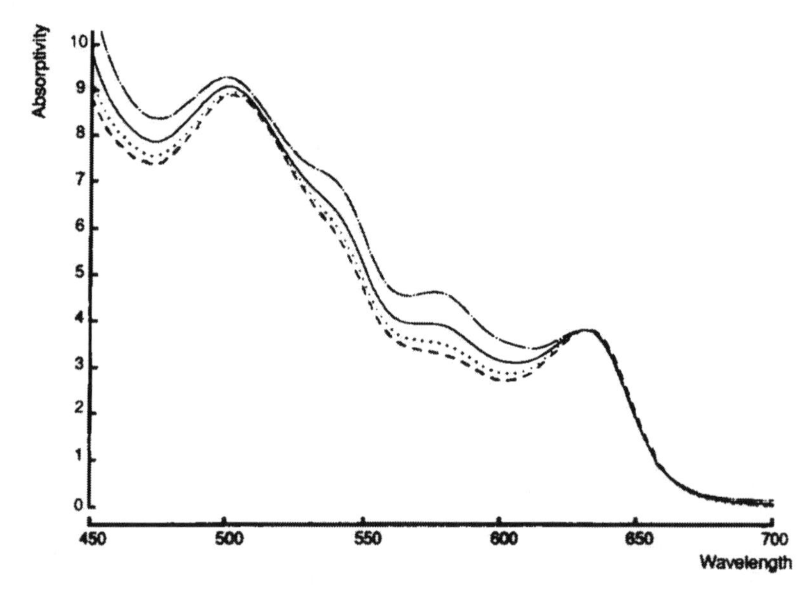

Figure 8.2. Absorption spectra of Hi of adult human haemoglobin at various pH values.

Absorptivity in $L \cdot mmol^{-1} \cdot cm^{-1}$, wavelength in nm. The solid line represents the standard Hi spectrum (pH ≈ 7.21). The three other spectra of human Hi are obtained at pH = 7.54 (·——·), pH = 7.06 (· · ·), and pH = 6.80 (- - - -). For easy comparison the absorptivity at 630 nm has been normalised at 3.895 $L \cdot mmol^{-1} \cdot cm^{-1}$, the absorptivity at 630 nm of the standard Hi spectrum. HP8450A.

Table 8.3.
Mean and standard error of absorptivities ($L \cdot mmol^{-1} \cdot cm^{-1}$) of various derivatives of adult human haemoglobin at selected wavelengths

λ (nm)	ε_{HHb} $N = 6$	*SEM*	λ (nm)	ε_{SHb} $N = 6$	*SEM*
476	3.298	0.013	620	20.91	0.034
478	3.302	0.012	622	21.21	0.078
552	13.21	0.044	624	20.73	0.128
554	13.35	0.045			
556	13.36	0.045			
λ (nm)	ε_{O_2Hb} $N = 6$	*SEM*	λ (nm)	ε_{Hi} $N = 15$	*SEM*
508	4.882	0.007	498	9.088	0.028
540	14.32	0.027	500	9.106	0.029
542	14.52	0.023	502	9.098	0.028
544	14.34	0.022	630	3.895	0.021
560	8.767	0.015	632	3.896	0.021
576	15.26	0.034	634	3.857	0.020
578	15.36	0.027			
λ (nm)	ε_{COHb} $N = 8$	*SEM*	λ (nm)	ε_{HiCN} $N = 11$	*SEM*
496	5.225	0.017	502	6.962	0.016
538	14.30	0.039	504	6.945	0.015
540	14.27	0.036	506	7.011	0.015
542	14.04	0.036	540	11.05	0.020
554	11.63	0.029	542	11.08	0.017
568	14.43	0.037	544	11.06	0.019
570	14.46	0.039			

Standard procedure; HP8450A.
SEM is standard error of the mean; *N* is number of specimens.

in Fig. 8.2. When fresh human whole blood is haemolysed, a solution with a pH of about 7.21 is obtained. Therefore, the mean of the Hi spectra of 15 specimens of human haemoglobin measured at pH values between 7.14 and 7.30 has been chosen as the standard Hi spectrum. This range of pH values appears to be near the acid end of the titration curve: increasing the pH has a stronger effect on the absorption spectrum than a further decrease in pH.

Table 8.4 presents the absorptivities of the common derivatives (HHb, O_2Hb, COHb and Hi) of HbA for selected wavelengths in the spectral range of 600–1000 nm, measured with an Optica CF4 grating spectrophotometer. Wavelength selection was necessary to complete this series of manual measurements within a reasonable time. Wavelengths of primary importance for biomedical application have been chosen preferentially.

Table 8.4.
Absorptivities ($L \cdot mmol^{-1} \cdot cm^{-1}$) of the common derivatives of adult human haemoglobin at selected wavelengths between 600 and 1000 nm

λ (nm)	ε_{HHb} $N = 6$	ε_{O_2Hb} $N = 6$	ε_{COHb} $N = 6$	ε_{Hi} $N = 7$
600		0.879	1.050	3.158
630	1.058	0.115	0.187	3.914
660	0.814	0.080	0.061	0.819
680	0.610	0.085	0.033	0.258
700	0.437	0.088	0.021	0.204
750	0.387	0.136	0.013	0.251
775	0.298	0.169	0.011	0.317
800	0.206	0.200	0.010	0.399
805	0.204	0.207	0.009	0.415
840	0.190	0.248	0.007	0.521
845	0.192	0.253	0.006	0.533
880	0.200	0.284	0.005	0.611
904	0.213	0.297	0.003	0.664
920	0.208	0.299	0.003	0.696
940	0.183	0.294	0.001	0.729
960	0.136	0.283	0.000	0.747
1000	0.057	0.251	0.000	0.760

Standard procedure; Optica CF4.
Absorbance measured with $l = 0.100$ cm (Hi at 600 and 630 nm), $l = 0.200$ cm (HHb and O_2Hb; COHb at 600 nm; Hi > 630 nm), $l = 1.000$ cm (COHb > 600 nm). N is number of specimens. $Na_2S_2O_4$ used to prevent reoxygenation of HHb. pH range (Hi) = 7.14–7.30. In the wavelength range 630–1000 nm a red sensitive photomultiplier tube and a red filter have been used.

ABSORPTIVITY OF HbA: SFH PROCEDURE

To exactly determine the maximum and minimum values in the absorption spectra of HHb, O_2Hb, COHb and Hi of adult human haemoglobin prepared with the SFH procedure, measurements have been made with an Optica CF4 spectrophotometer using a spectral bandwidth of only 0.06 nm. The absorbance was read at 0.5 nm intervals. In the measurements of ε_{HHb}, no $Na_2S_2O_4$ was used to prevent reoxygenation during filling of the cuvettes. Table 8.5 shows the results for HHb and O_2Hb, Table 8.6 the results for COHb and Hi. The mean value of the results obtained for the indicated number of specimens and the standard error are given for the absorptivity values.

Some of the data shown in Tables 8.5 and 8.6 have been used in a comparison of the haemoglobin absorption spectra of dog and man [472]. In contrast with the presentation in Tables 8.5 and 8.6, Table 2 of ref. [472] gives the mean values of the wavelengths found for the maximum and minimum

Table 8.5.
Absorption maxima and minima of HHb and O_2Hb of human adult haemoglobin

λ (nm)	ε_{HHb} $N = 5$	*SEM*	λ (nm)	ε_{O_2Hb} $N = 5$	*SEM*
475	3.291	0.019	508	4.799	0.013
475.5	3.288	0.020	508.5	4.798	0.013
476	3.285	0.019	509	4.799	0.013
476.5	3.281	0.018	509.5	4.806	0.011
477	3.283	0.017	510	4.811	0.011
477.5	3.284	0.015			
478	3.289	0.016	540	14.28	0.034
			540.5	14.34	0.027
552	12.76	0.029	541	14.41	0.043
553	12.81	0.036	541.5	14.43	0.048
554	12.84	0.041	542	14.44	0.046
554.5	12.83	0.040	542.5	14.43	0.045
555	12.82	0.043	543	14.37	0.049
555.5	12.80	0.041	543.5	14.26	0.037
556	12.79	0.043	544	14.19	0.028
			560	8.554	0.024
			560.5	8.546	0.025
			561	8.554	0.027
			576	15.31	0.044
			576.5	15.39	0.043
			577	15.40	0.041
			577.5	15.37	0.042
			578	15.20	0.038

SFH procedure; Optica CF4.
Absorptivity in $L \cdot mmol^{-1} \cdot cm^{-1}$; *N* is number of specimens; *SEM* is standard error of the mean.

points in the absorption spectra of the specimens, and the corresponding values of the absorptivity. This yielded for the α- and β-peaks of O_2Hb 576.9 and 541.9 nm, and for the α- and β-peaks of COHb 568.7 and 538.8 nm. This is not significantly different from the corresponding values in Tables 8.5 and 8.6. The same holds for the other maxima and minima given in these tables.

Table 8.7 shows the absorptivities of HHb, O_2Hb, COHb and Hi of HbA prepared with the SFH procedure for the spectral range of 450–800 nm measured with an HP8450A diode array spectrophotometer in 0.100-cm cuvettes with plan parallel glass inserts. As these measurements were made with a constant small lightpath length of 0.013 cm, the absorbance was often too low for accurate measurement. When the absorbance was < 0.050, no absorptivity value is given.

Table 8.6.
Absorption maxima and minima of COHb and Hi of adult human haemoglobin

λ (nm)	ε_{COHb} $N = 5$	*SEM*	λ (nm)	ε_{Hi} $N = 5$	*SEM*
495	5.090	0.018	498	8.997	0.030
495.5	5.094	0.019	499	9.015	0.032
496	5.093	0.017	499.5	9.016	0.027
			500	9.023	0.031
538	14.16	0.017	500.5	9.021	0.028
538.5	14.17	0.021	501	9.019	0.028
539	14.18	0.027	501.5	9.016	0.029
539.5	14.15	0.025	502	9.008	0.030
540	14.10	0.037			
542	13.80	0.031	630	3.858	0.023
			630.5	3.860	0.023
554	11.28	0.019	631	3.861	0.022
554.5	11.28	0.022	631.5	3.861	0.024
555	11.31	0.033	632	3.852	0.022
			632.5	3.845	0.024
568	14.29	0.030	633	3.834	0.024
568.5	14.32	0.024	634	3.803	0.023
569	14.31	0.026	640	3.275	0.022
569.5	14.31	0.030	650	1.811	0.016
570	14.24	0.020			

SFH procedure; Optica CF4.

Absorptivity in $L \cdot mmol^{-1} \cdot cm^{-1}$; *N* is number of specimens; *SEM* is standard error of the mean.

The absorptivities of the derivatives of HbA as determined with the SFH procedure have been compared with those of the standard procedure by means of Student's *t*-test for unpaired samples. The results are consolidated in Table 8.8. Figure 8.3 shows the absorption spectra of HHb, O_2Hb, COHb and Hi of HbA prepared with the two procedures for the spectral range of 450–800 nm measured with an HP8450A.

Figure 8.3 and Table 8.8 show that the absorptivities measured for the haemoglobin derivatives determined with the SFH procedure are generally somewhat lower than those of haemoglobin derivatives prepared with the standard procedure. At several wavelengths the difference is statistically significant. This is obviously due to the fact that the toluene erythrolysis used in this procedure thoroughly removes all cell debris from the haemoglobin solution, as described in Chapter 5.

The reproducibility of the measurements of the absorptivities of O_2Hb prepared with the standard procedure is better than that found for O_2Hb prepared with the SFH procedure. For the β-peak of O_2Hb and for the absorptivity at 508 nm of O_2Hb the coefficient of variation is 0.4% when the standard

Table 8.7.
Absorptivities ($L \cdot mmol^{-1} \cdot cm^{-1}$) of the common derivatives of adult human haemoglobin

λ (nm)	ε_{HHb} $N = 11$	ε_{O_2Hb} $N = 11$	ε_{COHb} $N = 11$	ε_{Hi} $N = 11$
450	13.02	15.97	10.04	8.943
452	10.30	14.71	9.345	8.684
454	8.352	13.63	8.766	8.485
456	6.931	12.66	8.275	8.329
458	5.893	11.81	7.854	8.203
460	5.125	11.05	7.487	8.096
462	4.562	10.38	7.170	8.001
464	4.146	9.772	6.888	7.916
466	3.847	9.238	6.642	7.847
468	3.633	8.750	6.419	7.786
470	3.489	8.322	6.224	7.751
472	3.397	7.927	6.049	7.737
474	3.346	7.575	5.898	7.753
476	3.323	7.250	5.764	7.793
478	3.325	6.960	5.655	7.868
480	3.342	6.689	5.562	7.962
482	3.375	6.545	5.484	8.086
484	3.419	6.240	5.411	8.220
486	3.478	6.057	5.347	8.366
488	3.550	5.887	5.288	8.508
490	3.638	5.740	5.242	8.655
492	3.740	5.604	5.209	8.782
494	3.858	5.485	5.191	8.901
496	3.992	5.374	5.187	8.988
498	4.145	5.276	5.207	9.057
500	4.312	5.184	5.244	9.078
502	4.496	5.101	5.309	9.064
504	4.693	5.023	5.396	9.001
506	4.899	4.962	5.521	8.906
508	5.106	4.925	5.681	8.764
510	5.319	4.923	5.897	8.608
512	5.523	4.971	6.174	8.417
514	5.735	5.090	6.542	8.224
516	5.942	5.291	7.004	8.007
518	6.165	5.605	7.578	7.803
520	6.397	6.046	8.242	7.589
522	6.659	6.638	8.998	7.391
524	6.936	7.381	9.795	7.196
526	7.252	8.285	10.63	7.029
528	7.592	9.302	11.45	6.875
530	7.974	10.39	12.24	6.746
532	8.380	11.44	12.93	6.627
534	8.840	12.39	13.55	6.523

Table 8.7.
(Continued)

λ (nm)	ε_{HHb} $N = 11$	ε_{O_2Hb} $N = 11$	ε_{COHb} $N = 11$	ε_{Hi} $N = 11$
536	9.324	13.19	13.98	6.410
538	9.863	13.84	14.22	6.288
540	10.41	14.24	14.19	6.139
542	10.97	14.41	13.92	5.960
544	11.49	14.21	13.46	5.742
546	11.98	13.69	12.93	5.491
548	12.37	12.84	12.40	5.220
550	12.69	11.87	11.94	4.942
552	12.89	10.90	11.61	4.676
554	13.00	10.06	11.45	4.435
556	12.99	9.401	11.47	4.231
558	12.90	8.950	11.68	4.069
560	12.72	8.719	12.07	3.950
562	12.52	8.736	12.63	3.870
564	12.24	9.022	13.26	3.823
566	11.93	9.603	13.88	3.804
568	11.57	10.49	14.27	3.804
570	11.20	11.66	14.30	3.814
572	10.78	13.00	13.82	3.830
574	10.36	14.28	12.88	3.844
576	9.924	15.10	11.57	3.851
578	9.493	15.13	10.04	3.845
580	9.059	14.14	8.447	3.823
582	8.628	12.30	6.924	3.776
584	8.184	10.01	5.570	3.712
586	7.714	7.718	4.439	3.632
588	7.200	5.727	3.535	3.544
590	6.635	4.156	2.829	3.456
592	6.028	3.001	2.288	3.372
594	5.396	2.185	1.872	3.299
596	4.759	1.624	1.551	3.236
598	4.146	1.240	1.301	3.187
600	3.587	0.977	1.106	3.153
602	3.093	0.790	0.949	3.131
604	2.678	0.654	0.821	3.120
606	2.340	0.552	0.717	3.119
608	2.072	0.474	0.631	3.128
610	1.859		0.559	3.147
612	1.691		0.498	3.182
614	1.555			3.234
616	1.445			3.304
618	1.355			3.391
620	1.281			3.490
622	1.217			3.588

Table 8.7.
(Continued)

λ (nm)	ε_{HHb} $N = 11$	ε_{O_2Hb} $N = 11$	ε_{COHb} $N = 11$	ε_{Hi} $N = 11$
624	1.161			3.679
626	1.113			3.762
628	1.071			3.827
630	1.035			3.868
632	1.002			3.877
634	0.973			3.845
636	0.947			3.759
638	0.924			3.615
640	0.904			3.412
642	0.886			3.158
644	0.869			2.869
646	0.854			2.561
648	0.840			2.250
650	0.827			1.946
652	0.815			1.665
654	0.803			1.416
656	0.786			1.210
658	0.772			0.973
660	0.764			0.831
662	0.751			0.700
664	0.736			0.589
666	0.719			0.498
668	0.701			
670	0.681			
672	0.660			
674	0.638			
676	0.618			
678	0.596			
680	0.575			
682	0.554			
684	0.534			
686	0.514			
688	0.495			
690	0.477			
692	0.460			

λ = 450–800 nm. SFH procedure; HP8450A.

Absorbance measured with l = 0.010–0.016 cm. N is number of specimens. pH range (Hi) = 7.1–7.2.

procedure is used, whereas with the SFH procedure the corresponding values are 0.9 and 0.8%, respectively. It may be that during tonometry with a high p_{O_2} deterioration of a small amount of haemoglobin occurs more readily when haemoglobin solutions are equilibrated with the gas mixture than when

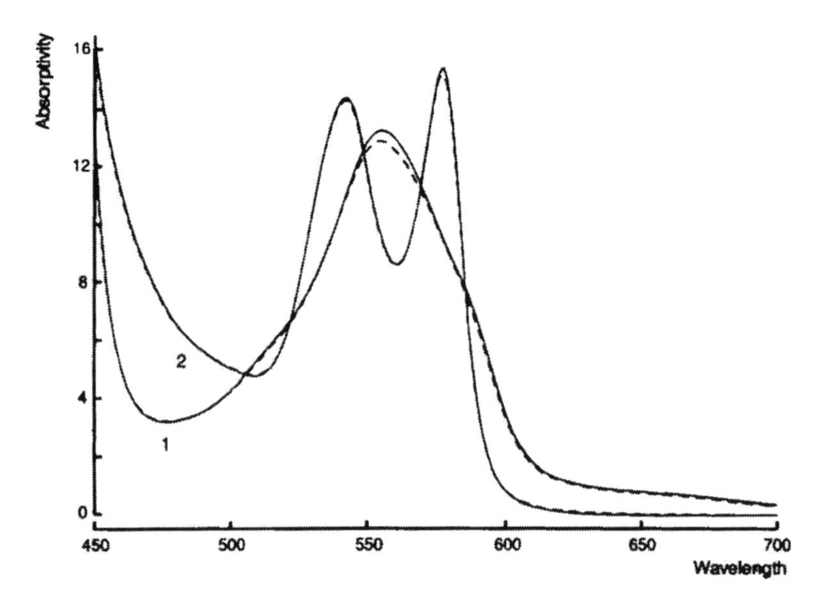

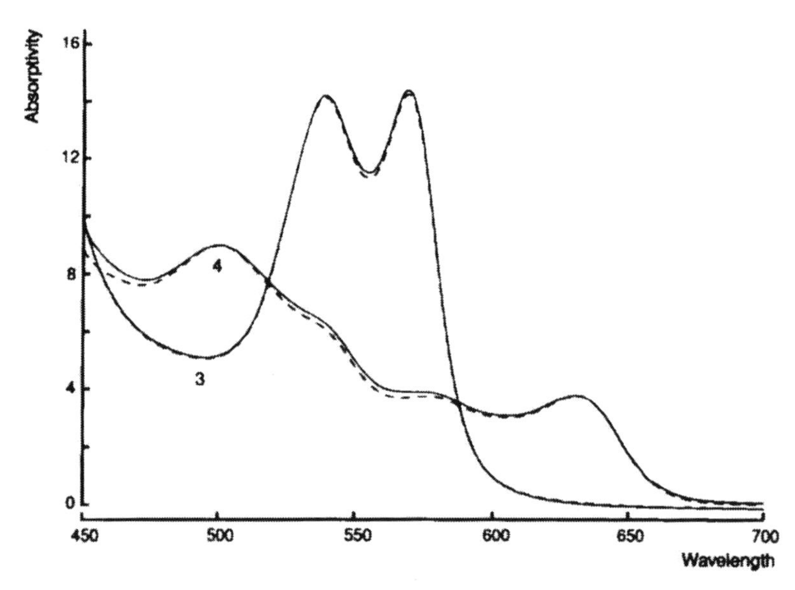

Figure 8.3. Absorption spectra of the common derivatives of HbA.

Absorptivity in $L \cdot mmol^{-1} \cdot cm^{-1}$, wavelength in nm. Solid lines: standard procedure; dashed lines SFH procedure. Upper panel: HHb (1) and O_2Hb (2); lower panel: COHb (3) and Hi (4). $l = 0.010–0.016$ cm. HP8450A.

haemoglobin remains in the protective environment of the red blood cells during tonometry. For Hi, where in the preparation of the samples tonometry plays no role, the situation is the other way round: for the absorptivity at

Table 8.8.
Comparison of absorptivities of the common derivatives of adult human haemoglobin as prepared with the standard procedure (S) and with SFH procedure (T)

λ (nm)	$\varepsilon_{HHb}(S)$ $N = 6$	*SEM*	$\varepsilon_{HHb}(T)$ $N = 11$	*SEM*	*p*
476	3.298	0.013	3.323	0.008	
478	3.302	0.012	3.325	0.007	
552	13.21	0.044	12.89	0.030	< 0.001
554	13.35	0.045	13.00	0.029	< 0.001
556	13.36	0.045	12.99	0.031	< 0.001
λ (nm)	$\varepsilon_{O_2Hb}(S)$ $N = 6$	*SEM*	$\varepsilon_{O_2Hb}(T)$ $N = 11$	*SEM*	*p*
508	4.882	0.007	4.925	0.012	
540	14.32	0.027	14.24	0.043	
542	14.52	0.023	14.41	0.039	
544	14.34	0.022	14.21	0.042	
560	8.767	0.015	8.719	0.026	
576	15.26	0.034	15.10	0.053	
578	15.36	0.027	15.13	0.054	< 0.02
λ (nm)	$\varepsilon_{COHb}(S)$ $N = 8$	*SEM*	$\varepsilon_{COHb}(T)$ $N = 11$	*SEM*	*p*
496	5.225	0.017	5.187	0.012	
538	14.30	0.039	14.22	0.034	
540	14.27	0.036	14.19	0.030	
542	14.04	0.036	13.92	0.032	< 0.02
554	11.63	0.029	11.45	0.025	< 0.001
568	14.43	0.037	14.27	0.037	< 0.02
570	14.46	0.039	14.30	0.037	< 0.002
λ (nm)	$\varepsilon_{Hi}(S)$ $N = 8$	*SEM*	$\varepsilon_{Hi}(T)$ $N = 11$	*SEM*	*p*
498	9.088	0.028	9.057	0.010	
500	9.106	0.029	9.078	0.010	
502	9.098	0.028	9.064	0.011	
630	3.895	0.021	3.868	0.006	
632	3.896	0.021	3.877	0.006	
634	3.857	0.020	3.845	0.006	

HP8450A.
Absorptivity in $L \cdot mmol^{-1} \cdot cm^{-1}$; *N* is number of specimens; *SEM* is the standard error of the mean. *p* values of Student's *t*-test for unpaired samples given only when $p < 0.05$.

630 nm the coefficient of variation is only 0.5% when the samples have been prepared with the SFH procedure against 2.1% when the standard procedure has been used.

ABSORPTIVITY OF HbF: STANDARD PROCEDURE

Table 8.9 shows the absorptivities of HHb, O_2Hb, COHb and Hi of human foetal haemoglobin prepared with the standard procedure for the spectral range of 450–630 nm measured with an HP8450A in 0.100-cm cuvettes with 0.084 to 0.090-cm plan parallel glass inserts. No absorptivity values have been given when the measured absorbance was < 0.050. Table 8.10

Table 8.9.
Absorptivities ($L \cdot mmol^{-1} \cdot cm^{-1}$) of the common derivatives of foetal human haemoglobin

λ (nm)	ε_{HHb} $N = 5$	ε_{O_2Hb} $N = 5$	ε_{COHb} $N = 5$	ε_{Hi} $N = 5$
450	12.97	15.82	9.984	10.34
452	10.16	14.57	9.314	9.918
454	8.164	13.50	8.762	9.560
456	6.726	12.54	8.291	9.266
458	5.675	11.71	7.879	9.011
460	4.918	10.96	7.527	8.794
462	4.351	10.29	7.218	8.611
464	3.944	9.691	6.947	8.440
466	3.650	9.162	6.702	8.311
468	3.450	8.676	6.486	8.201
470	3.314	8.246	6.298	8.132
472	3.229	7.854	6.124	8.094
474	3.188	7.497	5.968	8.087
476	3.178	7.165	5.830	8.133
478	3.182	6.877	5.722	8.197
480	3.211	6.605	5.627	8.287
482	3.246	6.368	5.546	8.420
484	3.297	6.147	5.464	8.552
486	3.354	5.961	5.395	8.690
488	3.431	5.791	5.331	8.843
490	3.520	5.648	5.286	8.987
492	3.627	5.507	5.243	9.123
494	3.755	5.392	5.231	9.251
496	3.904	5.278	5.217	9.345
498	4.067	5.177	5.237	9.411
500	4.249	5.082	5.276	9.434
502	4.451	5.001	5.345	9.437
504	4.674	4.927	5.429	9.392
506	4.904	4.868	5.556	9.319
508	5.133	4.831	5.720	9.203
510	5.362	4.834	5.944	9.053
512	5.580	4.889	6.229	8.890
514	5.810	5.019	6.618	8.719
516	6.020	5.238	7.094	8.526

Table 8.9.
(Continued)

λ (nm)	ε_{HHb} $N = 5$	ε_{O_2Hb} $N = 5$	ε_{COHb} $N = 5$	ε_{Hi} $N = 5$
518	6.260	5.566	7.693	8.336
520	6.489	6.028	8.374	8.138
522	6.753	6.646	9.155	7.945
524	7.029	7.420	9.966	7.745
526	7.339	8.372	10.84	7.569
528	7.683	9.425	11.67	7.389
530	8.064	10.54	12.48	7.233
532	8.471	11.62	13.18	7.087
534	8.948	12.62	13.82	6.946
536	9.455	13.42	14.24	6.806
538	10.02	14.07	14.47	6.642
540	10.60	14.48	14.40	6.472
542	11.21	14.62	14.12	6.262
544	11.78	14.38	13.60	6.026
546	12.31	13.79	13.05	5.767
548	12.77	12.87	12.48	5.493
550	13.14	11.84	12.02	5.217
552	13.39	10.82	11.68	4.958
554	13.52	9.946	11.52	4.722
556	13.53	9.271	11.55	4.528
558	13.44	8.814	11.77	4.373
560	13.26	8.599	12.18	4.254
562	13.02	8.625	12.74	4.161
564	12.71	8.938	13.39	4.104
566	12.36	9.549	14.01	4.059
568	11.96	10.49	14.41	4.021
570	11.52	11.72	14.42	3.979
572	11.05	13.14	13.93	3.942
574	10.57	14.48	12.99	3.905
576	10.10	15.38	11.66	3.858
578	9.646	15.45	10.09	3.808
580	9.216	14.47	8.454	3.744
582	8.811	12.55	6.884	3.679
584	8.408	10.12	5.500	3.603
586	7.973	7.684	4.359	3.522
588	7.481	5.604	3.456	3.435
590	6.920	4.004	2.761	3.353
592	6.296	2.853	2.228	3.273
594	5.633	2.058	1.825	3.204
596	4.959	1.515	1.512	3.143
598	4.303	1.146	1.271	3.094
600	3.713	0.896	1.082	3.063
602	3.188	0.719	0.931	3.049
604	2.748	0.589	0.808	3.047

Table 8.9.
(Continued)

λ (nm)	ε_{HHb} $N = 5$	ε_{O_2Hb} $N = 5$	ε_{COHb} $N = 5$	ε_{Hi} $N = 5$
606	2.392	0.494	0.709	3.066
608	2.109		0.625	3.094
610	1.884		0.556	3.145
612	1.708		0.497	3.206
614	1.567		0.447	3.289
616	1.455			3.392
618	1.360			3.490
620	1.279			3.595
622	1.212			3.710
624	1.157			3.813
626	1.108			3.903
628	1.064			3.972
630	1.026			4.021

λ = 450–630 nm. Standard procedure; HP8450A.

Absorbance measured with l = 0.010–0.016 cm. N is number of specimens. $Na_2S_2O_4$ used to prevent reoxygenation of HHb. pH range (Hi) = 7.01–7.15.

Table 8.10.
Absorptivities ($L \cdot mmol^{-1} \cdot cm^{-1}$) of the common derivatives of foetal human haemoglobin

λ (nm)	ε_{HHb} $N = 5$	ε_{O_2Hb} $N = 5$	ε_{COHb} $N = 4$	ε_{Hi} $N = 5$
600		0.883		3.031
602		0.714		3.019
604	2.62	0.590		3.012
606	2.350	0.497		3.026
608	2.090	0.424		3.052
610	1.891	0.366		3.097
612	1.720	0.320		3.149
614	1.589	0.282		3.221
616	1.478	0.251	0.389	3.305
618	1.386	0.224	0.354	3.404
620	1.308	0.201	0.323	3.501
622	1.243	0.181	0.295	3.609
624	1.185	0.164	0.270	3.707
626	1.136	0.149	0.246	3.792
628	1.092	0.136	0.225	3.860
630	1.054	0.124	0.206	3.898
632	1.020	0.115	0.188	3.909
634	0.991	0.107	0.172	3.878

Table 8.10.
(Continued)

λ (nm)	ε_{HHb} $N = 5$	ε_{O_2Hb} $N = 5$	ε_{COHb} $N = 4$	ε_{Hi} $N = 5$
636	0.965	0.100	0.158	3.796
638	0.944	0.095	0.145	3.661
640	0.924	0.092	0.134	3.469
642	0.909	0.090	0.125	3.232
644	0.896	0.089	0.117	2.961
646	0.885	0.088	0.110	2.677
648	0.875	0.089	0.103	2.386
650	0.868	0.090	0.098	2.101
652	0.859	0.091	0.093	1.831
654	0.852	0.092	0.088	1.592
656	0.841	0.090	0.084	1.392
658	0.828	0.092	0.079	1.153
660	0.822	0.095	0.076	1.025
662	0.809	0.095	0.072	0.898
664	0.794	0.096	0.069	0.788
666	0.778	0.096	0.066	0.697
668	0.759	0.096	0.063	0.622
670	0.739	0.097	0.060	0.561
672	0.718	0.097	0.057	0.511
674	0.696	0.097	0.055	0.471
676	0.674	0.097	0.053	0.439
678	0.652	0.097	0.050	0.413
680	0.631	0.098	0.049	0.392
682	0.610	0.098	0.047	0.375
684	0.589	0.098	0.045	0.361
686	0.569	0.098	0.044	0.350
688	0.550	0.098	0.042	0.340
690	0.532	0.098	0.041	0.332
692	0.515	0.098	0.040	0.326
694	0.500	0.099	0.039	0.320
696	0.485	0.099	0.038	0.316
698	0.472	0.100	0.037	0.312
700	0.460	0.101	0.036	0.309
702	0.449	0.102	0.035	0.305
704	0.438	0.103	0.034	0.303
706	0.427	0.105	0.034	0.301
708	0.417	0.106	0.033	0.299
710	0.408	0.107	0.033	0.298
712	0.398	0.109	0.032	0.297
714	0.389	0.110	0.032	0.295
716	0.380	0.111	0.031	0.295
718	0.371	0.113	0.031	0.295
720	0.362	0.114	0.031	0.294
722	0.354	0.116	0.030	0.294

Table 8.10.
(Continued)

λ (nm)	ε_{HHb} $N = 5$	ε_{O_2Hb} $N = 5$	ε_{COHb} $N = 4$	ε_{Hi} $N = 5$
724	0.347	0.117	0.030	0.295
726	0.340	0.119	0.030	0.295
728	0.335	0.121	0.029	0.296
730	0.332	0.123	0.029	0.297
732	0.329	0.124	0.029	0.299
734	0.328	0.126	0.029	0.300
736	0.330	0.128	0.028	0.302
738	0.333	0.130	0.028	0.304
740	0.338	0.131	0.028	0.306
742	0.344	0.133	0.027	0.308
744	0.352	0.135	0.027	0.311
746	0.362	0.137	0.027	0.314
748	0.373	0.139	0.027	0.317
750	0.385	0.141	0.027	0.320
752	0.397	0.143	0.026	0.323
754	0.408	0.146	0.026	0.326
756	0.417	0.148	0.026	0.330
758	0.422	0.150	0.026	0.334
760	0.423	0.153	0.025	0.338
762	0.418	0.155	0.025	0.342
764	0.409	0.158	0.025	0.346
766	0.396	0.160	0.025	0.350
768	0.380	0.162	0.025	0.354
770	0.362	0.165	0.024	0.359
772	0.344	0.167	0.024	0.364
774	0.327	0.169	0.024	0.368
776	0.311	0.172	0.024	0.373
778	0.296	0.174	0.023	0.379
780	0.283	0.177	0.023	0.384
782	0.271	0.179	0.023	0.389
784	0.261	0.181	0.023	0.394
786	0.252	0.184	0.023	0.400
788	0.244	0.186	0.022	0.405
790	0.238	0.189	0.022	0.411
792	0.232	0.191	0.022	0.416
794	0.227	0.193	0.022	0.422
796	0.222	0.196	0.022	0.428
798	0.219	0.198	0.021	0.434
800	0.215	0.200	0.021	0.439

λ = 600–800 nm. Standard procedure; HP8450A.

Absorbance measured with l = 0.100 cm (Hi), l = 0.200 cm (HHb and O_2Hb), l = 1.000 cm (COHb). N is number of specimens. $Na_2S_2O_4$ used to prevent reoxygenation of HHb. pH range (Hi) = 7.01–7.15.

Table 8.11.
Comparison of absorptivities of the common derivatives of adult (HbA) and foetal (HbF) human haemoglobin

λ (nm)	ε_{HHbA} $N = 6$	*SEM*	ε_{HHbF} $N = 5$	*SEM*	*p*
476	3.298	0.013	3.178	0.015	< 0.001
478	3.302	0.012	3.182	0.015	< 0.001
552	13.21	0.044	13.39	0.050	< 0.02
554	13.35	0.045	13.52	0.043	< 0.02
556	13.36	0.045	13.53	0.050	< 0.02
λ (nm)	ε_{O_2HbA} $N = 6$	*SEM*	ε_{O_2HbF} $N = 5$	*SEM*	*p*
508	4.882	0.007	4.831	0.035	
540	14.32	0.027	14.48	0.033	< 0.01
542	14.52	0.023	14.62	0.036	< 0.01
544	14.34	0.022	14.38	0.038	
560	8.767	0.015	8.599	0.043	< 0.01
576	15.26	0.034	15.38	0.037	0.05
578	15.36	0.027	15.45	0.034	< 0.05
λ (nm)	ε_{COHbA} $N = 8$	*SEM*	ε_{COHbF} $N = 5$	*SEM*	*p*
496	5.225	0.017	5.217	0.019	
538	14.30	0.039	14.47	0.029	< 0.02
540	14.27	0.036	14.40	0.031	< 0.05
542	14.04	0.036	14.12	0.032	
554	11.63	0.029	11.52	0.020	< 0.02
568	14.43	0.037	14.41	0.027	
570	14.46	0.039	14.42	0.031	
λ (nm)	ε_{HiA} $N = 15$	*SEM*	ε_{HiF} $N = 5$	*SEM*	*p*
498	9.088	0.028	9.411	0.036	< 0.001
500	9.106	0.029	9.434	0.039	< 0.001
502	9.098	0.028	9.437	0.041	< 0.001
630	3.895	0.021	4.021	0.030	< 0.01
632	3.896	0.021	4.029	0.029	< 0.01
634	3.857	0.020	4.000	0.028	< 0.002

Standard procedure; HP8450A.

Absorptivity in $L \cdot mmol^{-1} \cdot cm^{-1}$; *N* is number of specimens; *SEM* is standard error of the mean. *p* values are from Student's *t*-test for unpaired samples.

shows the absorptivities for the spectral range of 600–800 nm measured with the same instrument; the lightpath was adjusted so that the measured absorbance was > 0.050. Absorptivity values have not been given when the

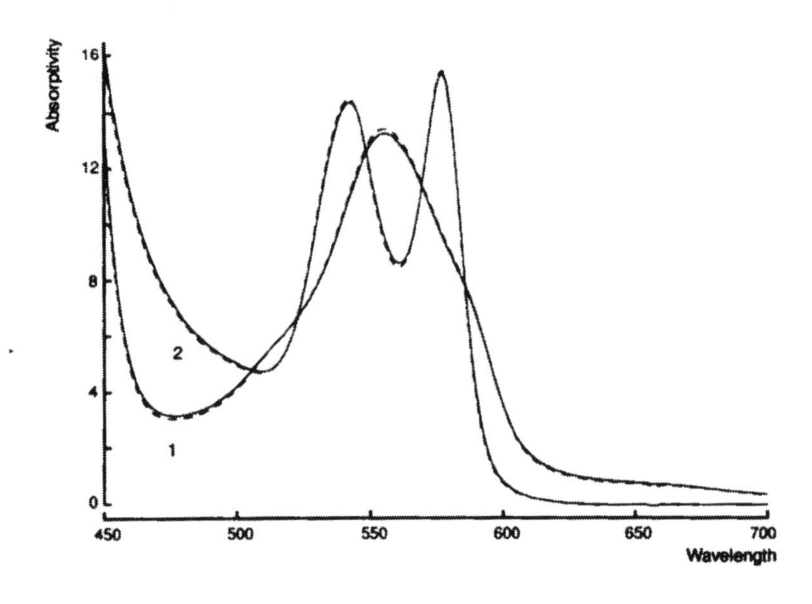

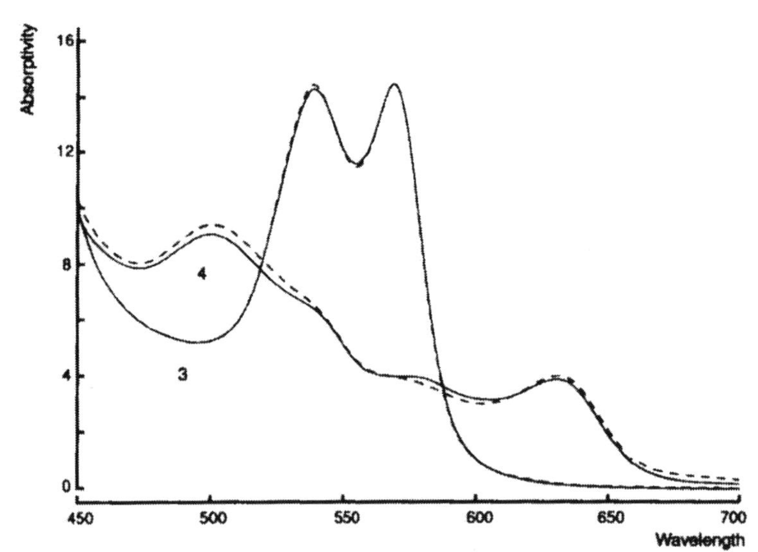

Figure 8.4a. Absorption spectra of the common derivatives of HbF in comparison with those of HbA.

Absorptivity in $L \cdot mmol^{-1} \cdot cm^{-1}$, wavelength in nm. Solid lines: HbA; dashed lines: HbF. Upper panel: HHb (1) and O_2Hb (2); lower panel: COHb (3) and Hi (4). Absorbance measured with $l = 0.010$–0.016 cm. HP8450A. pH (foetal Hi) ≈ 7.08; pH (adult Hi) ≈ 7.21.

spectrophotometer rejected the measurements because the absorbance of the samples was too high. This happened occasionally in measurements with $l = 0.200$ cm between 600 and 650 nm. The spectra represented by the

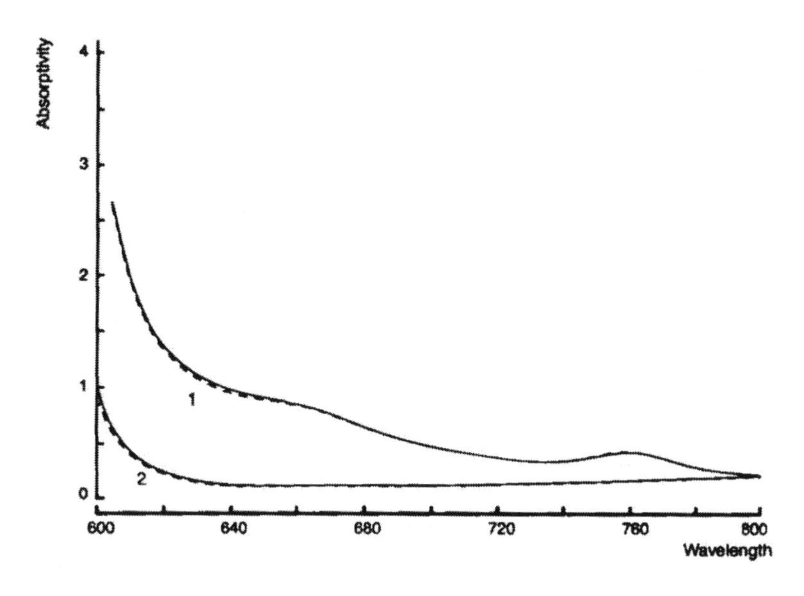

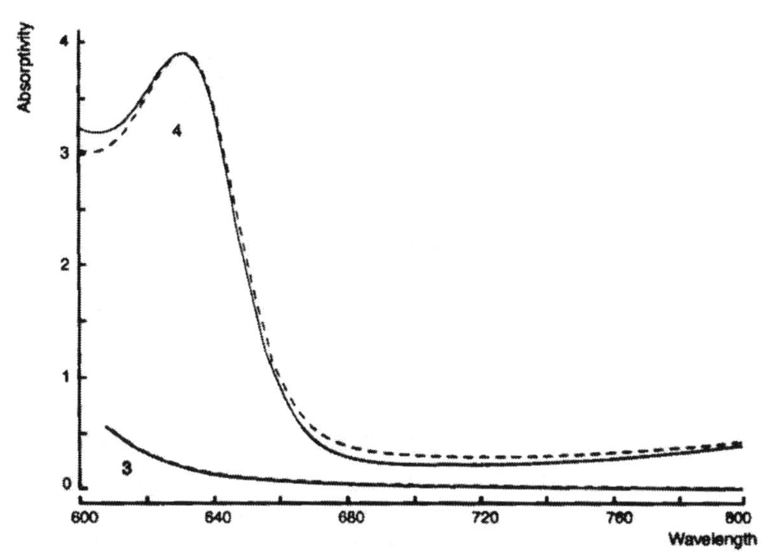

Figure 8.4b. Absorption spectra of the common derivatives of HbF in comparison with those of HbA.

Absorptivity in $L \cdot mmol^{-1} \cdot cm^{-1}$, wavelength in nm. Solid lines: HbA; dashed lines: HbF. Upper panel: HHb (1) and O_2Hb (2); lower panel: COHb (3) and Hi (4). Absorbance measured with $l = 0.100$ cm (Hi); $l = 0.200$ cm (HHb and O_2Hb); $l = 1.000$ cm (COHb). HP8450A. pH (foetal Hi) ≈ 7.08; pH (adult Hi) ≈ 7.21.

absorptivities of Tables 8.9 and 8.10 are designated as the **standard spectra** of foetal human haemoglobin.

Table 8.12.
Absorptivities ($L \cdot mmol^{-1} \cdot cm^{-1}$) of the common derivatives of foetal human haemoglobin at selected wavelengths between 600 and 1000 nm

λ (nm)	ε_{HHb} $N = 5$	ε_{O_2Hb} $N = 5$	ε_{COHb} $N = 4$	ε_{Hi} $N = 3$
600		0.799		3.073
630	1.046	0.100	0.183	3.953
660	0.826	0.074	0.065	0.866
680	0.618	0.079	0.038	0.286
700	0.450	0.083	0.027	0.228
750	0.397	0.129	0.020	0.259
775	0.289	0.157	0.017	0.322
800	0.206	0.189	0.015	0.400
805	0.196	0.193	0.014	0.412
840	0.183	0.234	0.012	0.525
845	0.185	0.239	0.012	0.538
880	0.194	0.270	0.009	0.623
904	0.201	0.281	0.008	0.683
920	0.197	0.285	0.006	0.719
940	0.167	0.284	0.004	0.751
960	0.119	0.274	0.003	0.770
1000	0.044	0.246	0.000	0.804

Standard procedure; Optica CF4.
Absorbance measured with $l = 0.100$ cm (Hi); $l = 0.200$ cm (HHb and O_2Hb); $l = 1.000$ cm (COHb). N is number of specimens. $Na_2S_2O_4$ used to prevent reoxygenation of HHb. In the wavelength range 630–1000 nm a red sensitive photomultiplier tube and a red filter have been used. pH range (Hi) = 7.01–7.15.

Figure 8.4 shows the absorption spectra of HHb, O_2Hb, COHb and Hi of HbF in comparison with those of HbA for the spectral ranges of 450–700 nm (Table 8.9) and 600–800 nm (Table 8.10) measured with an HP8450A. In Fig. 8.4a, the spectral absorption curves run to 700 nm, but for $\lambda > 600$ nm part of the measurements resulted in $A < 0.050$; these values have not been given in Table 8.9. The absorptivities of the derivatives of HbF have been compared with those of HbA determined with the standard procedure, by means of Student's t-test for unpaired samples. The results are presented in Table 8.11.

Figure 8.4 and Table 8.11 show no differences in the position of the maxima and minima in the absorption spectra of the common derivatives of HbA and HbF. As the measurements have been made with a diode array spectrophotometer, which has a limited spectral resolution, this does not rule out that slight differences in the position of the maxima and minima do exist. More important are the distinct differences in the absorptivities between HbF and HbA of all four the common derivatives.

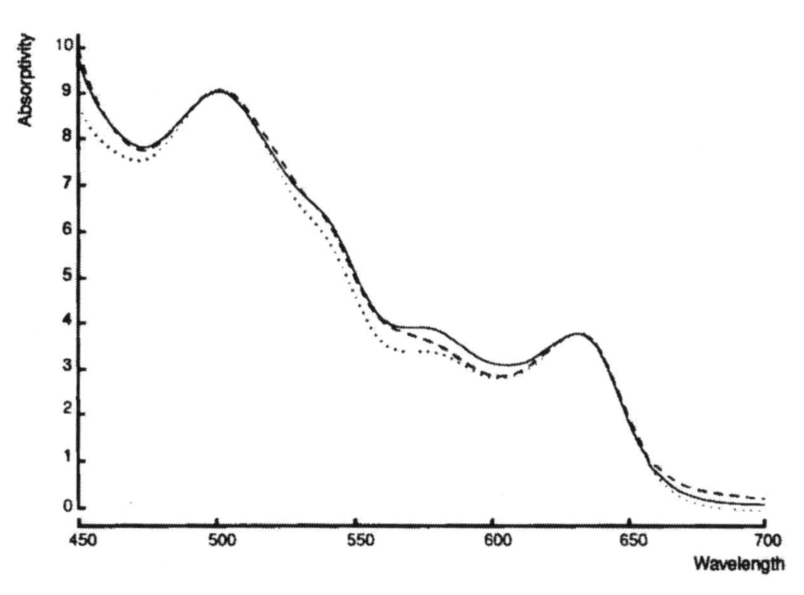

Figure 8.5. Absorption spectrum of Hi of foetal human haemoglobin at pH = 7.07 (- - - -) in comparison with the standard Hi spectrum of adult human haemoglobin (pH ≈ 7.21) (——) and an adult human Hi spectrum at pH = 7.03 (· · ·).

Absorptivity in $L \cdot mmol^{-1} \cdot cm^{-1}$, wavelength in nm. For easy comparison the absorptivity at 630 nm has been normalised at 3.895 $L \cdot mmol^{-1} \cdot cm^{-1}$, the absorptivity at 630 nm of the standard Hi spectrum. HP8450A.

Figure 8.5 presents an absorption spectrum of foetal human Hi at a pH value of 7.07 between two spectra of adult Hi at different values of pH. As the pH of the solution of foetal Hi is between the two pH values of the spectra of adult Hi, these data demonstrate that the spectral difference between foetal and adult Hi is, at least partly, not caused by differences in the pH values of the solutions.

Table 8.12 presents the absorptivities of HHb, O_2Hb, COHb and Hi of foetal human haemoglobin at selected wavelengths between 600 and 1000 nm, measured with an Optica CF4. Wavelength selection was necessary to complete this series of manual measurements within a reasonable time. Wavelengths of primary importance for biomedical application have been chosen preferentially.

COMMENTS

In the determination of the absorption spectra of human haemoglobin various methods have been used. Different spectrophotometers were used, as well as different procedures to prepare pure and clear solutions of the various derivatives. The influence of the type of spectrophotometer on the results is largely predictable. A diode array spectrophotometer, for instance, scanning the visible spectrum in steps of 2 nm, will obviously be unsuited to exactly represent a spectrum with very sharp absorption maxima and minima. Effects of the technique of measurement are further discussed in Chapter 15.

The influence of the way the samples are prepared for the measurement is less predictable; the standard procedure used for most of the data presented in this chapter only gradually evolved by much trial and error. A case in point is the use of sodium dithionite ($Na_2S_2O_4$) in the preparation and handling of HHb solutions (p. 43). Ideally, any contact with air should be avoided after all oxygen has been removed from the erythrocyte suspension by tonometry with N_2/CO_2, and no $Na_2S_2O_4$ should be added.

Table 8.13 shows two series of measurements of ε_{HHb}: series (1) is identical with the corresponding values given in Tables 8.1 and 8.2, where reoxygenation has been prevented by the addition of a trace of $Na_2S_2O_4$ after deoxygenation by tonometry; the data of series (2) were obtained by the same procedure with the sole exception that no $Na_2S_2O_4$ was added. In the second series the maximum of HHb around 555 nm is clearly lower, whereas the minimum absorptivity near 478 nm is higher than in series (1). This demonstrates that the samples underlying series (2) contain a little

Table 8.13.
Absorptivity of HHb of adult human haemoglobin measured with (1) and without (2) addition of $Na_2S_2O_4$

λ (nm)	$\varepsilon_{HHb}(1)$ $N = 6$	$\varepsilon_{HHb}(2)$ $N = 5$
450	13.48	13.46
478	3.302	3.446
480	3.322	3.454
530	8.010	8.057
540	10.47	10.56
552	13.21	12.99
554	13.35	13.08
556	13.36	13.06
558	13.29	12.96
560	13.11	12.78
570	11.42	11.33
630	1.082	1.019
	$N = 6$	$N = 6$
610	1.753	1.844
630	1.051	1.041
660	0.815	0.795
700	0.447	0.440
750	0.388	0.362
800	0.215	0.210

Standard procedure; HP8450A.

O_2Hb (compare HHb and O_2Hb spectra in Fig. 8.1). A similar difference is discernible in the absorptivity values of HHb in Table 8.8. In the SFH procedure no $Na_2S_2O_4$ is used; although the absorptivities obtained with the SFH procedure are generally somewhat lower (through more thorough clearing, as described), in the minimum of HHb at 476 nm the absorptivity of HHb is higher than the corresponding value obtained with the standard procedure. These results justify the procedure in which a trace of $Na_2S_2O_4$ is used, in spite of the risk of some distortion of the HHb spectrum in the red and near infrared region.

That in the standard HHb spectrum of human HbA some $Na_2S_2O_4$-induced distorsion may also be present is suggested by a comparison of the absorptivities of HHb and O_2Hb at 800 nm. Table 8.2 gives $\varepsilon_{HHb}(800) = 0.215$ and $\varepsilon_{O_2Hb}(800) = 0.208\ L \cdot mmol^{-1} \cdot cm^{-1}$; this suggests an isosbestic point slightly above 800 nm, the formerly established value for HHb and O_2Hb [272]. Taking the value for ε_{HHb} as obtained without $Na_2S_2O_4$ ($0.210\ L \cdot mmol^{-1} \cdot cm^{-1}$; Table 8.13), gives 800 nm for the isosbestic wavelength (compare p. 144).

The appreciable influence of the pH on the absorption spectrum of Hi through the binding of OH^- ions to the ferric iron (pp. 12; 52), as shown in Fig. 8.2, has to be taken into account in spectrophotometric haemoglobin analysis when more than traces of Hi may be present in the blood specimen, as well as in comparing Hi absorption spectra from different animal species. In the latter application it does not suffice to compare these spectra at similar pH, because of the difference in OH^- affinity of Hi between animals.

The choice of a pH value for the standard Hi spectrum, *i.e.* the Hi spectrum which is to be used, in combination with the spectra of the other common haemoglobin derivatives, in various applications as described in Chapters 16–20, is a matter of opinion. For practical reasons we have chosen for this purpose a Hi spectrum at a pH within the range of pH values which may be expected to prevail in fresh haemolysed whole blood. This rather imprecise definition is acceptable, because in this pH range the effect of a change in pH is only limited. As shown in Fig. 8.2, the influence of pH increases considerably, when the solution becomes more alkaline.

This is also shown in Table 8.14. This table is the result of multicomponent analysis as described in Chapter 17, when pure Hi solutions are subjected to a procedure for analysing the system HHb/O_2Hb/COHb/Hi/SHb. Ideally, this procedure should yield $F_{Hi} = 100\%$. For the specimens with a pH within the range of pH values of the 15 specimens underlying the standard Hi spectrum of human HbA (Table 8.1), pH = 7.14–7.30, the expectation comes true to a reasonable extent, but outside this range the error increases rapidly, especially when the pH deviates to the alkaline side. In Table 11.5 the effect is still more clearly shown; this table gives the result of multicomponent analysis of Hi solutions of a *single* bovine haemoglobin specimen at different pH values.

Table 8.14.
Hi fractions as determined by MCA for pure haemiglobin solutions of adult human haemoglobin at various pH values

pH	F_{Hi} (%)
7.07	103.2
7.15	100.3
7.18	100.6
7.21	99.2
7.24	100.4
7.27	98.7
7.28	100.6
7.30	100.6
7.36	96.9
7.54	92.5
7.63	91.7

MCA = multicomponent analysis in the system HHb/O_2Hb/COHb/Hi/SHb. The standard Hi spectrum in the calculation was based on Hi spectra of 15 specimens of adult human haemoglobin recorded at pH values of 7.14–7.30. HP8450A.

Figure 8.4a shows that there are several differences in the absorptivity of the common haemoglobin derivatives between human HbA and HbF. In the HHb spectrum, the maximum near 554 nm is higher and the minimum near 476 nm is lower in HbF than in HbA. The α- and β-peak of foetal O_2Hb are higher and the minimum near 560 nm is deeper in HbF than in HbA. Table 8.11 shows that these differences are statistically significant. Most distinct in the COHb spectrum is the higher β-peak of HbF, while the minimum near 554 nm is lower in comparison with HbA. The most conspicuous differences between HbF and HbA are in the Hi spectra, as shown in the left lower panel of Fig. 8.4a. In the spectral region above 600 nm, there are no differences in the absorptivity of HHb, O_2Hb and COHb between foetal and adult human haemoglobin. Fogh-Andersen *et al.* [118] have studied the spectral differences between HbA and HbF for HHb, O_2Hb and COHb in the spectral range of 450–700 nm; their results are in fair agreement with those of Fig. 8.4 and Table 8.11.

In comparing the spectra of Hi of foetal and adult human haemoglobin, any observed difference may either be due to a difference in pH or result from a structural difference between the haemoglobin molecules. If there is only a difference in OH^- affinity between the two haemiglobins, there will be a pH value at which the two Hi spectra are identical. The absorption spectra shown in Fig. 8.5 make it very improbable that for foetal and adult Hi such an absorption spectrum exists. Consequently, it can be inferred that there is an intrinsic difference between foetal and adult haemiglobin. The practical significance of this difference is shown by reports of a clinically significant overestimation of F_{Hi} in neonatal blood by some types of instruments for multicomponent analysis of haemoglobin [246, 369].

In the determination of our two early series of absorptivities of human haemoglobin derivatives [7, 211], the erythrocytes of fresh human blood anticoagulated with heparin were washed three times with isotonic saline solution, then lysed by adding a 7 mL/L Sterox SE solution to the packed cells. The erythrolysate was filtered through a folded paper filter and stored in a glass syringe. O_2Hb and COHb were prepared by tonometry of the erythrolysate in an open cylindrical tonometer for 10–15 min with water-saturated O_2 and CO, respectively. HHb was prepared through addition of 4 mg/mL sodium dithionite to the oxygenated erythrolysate. Hi was prepared by adding powdered $K_3Fe(CN)_6$ to an end molar ratio $Hb/K_3Fe(CN)_6$ of 1 : 8. The absorbance was measured with a Beckman DU spectrophotometer [210] and with Optica CF4 and CF4DR grating spectrophotometers [7] (lightpath 0.013 cm). The use of $Na_2S_2O_4$ for the deoxygenation of a O_2Hb solution resulted in a solution of HHb with a slightly distorted spectrum. This was most clearly seen in the position of the isosbestic point of HHb and O_2Hb in the near infrared, which had shifted to 815 nm. $\varepsilon_{O_2Hb}(542)$ values obtained in these two investigations are included in Table 8.15.

In a study concerning the validity of Lambert–Beer's law for haemoglobin solutions [481], a stroma-free haemoglobin (SFH) solution was prepared as described in Chapter 5, but with only half the volume of water. The clear haemoglobin solution was diluted to $ct_{Hb} \approx 10$ mmol/L using 50 mmol/L Tris-HCl buffer with pH 7.2. O_2Hb was prepared by tonometry of a SFH solution with a water-saturated O_2/CO_2 mixture (94.4/5.6%). The absorbance

Table 8.15.
$\varepsilon_{O_2Hb}(542)$ of human HbA as prepared and measured by different procedures

Ref.	$\varepsilon_{O_2Hb}(542)$	Preparation	Spectrophotometer
Drabkin [94]	14.68	Crystalline Hb	Visual
van Kampen *et al.* [211]	14.37	Erythrolysate	Beckman DU
van Assendelft [7]	14.37	Erythrolysate	Optica CF4
Zijlstra *et al.* [481]	14.45	SFH in Tris buffer	Optica CF4
	14.41	SFH in Tris buffer	HP8450A
Zwart *et al.* [444, 447]	14.62	Haemolysate	HP8450A
Zijlstra *et al.* [480]	14.52	Standard procedure	HP8450A
Table 8.1/8.3	14.52	Standard procedure	HP8450A
Table 8.5	14.44	SFH procedure	Optica CF4
Table 8.7	14.41	SFH procedure	HP8450A

Absorptivity in $L \cdot mmol^{-1} \cdot cm^{-1}$; SFH is stroma-free haemoglobin; Tris is tris(hydroxymethyl)aminomethane.

was measured with Optica CF4 and HP8450A spectrophotometers. The values obtained for $\varepsilon_{O_2Hb}(542)$ are included in Table 8.15.

For a series of absorptivity values of haemoglobin derivatives to be used in multicomponent analysis of human blood [444, 447], freshly drawn heparinised human blood was introduced directly into tonometers similar to the one described in Chapter 6 and equilibrated with N_2/CO_2 (94.4/5.6%), O_2/CO_2 (94.4/5.6%) and CO/N_2 (5.0/95.0%), to prepare HHb, O_2Hb and COHb. The samples were drawn anaerobically from the tonometers into glass syringes of which the dead space contained a Sterox SE solution for haemolysis. From the syringes, 0.100-cm cuvettes were filled through a cotton wool filter and 0.0906-cm glass inserts were then introduced into the cuvettes, leaving a lightpath of 0.0094 cm. Hi solutions were prepared by 51-fold dilution of a blood sample with a phosphate buffer containing 3.04 mmol $K_3Fe(CN)_6$ and 0.5 mL Sterox SE per L. This yielded Hi solutions with pH $\sim$ 7.20, which were measured with $l = 0.500$ cm. Absorbances were measured with an HP8450A.

The absorptivities obtained in this investigation are summarised in the appendix of ref. [447]. The numbers of specimens (N) and of samples (n) were, for O_2Hb, $N = 21$, $n = 105$; for COHb, $N = 9$, $n = 45$; for HHb, $N = 10$, $n = 50$; and for Hi, $N = 16$, $n = 80$. The results are in excellent agreement with those of Table 8.1. The absorptivities are generally slightly higher than those of Table 8.1, because the samples have been less thoroughly clarified. This is also reflected by the values of $\varepsilon_{O_2Hb}(542)$ shown in Table 8.15. The difference, however, is very small, showing the effectiveness of the cotton wool filtration. Only at the maximum in the absorption spectrum of HHb at 554–556 nm is the absorptivity found in this investigation *lower* than the value of Table 8.1: $\varepsilon_{HHb}(556) = 13.29\ L \cdot mmol^{-1} \cdot cm^{-1}$ (ref. [447]) *vs* $13.36\ L \cdot mmol^{-1} \cdot cm^{-1}$ (Table 8.1). This is most probably caused by slight reoxygenation, because no $Na_2S_2O_4$ had been added.

The absorptivity values of the common derivatives of HbA and HbF published previously [480] correspond with parts of Tables 8.1, 8.2 and 8.4 for HbA and parts of Tables 8.9, 8.10 and 8.12 for HbF, with the exception of the absorptivities of the Hi derivative of HbA, for which the tables include new data measured with stricter control of the pH. The $\varepsilon_{O_2Hb}(542)$ values of these investigations have also been added to Table 8.15.

The data of Table 8.15 suggest that the measured absorptivity of O_2Hb decreases the more the haemoglobin solutions have been cleared. This is in agreement with the data of Table 8.8, showing that the absorptivities measured in samples prepared with the standard procedure are generally slightly higher than those found in samples prepared with the SFH procedure. The value obtained for the haemolysate is the highest. The slightly lower value obtained by van Kampen *et al.* [211] and by van Assendelft [7] may be due to the short equilibration time and the use of an open tonometer.

The SHb spectrum shown in Fig. 8.1 has been determined as described in Chapter 7 and the absorptivities have been calculated with equation 7.4. In this calculation F_{SHb} is ~ 20% and has been determined with the help of titrimetric oxygen measurements as described in Chapter 6. It may therefore be that the absorption minima in the SHb spectrum near 540 and 580 nm (partly) result from overestimation of the O_2Hb fraction in the SHb/O_2Hb mixture. The same may hold for the SHb data in the appendix of ref. [447]. In an earlier study [103], SHb spectra had been determined with an Aminco Chance split beam spectrophotometer. The sample cuvette (l = 0.50 cm) was filled with an SHb/O_2Hb mixture diluted with Tris buffer solution (pH = 8.0), the reference cuvette with a pure O_2Hb solution with the same concentration. As the output signal of this spectrophotometer represents the difference between the absorbances of the two cuvettes, the absorption spectrum of pure SHb is obtained directly. The SHb spectrum thus obtained, also shown in Figs 10 and 17 of ref. [210], is clearly flatter between 500 and 600 nm.

SPECTRAL CHARACTERISTICS OF MODIFIED HUMAN HAEMOGLOBIN

Although potentially toxic [418], stroma-free haemoglobin (SFH) solutions have long been recognised as promising substitutes for blood because they combine oxygen carrying capability with colloid-osmotic activity and low viscosity. However, a too high oxygen affinity and a too short retention time limit application of simple SFH solutions to short-time physiological experiments [233, 404, 405]. Clinical application in humans requires a lower oxygen affinity and a longer retention time. A decrease in oxygen affinity can be effected by covalent binding of organic phosphate compounds such as pyridoxal-5′-phosphate [345]; the retention time can be increased by stabilising the haemoglobin tetramer through cross-linking [306] and by polymerisation [133]. Cross-linking is effective because most of the loss of SFH is through dimeric renal excretion following dissociation of the tetramer. Polymerisation also decreases the colloid-osmotic activity thus allowing an increase of the oxygen capacity of the solution through increasing the mass concentration of haemoglobin.

When a modified haemoglobin solution is used for infusion in patients, errors in optical measurements on blood samples or *in vivo* would result if spectral differences should exist between modified and unmodified haemoglobin. It is therefore desirable to determine the absorption spectra of the common derivatives of a modified haemoglobin before it is used in patients in the care of whom methods like pulse oximetry and multicomponent analysis of haemoglobin derivatives are used.

Table 8.16.
Absorptivity ($L \cdot mmol^{-1} \cdot cm^{-1}$) at various wavelengths of the common derivatives of Hb-NFPLP in comparison with the corresponding derivatives of adult human haemoglobin

λ (nm)	ε_{HHb}		λ (nm)	ε_{COHb}	
	M	U		M	U
478	3.18	3.30	496	5.15	5.23
506	4.78	4.95	538	14.23	14.30
554	13.05	13.35	554	11.67	11.63
			568	14.42	14.43
	ε_{O_2Hb}			ε_{Hi}	
	M	U		M	U
508	4.74	4.88	502	9.06	9.10
524	7.21	7.32	524	7.21	7.37
542	14.42	14.52	602	3.07	3.20
560	8.72	8.77	632	3.87	3.90
578	15.36	15.36			

Standard procedure. HP8450A.
M (modified) is Hb-NFPLP; data from ref. [307]. U (unmodified) is human HbA; data from Tables 8.1 and 8.2.

In anticipation of possible clinical application, spectral properties have been determined of the HHb, O_2Hb, COHb, Hi and HiCN derivatives of adult human haemoglobin modified by coupling to 2-nor-2-formylpyridoxal 5′-phosphate (NFPLP) [307]. Human haemoglobin cross-linked with NFPLP between the β-chains has a right-shifted oxygen dissociation curve, and a normal oxygen capacity of 1.4 mL/g [307]. It has been shown that the lower oxygen affinity of Hb-NFPLP, as compared to that of unmodified haemoglobin, enhances oxygen delivery to the tissues [305], and that the retention time has increased by a factor 3, due to prevention of renal excretion [32].

The absorptivity at 540 nm of the HiCN derivative of Hb-NFPLP was determined by the methods described in Chapter 5, using an Optica CF4 grating spectrophotometer. A mean value of 11.01 $L \cdot mmol^{-1} \cdot cm^{-1}$ was found for $\varepsilon_{HiCN\text{-}NFPLP}(540)$ with a standard error of 0.02 $L \cdot mmol^{-1} \cdot cm^{-1}$ ($N = 4$). This value does not differ from the internationally established value for HiCN (11.0 $L \cdot mmol^{-1} \cdot cm^{-1}$; Chapter 4) demonstrating that the total haemoglobin concentration of blood containing Hb-NFPLP can be reliably determined by the HiCN method (Chapter 16). $\varepsilon_{HiCN\text{-}NFPLP}(540) = 11.0\ L \cdot mmol^{-1} \cdot cm^{-1}$ has also been used as the anchor value in the determination of the absorptivities of the common derivatives of Hb-NFPLP.

Samples of the HHb, O_2Hb, COHb, and Hi derivatives of Hb-NFPLP were prepared using, as far as possible, the procedures described in Chapter 6.

Since the stock preparation already was a haemoglobin solution, no detergent was introduced into the tonometers at the end of the equilibration procedure. No sodium dithionite was added to prevent reoxygenation of the HHb derivative. The absorbance in the range of 480–680 nm was measured with an HP8450A, with a lightpath length of 0.013 cm (0.100-cm cuvettes with 0.087 cm glass inserts). Correction for contaminating Hi in the HHb, O_2Hb, and COHb samples of Hb-NFPLP was made according to equation 7.2.

Table 8.16 compares the absorptivities at wavelengths near the absorption maxima and minima obtained for the common derivatives of Hb-NFPLP with those of human HbA prepared with the standard procedure (Tables 8.1 and 8.2). In spite of the differences in sample preparation, there is good agreement between the absorptivities of Hb-NFPLP and those of unmodified adult human haemoglobin. The lower absorption maximum of HHb-NFPLP in comparison with that of unmodified HHb is probably due to slight reoxygenation since no $Na_2S_2O_4$ had been added. These data demonstrate that cross-linking human HbA with NFPLP has very little, if any, influence on the absorption spectra of the common derivatives. For the complete spectra of Hb-NFPLP, see ref. [307].

CHAPTER 9

ABSORPTION SPECTRA OF DOG HAEMOGLOBIN

Because the dog was frequently used in our laboratory as an experimental animal in blood gas physiology [168, 238, 324, 452] and in circulation research involving exchange transfusions with stroma-free haemoglobin solution [404, 405], a study of the light absorption spectra of dog haemoglobin was deemed necessary after differences between human foetal and human adult haemoglobin had been found [177, 446]. Dog haemoglobin thus became the first non-human haemoglobin of which we made a study of this kind [472].

The first step in this investigation of course was to determine the absorptivity of HiCN at 540 nm, $\varepsilon_{HiCN}(540)$, the anchor value for the whole of haemoglobin spectrophotometry, as described in Chapter 5. This yielded a value of 10.96 $L \cdot mmol^{-1} \cdot cm^{-1}$ for $\varepsilon_{HiCN}(540)$ (Tables 5.1 and 5.2), which is within the range of values underlying the internationally accepted value of 11.0 $L \cdot mmol^{-1} \cdot cm^{-1}$. Hence, the latter value was used in the calculation of the absorptivities of the haemoglobin derivatives of the dog.

At the time of this investigation the standard procedure (see p. 55) had not yet been developed. Therefore the obvious approach was to first prepare a stroma-free haemoglobin solution, as described in Chapter 5, then to use this stock solution to prepare pure solutions of the common haemoglobin derivatives. This procedure was similar to the SFH procedure, which also has been applied to human blood as described in Chapter 8. The results of a comparative study of the standard procedure and the SFH procedure are presented in Table 8.8 and Fig. 8.3.

Blood from healthy dogs (*Canis familiaris*) was obtained by venipuncture and collected in flasks containing 100 USP units of sodium heparin per mL blood. When measurements of $\varepsilon_{HiCN}(540)$ were required, iron free flasks were used. Solutions containing HHb, O_2Hb, COHb and Hi were prepared from the stock solution using procedures similar to those described in Chapter 6. As the stock solution was an erythrolysate, no non-ionic detergent solution was introduced into the tonometers near the end of the equilibration period, neither was $Na_2S_2O_4$ added to the deoxygenated haemoglobin solutions before filling the cuvettes. HiCN solutions were prepared as described in Chapter 5. The absorption spectra were determined using an HP8450A diode array spectrophotometer and an Optica CF4 grating spectrophotometer as described in Chapter 7.

Table 9.1 shows the absorptivities of HHb, O_2Hb, COHb, Hi and HiCN of dog haemoglobin for the spectral range of 450–700 nm measured with an HP8450A diode array spectrophotometer in 0.100-cm cuvettes with plan parallel glass inserts, with the exception of HiCN of which a diluted solution was measured in a 1.000-cm cuvette. Part of these data have been published earlier [472]. Since, for obtaining accurate results, the measured absorbance should not be below 0.050, only values which have been measured with $A \geqslant 0.050$ have been included in the table. The absorptivities of the derivatives of canine haemoglobin have been compared with those of adult human haemoglobin by means of Student's t-test for unpaired samples. The results at the wavelengths near the principal absorption maxima and minima in the spectra of the common derivatives are presented in Table 9.2.

In Fig. 9.1 the absorption spectra of the common derivatives of dog haemoglobin are shown in comparison with those of human HbA. The spectra have been recorded in a continuous run over the indicated spectral range with a constant small lightpath. For some haemoglobin derivatives the measured absorbance at $\lambda > 600$ nm was < 0.050. These values have not been given in Table 9.1. The samples measured have been prepared with the SFH procedure, for both canine and human haemoglobin. There appears to be little difference between the absorption spectra of the two haemoglobins. The maximum in the HHb spectrum is a little lower in the dog. In the O_2Hb spectrum the α-peak is lower in the dog, while at the β-peak and in the minimum near 508 nm the absorptivity of canine O_2Hb slightly exceeds that of human O_2Hb. Also in the COHb spectrum, the α-peak seems a little lower, the β-peak slightly higher in the dog, but most distinct is the deeper trough in the dog COHb spectrum at 554 nm. Table 9.2 shows that most of these minute differences are statistically significant. The most conspicious differences between canine and human haemoglobin, however, are in the Hi spectra, as clearly demonstrated in Fig. 9.1 and Table 9.2. These differences should be considered while taking any differences in pH of the solutions into account.

Figure 9.1 does not show any difference in the position of the principal absorption maxima and minima in the spectra of the haemoglobin derivatives between the haemoglobins of dog and man. However, since these spectra have been recorded with a diode array spectrophotometer that scans the spectrum at 2 nm intervals, it cannot be ruled out, based on these data, that small differences in the position of the absorption maxima do exist. Therefore, the maxima and minima in the spectra of these haemoglobin derivatives have been exactly located with the aid of an Optica CF4 grating spectrophotometer.

For this purpose, measurements were made at intervals of 0.5 nm, with a spectral bandwidth of only 0.06 nm. The results are consolidated in

Table 9.1.
Absorptivities ($L \cdot mmol^{-1} \cdot cm^{-1}$) of the common derivatives and of HiCN of dog haemoglobin

λ (nm)	ε_{HHb} $N = 8$	ε_{O_2Hb} $N = 11$	ε_{COHb} $N = 7$	ε_{Hi} $N = 10$	ε_{HiCN} $N = 5$
450	12.85	15.82	10.02	9.116	19.87
452	10.15	14.57	9.334	8.803	18.61
454	8.233	13.50	8.769	8.561	17.37
456	6.840	12.54	8.276	8.375	16.16
458	5.829	11.71	7.862	8.227	14.99
460	5.086	10.96	7.495	8.102	13.92
462	4.538	10.30	7.180	8.000	12.98
464	4.139	9.708	6.899	7.909	12.17
466	3.852	9.183	6.655	7.838	11.47
468	3.648	8.706	6.432	7.780	10.91
470	3.512	8.282	6.234	7.747	10.46
472	3.425	7.896	6.056	7.738	10.07
474	3.377	7.550	5.906	7.763	9.767
476	3.359	7.231	5.767	7.820	9.512
478	3.360	6.945	5.656	7.906	9.272
480	3.381	6.685	5.557	8.023	9.056
482	3.415	6.453	5.474	8.164	8.822
484	3.462	6.247	5.393	8.322	8.564
486	3.521	6.070	5.331	8.490	8.313
488	3.595	5.906	5.267	8.663	8.043
490	3.685	5.763	5.219	8.834	7.776
492	3.788	5.633	5.182	8.991	7.519
494	3.909	5.520	5.164	9.133	7.313
496	4.044	5.412	5.160	9.245	7.129
498	4.198	5.317	5.182	9.327	7.001
500	4.368	5.229	5.218	9.363	6.888
502	4.550	5.149	5.283	9.358	6.855
504	4.745	5.074	5.370	9.299	6.862
506	4.952	5.018	5.497	9.194	6.925
508	5.160	4.986	5.660	9.049	7.050
510	5.369	4.990	5.882	8.875	7.214
512	5.567	5.046	6.164	8.667	7.440
514	5.775	5.174	6.546	8.445	7.713
516	5.978	5.388	7.021	8.206	8.031
518	6.195	5.718	7.616	7.960	8.393
520	6.424	6.177	8.300	7.715	8.765
522	6.676	6.785	9.076	7.475	9.174
524	6.951	7.552	9.889	7.242	9.569
526	7.261	8.473	10.74	7.031	9.932
528	7.595	9.496	11.55	6.835	10.24
530	7.966	10.57	12.34	6.662	10.50
532	8.373	11.61	13.04	6.501	10.70

Table 9.1.
(Continued)

λ (nm)	ε_{HHb} $N = 8$	ε_{O_2Hb} $N = 11$	ε_{COHb} $N = 7$	ε_{Hi} $N = 10$	ε_{HiCN} $N = 5$
534	8.821	12.55	13.66	6.360	10.83
536	9.306	13.33	14.07	6.213	10.93
538	9.840	13.95	14.27	6.059	10.96
540	10.39	14.33	14.20	5.887	11.00
542	10.95	14.45	13.90	5.687	11.00
544	11.47	14.21	13.40	5.458	10.99
546	11.94	13.62	12.85	5.202	10.92
548	12.33	12.74	12.28	4.929	10.83
550	12.64	11.76	11.83	4.655	10.68
552	12.83	10.79	11.49	4.392	10.49
554	12.92	9.968	11.33	4.154	10.25
556	12.90	9.315	11.35	3.949	9.976
558	12.79	8.875	11.58	3.784	9.682
560	12.60	8.666	11.98	3.657	9.373
562	12.37	8.710	12.56	3.569	9.071
564	12.10	9.027	13.21	3.510	8.760
566	11.78	9.640	13.82	3.476	8.458
568	11.42	10.56	14.19	3.463	8.144
570	11.04	11.75	14.16	3.461	7.834
572	10.64	13.08	13.62	3.466	7.510
574	10.22	14.30	12.63	3.473	7.173
576	9.792	15.02	11.31	3.477	6.814
578	9.365	14.93	9.824	3.474	6.448
580	8.939	13.91	8.284	3.460	6.081
582	8.517	12.12	6.812	3.432	5.708
584	8.084	9.928	5.501	3.391	5.352
586	7.624	7.715	4.394	3.338	5.002
588	7.111	5.756	3.499	3.279	4.661
590	6.550	4.213	2.811	3.220	4.329
592	5.927	3.052	2.273	3.161	3.996
594	5.283	2.237	1.865	3.109	3.676
596	4.639	1.676	1.551	3.066	3.357
598	4.027	1.288	1.308	3.032	3.049
600	3.479	1.024	1.117	3.014	2.755
602	3.003	0.835	0.964	3.007	2.486
604	2.610	0.698	0.840	3.013	2.225
606	2.293	0.596	0.740	3.032	1.987
608	2.041	0.517	0.658	3.062	1.780
610	1.841	0.455	0.589	3.105	1.592
612	1.683		0.531	3.162	1.424
614	1.556		0.481	3.238	1.291
616	1.455			3.330	1.175
618	1.369			3.438	1.065

Table 9.1.
(Continued)

λ (nm)	ε_{HHb} $N = 8$	ε_{O_2Hb} $N = 11$	ε_{COHb} $N = 7$	ε_{Hi} $N = 10$	ε_{HiCN} $N = 5$
620	1.297			3.555	0.971
622	1.235			3.672	0.881
624	1.183			3.784	0.802
626	1.138			3.884	0.740
628	1.098			3.965	0.681
630	1.062			4.021	0.633
632	1.031			4.042	0.593
634	1.002			4.019	0.556
636	0.977			3.938	0.525
638	0.953			3.796	0.495
640	0.933			3.593	0.467
642	0.912			3.336	
644	0.894			3.043	
646	0.876			2.731	
648	0.859			2.414	
650	0.843			2.105	
652	0.828			1.816	
654	0.811			1.555	
656	0.820			1.346	
658	0.805			1.108	
660	0.765			0.933	
662	0.751			0.791	
664	0.735			0.668	
666	0.717			0.568	
668	0.697			0.485	
670	0.676				
672	0.655				
674	0.633				
676	0.611				
678	0.589				
680	0.567				
682	0.547				
684	0.526				
686	0.506				
688	0.486				
690	0.469				
692	0.452				

λ = 450–700 nm. SFH procedure; HP8450A.

Absorbance measured with $l = 0.010–0.016$ cm and $l = 1.000$ cm (HiCN). N is number of specimens. pH (Hi) ≈ 7.15.

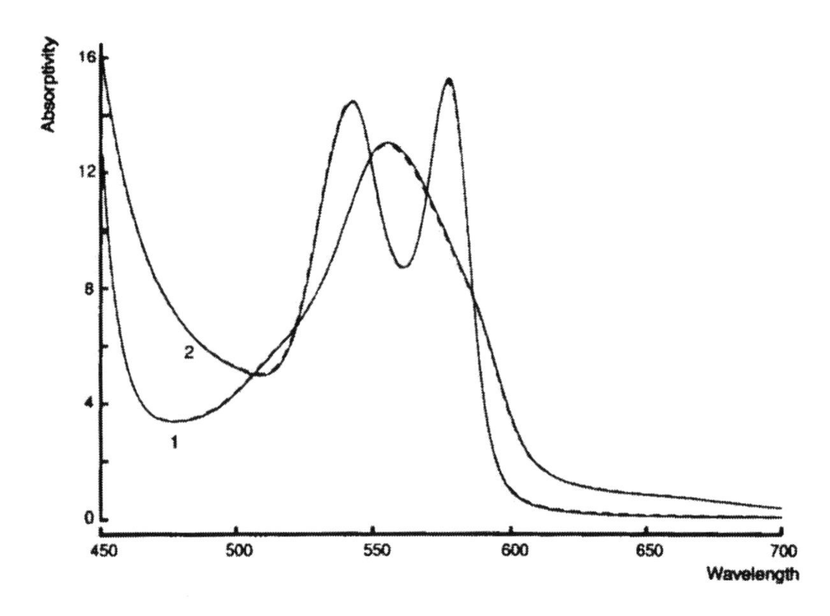

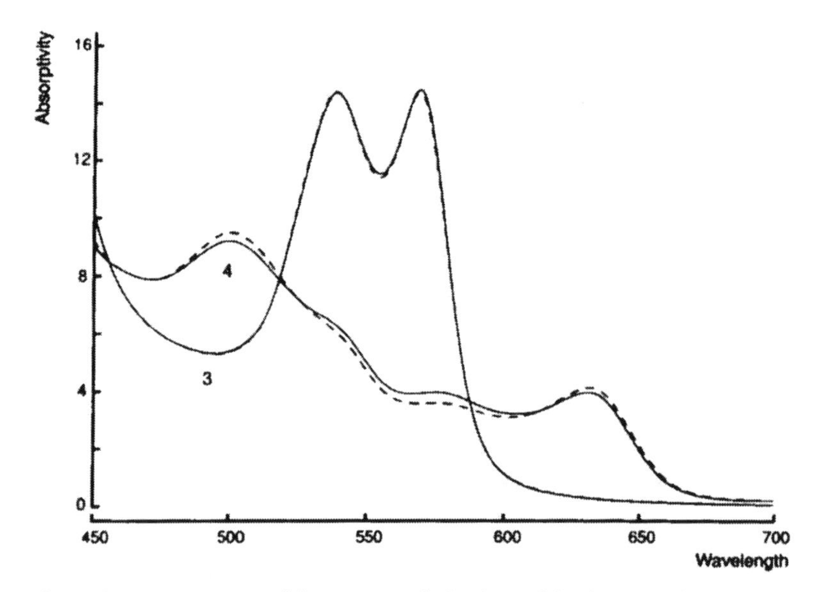

Figure 9.1. Absorption spectra of the common derivatives of dog haemoglobin in comparison with those of human HbA (SFH procedure).

Absorptivity in $L \cdot mmol^{-1} \cdot cm^{-1}$, wavelength in nm. Solid lines: human haemoglobin; dashed lines: dog haemoglobin. Upper panel: HHb (1) and O_2Hb (2); lower panel: COHb (3) and Hi (4). Absorbance measured with $l = 0.010–0.016$ cm. HP8450A. pH (Hi) ≈ 7.15.

Table 9.2.
Comparison of absorptivities of the common derivatives of human and dog haemoglobin

	Human		Dog		
λ (nm)	ε_{HHb} $N = 11$	*SEM*	ε_{HHb} $N = 8$	*SEM*	*p*
476	3.323	0.008	3.359	0.021	
478	3.325	0.007	3.360	0.020	
552	12.89	0.030	12.83	0.014	
554	13.00	0.029	12.92	0.015	< 0.05
556	12.99	0.031	12.90	0.014	< 0.05
λ (nm)	ε_{O_2Hb} $N = 11$	*SEM*	ε_{O_2Hb} $N = 11$	*SEM*	*p*
508	4.925	0.012	4.986	0.016	< 0.01
540	14.24	0.043	14.33	0.043	< 0.05
542	14.41	0.039	14.45	0.045	
560	8.719	0.026	8.667	0.030	
576	15.10	0.053	15.02	0.051	
578	15.13	0.054	14.93	0.053	< 0.01
λ (nm)	ε_{COHb} $N = 11$	*SEM*	ε_{COHb} $N = 7$	*SEM*	*p*
496	5.187	0.012	5.160	0.014	
538	14.22	0.034	14.27	0.040	
540	14.19	0.030	14.20	0.041	
554	11.45	0.025	11.33	0.029	< 0.01
568	14.27	0.037	14.19	0.037	
570	14.30	0.037	14.16	0.041	< 0.02
λ (nm)	ε_{Hi} $N = 11$	*SEM*	ε_{Hi} $N = 10$	*SEM*	*p*
498	9.057	0.010	9.327	0.043	< 0.001
500	9.078	0.010	9.363	0.043	< 0.001
502	9.064	0.011	9.358	0.044	< 0.001
630	3.868	0.006	4.021	0.021	< 0.001
632	3.877	0.006	4.042	0.022	< 0.001
634	3.845	0.006	4.019	0.023	< 0.001

SFH procedure; HP8450A.
Absorptivity in $L \cdot mmol^{-1} \cdot cm^{-1}$; *N* is number of specimens; *SEM* is standard error of the mean. *p* values of Student's *t*-test for unpaired samples given only when $p < 0.05$.

Tables 9.3 and 9.4. Part of these data have also been presented in Table 2 of ref. [472]. In this table the mean values of the wavelengths of maximum and minimum absorption in the spectra of the haemoglobin derivatives of each of the specimens are given. This yielded for the α- and β-peaks of O_2Hb 576.7 and 541.6 nm, and for the α- and β-peaks of COHb 568.6 and 538.4 nm. Tables 9.3 and 9.4 show the mean values of the absorptivities

Table 9.3.
Absorption maxima and minima in the spectra of HHb and O_2Hb of dog haemoglobin

λ (nm)	ε_{HHb}	*SEM*	*N*	λ (nm)	ε_{O_2Hb}	*SEM*	*N*
475	3.446	0.077	6	507	4.942	0.037	10
475.5	3.446	0.077	6	508	4.936	0.036	10
476	3.414	0.070	7	508.5	4.908	0.026	9
477	3.437	0.120	4	509	4.908	0.026	9
478	3.366	0.033	5	509.5	4.913	0.026	9
				510	4.962	0.037	9
552	12.95	0.077	7				
553	13.01	0.076	7				
553.5	13.01	0.077	7	540	14.42	0.065	9
554	13.04	0.077	7	541	14.46	0.065	10
554.5	13.03	0.078	7	541.5	14.47	0.064	10
555	13.02	0.081	7	542	14.47	0.064	10
556	13.00	0.093	6	542.5	14.44	0.067	9
				543	14.34	0.068	6
				559	8.587	0.043	11
				559.5	8.561	0.042	11
				560	8.545	0.042	11
				560.5	8.548	0.043	11
				561	8.569	0.043	11
				562	8.678	0.056	9
				575	15.01	0.083	11
				576	15.27	0.079	11
				576.5	15.32	0.078	11
				577	15.30	0.081	11
				577.5	15.26	0.096	9
				578	15.07	0.082	9

SFH procedure; Optica CF4.
Absorptivity in $L \cdot mmol^{-1} \cdot cm^{-1}$. Absorbance measured with $l = 0.015$ cm. *N* is number of specimens.

measured at the given wavelengths. As to the wavelengths of maximum and minimum absorption, the difference in location is negligible.

Comparing the data of Tables 9.3 and 9.4 with those of Tables 8.5 and 8.6 shows that there is indeed no difference in the position of the principal absorption maxima and minima in the spectra of the common derivatives between canine and human haemoglobin, perhaps with the single exception of the minimum in the O_2Hb spectrum, which is found in the dog at 560 nm, in man at 560.5 nm. In the calculation presented in ref. [472], this difference appeared to be statistically significant ($p < 0.05$).

The absorptivities of the derivatives of canine haemoglobin as measured with the Optica CF4 spectrophotometer have also been compared with those

Table 9.4.
Absorption maxima and minima in the spectra of COHb and Hi of dog haemoglobin

λ (nm)	ε_{COHb}	*SEM*	*N*	λ (nm)	ε_{Hi}	*SEM*	*N*
494	5.131	0.021	7	499	9.274	0.054	10
494.5	5.129	0.022	5	500	9.290	0.050	11
495	5.120	0.020	7				
495.5	5.119	0.020	7	630	3.988	0.026	10
496	5.123	0.021	7	630.5	3.995	0.030	9
496.5	5.118	0.028	5	631	3.999	0.027	10
497	5.124	0.020	7	631.5	3.998	0.027	10
				632	3.998	0.028	10
				633	4.000	0.023	9
537	14.34	0.083	6	634	3.976	0.030	7
537.5	14.41	0.076	6				
538	14.40	0.068	7				
538.5	14.39	0.065	7				
539	14.40	0.064	7				
540	14 37	0.050	6				
554	11.33	0.058	7				
554.5	11.33	0.057	7				
555	11.34	0.057	7				
568	14.40	0.068	7				
568.5	14.43	0.066	7				
569	14.42	0.060	7				
570	14.30	0.060	7				

SFH procedure; Optica CF4.
Absorptivity in $L \cdot mmol^{-1} \cdot cm^{-1}$. Absorbance measured with $l = 0.015$ cm. *N* is number of specimens.

of adult human haemoglobin measured with the same instrument. For this purpose data were taken from Tables 8.5 and 8.6 and Student's t-test for unpaired samples was used. The results are presented in Table 9.5.

Although the two sets of data given in Tables 9.2 and 9.5 generally agree, there are some remarkable differences. Firstly, the standard error is usually greater for the measurements with the Optica CF4. For a proper comparison of the data as obtained with the two spectrophotometers, the data of dog haemoglobin are the most suitable, because the number of measurements hardly differ between the two series. The greater spread in the measurements with the Optica CF4 is most probably due to the fact that it is a manual method in which the wavelength scale has to be adjusted, the potentiometer balanced and the absorbance read in each individual measurement, whereas the HP8450A automatically scans through the spectrum. Secondly, at wavelengths near narrow maxima the Optica CF4 measures a higher absorptivity, which is most clearly seen for the sharp α-peak of O_2Hb. This difference

Table 9.5.
Comparison of absorptivities of the common derivatives of dog and human haemoglobin

	Human		Dog			
λ (nm)	ε_{HHb} $N = 5$	*SEM*	ε_{HHb}	N	*SEM*	*p*
476	3.285	0.019	3.414	7	0.070	
478	3.289	0.016	3.366	5	0.033	
552	12.76	0.029	12.95	7	0.077	
554	12.84	0.041	13.04	7	0.077	
556	12.79	0.043	13.00	6	0.093	
λ (nm)	ε_{O_2Hb} $N = 5$	*SEM*	ε_{O_2Hb}	N	*SEM*	*p*
508	4.799	0.013	4.936	10	0.036	< 0.020
540	14.28	0.034	14.42	9	0.065	
542	14.44	0.046	14.47	10	0.064	
560	8.554	0.024	8.545	11	0.042	
576	15.31	0.044	15.27	11	0.079	
578	15.20	0.038	15.07	9	0.082	
λ (nm)	ε_{COHb} $N = 5$	*SEM*	ε_{COHb}	N	*SEM*	*p*
496	5.093	0.017	5.123	7	0.021	
538	14.16	0.017	14.40	7	0.068	< 0.020
540	14.10	0.037	14.37	6	0.050	< 0.010
554	11.28	0.019	11.33	7	0.058	
568	14.29	0.030	14.40	7	0.068	
570	14.24	0.020	14.30	7	0.060	
λ (nm)	ε_{Hi} $N = 5$	*SEM*	ε_{Hi}	N	*SEM*	*p*
499	9.015	0.032	9.274	10	0.054	< 0.010
500	9.023	0.031	9.290	11	0.050	< 0.010
630	3.858	0.023	3.988	10	0.026	< 0.010
632	3.852	0.022	3.998	10	0.028	< 0.010
634	3.803	0.023	3.976	7	0.030	< 0.002

SFH procedure; Optica CF4.
Absorptivity in $L \cdot mmol^{-1} \cdot cm^{-1}$; N is number of specimens; *SEM* is standard error of the mean. p values of Student's t-test for unpaired samples given only when $p < 0.05$.

originates from the considerable difference in spectral resolution between the two spectrophotometers (pp. 22; 48). These observations clearly show that for reliably demonstrating small differences between the absorption spectra of haemoglobin derivatives of different animal species a strict similarity of the applied techniques is necessary. This should be even more strict than has been realised in the present investigation of dog haemoglobin.

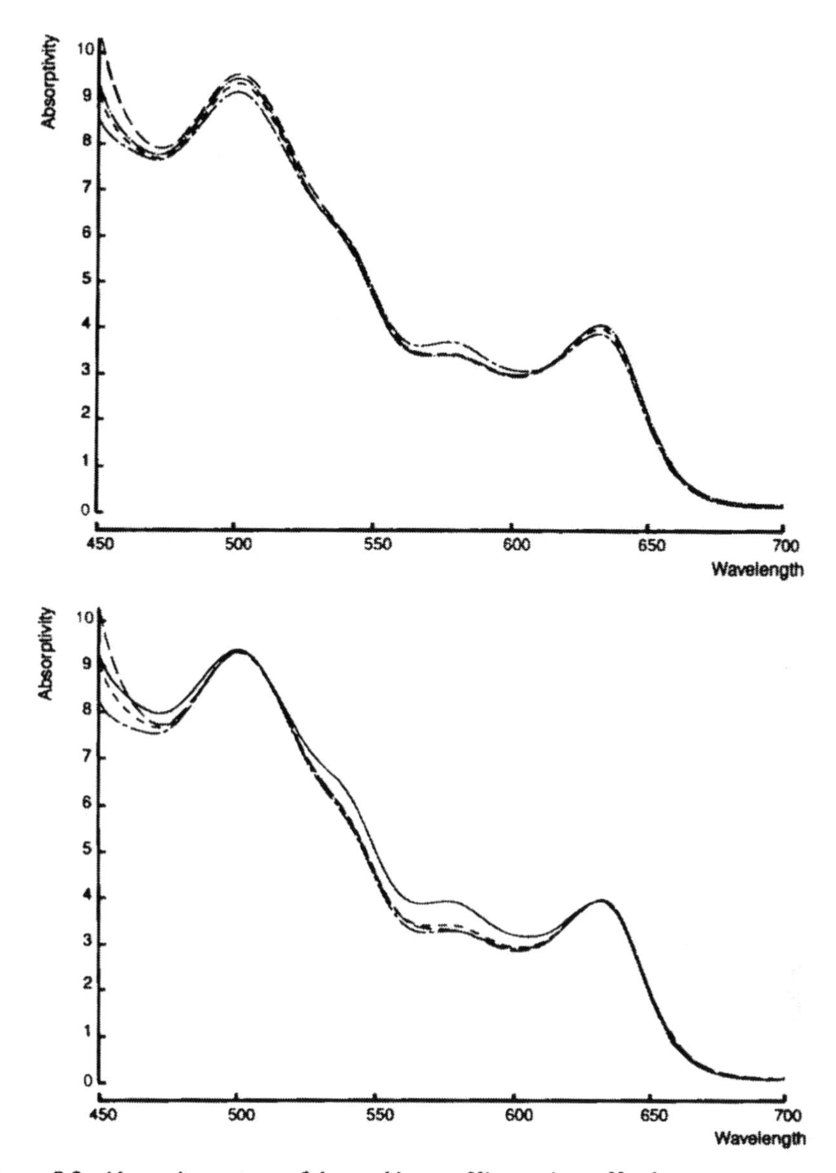

Figure 9.2. Absorption spectra of dog and human Hi at various pH values.

The upper panel shows the absorption spectra of dog Hi at pH = 7.05 (— —), 7.15 (·——·) and 7.35 (— - —). The mean dog Hi spectrum (Table 9.1), pH ≈ 7.15, has been added (- - - -). In the lower panel the dog Hi spectra at pH = 7.05 (— —) and 7.35 (— - —) are shown together with the mean dog Hi spectrum (- - - -), and the mean human Hi spectrum (SFH procedure; Table 8.7), pH range = 7.1–7.2 (——). The absorptivity at 630 nm has been normalised to 4.021 $L \cdot mmol^{-1} \cdot cm^{-1}$.

As to the absorptivity values of canine and human Hi, both Tables 9.2 and 9.5, as well as Fig. 9.1 show clear differences between dog and man.

This difference, however, should be interpreted in relation to the pH of the solution. In ref. [472], we concluded, from the similarity in shape of the two Hi spectra at similar pH, that the difference between canine and human Hi was due only to a different affinity for OH^- ions. In light of further investigation this conclusion does not seem to be warranted.

The upper panel of Fig. 9.2 shows three absorption spectra of dog haemiglobin at pH values between 7.05 and 7.35, and the mean dog Hi spectrum ($N = 10$; pH ≈ 7.15). There appears to be only a moderate influence of the differences in pH on shape and position of the dog Hi spectrum. The lower panel, in which the absorptivity curves have been normalised to $\varepsilon_{Hi}(630) = 4.021\ L \cdot mmol^{-1} \cdot cm^{-1}$, the actual value in the mean dog Hi spectrum, and in which the mean human Hi spectrum obtained with the SFH procedure ($N = 11$; pH range $= 7.1-7.2$) has been added, demonstrates that it is improbable that there is a pH at which the absorption spectra of canine and human Hi will coincide. Hence, there is an intrinsic difference between the Hi spectra of the two kinds of haemoglobin.

CHAPTER 10

ABSORPTION SPECTRA OF RAT HAEMOGLOBIN

Rat haemoglobin has been selected for spectrophotometric study because the rat (*Rattus rattus*) is among the most frequently used experimental animals in biomedical research. Preparation of haemoglobin stock solutions of rat blood by means of the standard procedure used for the blood of other animal species including man, proved to be impossible because of the limited solubility of rat haemoglobin in electrolyte-poor aqueous solutions. Therefore we had to use a different approach which resulted in stock solutions of a different total haemoglobin concentration, osmolality and pH than have been used in recording the haemoglobin spectra of the other animal species.

Blood was obtained from healthy male Wistar rats. Usually the blood of two rats, about 20 mL, was collected in an iron-free flask containing 0.5 mL heparin (2.5 mg), and 20 mL 9 g/L NaCl solution was added. When blood of only one rat was obtained, the volume of the diluents was adjusted accordingly. The mixture was subjected to centrifugation at 8000*g* for 10 min at 4°C. Washing of the erythrocytes with saline solution was repeated twice. The red cells were then weighed and for each gram of cells 2.75 mL distilled water was added. After mixing and waiting for erythrolysis, which took about 3 min to occur, 1.25 mL of a buffer solution and 750 mg NaCl were added per g of cells. The buffer solution was a mixture of $Na_2HPO_4 \cdot 2H_2O$ (0.50 mol/L) and KH_2PO_4 (0.50 mol/L) of pH = 7.17 at 25°C. After mixing thoroughly, the erythrolysate was subjected to centrifugation at 8000*g* for 15 min at 4°C.

The total haemoglobin concentration, ct_{Hb}, of the stock solution was about 2.5 mmol/L. The stock solution was used for the determination of the absorptivity of HiCN at 540 nm, $\varepsilon_{HiCN}(540)$, on the basis of determination of haemoglobin iron, as well as for the preparation of pure solutions of HHb, O_2Hb, COHb, and Hi for recording of the absorption spectra of these haemoglobin derivatives. When the stock solution was to be used for the determination of $\varepsilon_{HiCN}(540)$, all glassware had been made iron-free and all reagents were of analytical grade as described in Chapter 5.

In Chapter 5 it has been shown that the value of 10.97 $L \cdot mmol^{-1} \cdot cm^{-1}$ found for $\varepsilon_{HiCN}(540)$ of the rat (Tables 5.1 and 5.2) is within the range of values on which the internationally accepted value is based. Hence,

$\varepsilon_{HiCN}(540) = 11.0\ L \cdot mmol^{-1} \cdot cm^{-1}$ is also valid for rat haemoglobin. In this chapter the absorption spectra of HHb, O_2Hb, COHb and Hi of rat haemoglobin are presented. Some of these results have been reported earlier [477].

Starting with the haemoglobin stock solution prepared as described above, solutions of the four common haemoglobin derivatives were prepared, following, as far as possible, the procedures described in Chapter 6. As the stock solution was an erythrolysate, no non-ionic detergent solution was introduced into the tonometers near the end of the equilibration period, neither was any $Na_2S_2O_4$ added to the deoxygenated haemoglobin solutions before filling the cuvettes. The measurements were made as described in Chapter 7. As the complete procedure leading up to the absorptivities of the haemoglobin derivatives of rat haemoglobin was different from the standard procedure as well as from the SFH procedure used in preparing the haemoglobin solutions of the other animal species, this approach is called **Rat procedure**.

Table 10.1 presents the absorptivities of HHb, O_2Hb, COHb, Hi, and HiCN of rat haemoglobin for the spectral range of 450–630 nm measured with an HP8450A diode array spectrophotometer in 0.100-cm cuvettes with plan parallel glass inserts, with the exception of HiCN, of which a diluted solution was measured in a 1.000-cm cuvette. To obtain reliable results, the measured absorbance should not be less than 0.050 (p. 47). No absorptivity values have been given when the measured absorbance was < 0.050. Table 10.2 shows the absorptivities for the spectral range of 600–800 nm measured with the same instrument; the lightpath was adjusted so that the measured absorbance was > 0.050. Absorptivity values have not been given when the spectrophotometer rejected the measurements because the absorbance of the samples was too high. This happened occasionally in measurements with $l = 0.200$ cm between 600 and 650 nm. The pH of the Hi solutions was measured, but no attempt was made to bring the pH within the range of pH values that ensues when a fresh rat blood sample is haemolysed. The pH of the solutions of the haemoglobin derivatives was determined by the requirements of producing clear and stable solutions. The spectra represented by the absorptivities of Tables 10.1 and 10.2 are designated as the **standard spectra** of rat haemoglobin.

Figure 10.1 shows the absorption spectra of HHb, O_2Hb, COHb and Hi of rat haemoglobin in comparison with those of human HbA for the spectral ranges of 450–700 nm (Table 10.1) and 600–800 nm (Table 10.2). In Fig. 10.1a the spectral absorption curves run to 700 nm, but for $\lambda > 600$ nm part of the measurements have been made with $A < 0.050$; these values have not been given in Table 10.1.

The absorptivities of the derivatives of rat haemoglobin have been compared with those of adult human haemoglobin by means of Student's t-test for unpaired samples. Only p values < 0.05 are shown. The results at the

Table 10.1.
Absorptivities ($L \cdot mmol^{-1} \cdot cm^{-1}$) of the common derivatives and of HiCN of rat haemoglobin

λ (nm)	ε_{HHb} $N = 8$	ε_{O_2Hb} $N = 11$	ε_{COHb} $N = 8$	ε_{Hi} $N = 16$	ε_{HiCN} $N = 13$
450	13.92	16.37	9.991	9.630	20.08
452	10.99	15.10	9.319	9.269	18.83
454	8.923	14.00	8.766	8.979	17.60
456	7.420	13.02	8.296	8.732	16.38
458	6.330	12.15	7.897	8.524	15.23
460	5.518	11.38	7.549	8.339	14.16
462	4.926	10.69	7.245	8.171	13.21
464	4.482	10.07	6.988	8.016	12.40
466	4.162	9.521	6.759	7.881	11.71
468	3.930	9.024	6.545	7.764	11.12
470	3.774	8.578	6.363	7.678	10.65
472	3.667	8.171	6.196	7.617	10.25
474	3.605	7.812	6.055	7.593	9.933
476	3.576	7.480	5.930	7.608	9.648
478	3.564	7.177	5.817	7.658	9.410
480	3.573	6.895	5.737	7.741	9.149
482	3.598	6.653	5.655	7.853	8.929
484	3.636	6.432	5.585	7.985	8.655
486	3.693	6.238	5.523	8.136	8.405
488	3.759	6.069	5.467	8.292	8.121
490	3.844	5.910	5.428	8.449	7.867
492	3.941	5.769	5.392	8.599	7.614
494	4.055	5.640	5.371	8.732	7.389
496	4.187	5.525	5.377	8.846	7.207
498	4.336	5.421	5.399	8.935	7.067
500	4.502	5.322	5.442	8.991	6.979
502	4.682	5.233	5.505	9.011	6.923
504	4.878	5.150	5.603	8.995	6.927
506	5.078	5.080	5.733	8.941	6.992
508	5.288	5.037	5.889	8.859	7.100
510	5.496	5.037	6.120	8.753	7.258
512	5.701	5.072	6.397	8.628	7.463
514	5.907	5.188	6.773	8.493	7.738
516	6.115	5.388	7.236	8.346	8.036
518	6.331	5.698	7.804	8.191	8.383
520	6.571	6.151	8.472	8.026	8.745
522	6.828	6.751	9.223	7.862	9.139
524	7.118	7.514	10.01	7.687	9.517
526	7.446	8.440	10.85	7.512	9.876
528	7.800	9.493	11.67	7.339	10.19
530	8.194	10.62	12.47	7.173	10.45
532	8.628	11.72	13.18	7.015	10.65
534	9.106	12.71	13.82	6.858	10.81

Table 10.1.
(Continued)

λ (nm)	ε_{HHb} $N = 8$	ε_{O_2Hb} $N = 11$	ε_{COHb} $N = 8$	ε_{Hi} $N = 16$	ε_{HiCN} $N = 13$
536	9.620	13.55	14.27	6.698	10.90
538	10.17	14.21	14.51	6.530	10.97
540	10.75	14.65	14.50	6.342	11.00
542	11.33	14.81	14.26	6.125	11.02
544	11.88	14.62	13.82	5.881	11.01
546	12.37	14.07	13.29	5.615	10.98
548	12.77	13.20	12.75	5.335	10.90
550	13.09	12.18	12.29	5.058	10.76
552	13.29	11.17	11.94	4.799	10.57
554	13.38	10.29	11.76	4.572	10.33
556	13.35	9.598	11.75	4.385	10.06
558	13.23	9.117	11.94	4.241	9.763
560	13.03	8.868	12.30	4.132	9.454
562	12.79	8.874	12.81	4.053	9.133
564	12.50	9.157	13.40	3.992	8.818
566	12.18	9.754	13.97	3.942	8.500
568	11.83	10.68	14.34	3.891	8.193
570	11.46	11.91	14.37	3.838	7.882
572	11.08	13.35	13.95	3.782	7.556
574	10.68	14.75	13.09	3.726	7.227
576	10.27	15.72	11.84	3.667	6.876
578	9.860	15.87	10.34	3.605	6.524
580	9.430	14.94	8.722	3.536	6.162
582	8.984	13.02	7.151	3.460	5.803
584	8.512	10.55	5.749	3.377	5.446
586	8.009	8.077	4.577	3.289	5.100
588	7.460	5.943	3.639	3.199	4.760
590	6.861	4.286	2.910	3.110	4.428
592	6.218	3.083	2.350	3.026	4.096
594	5.547	2.244	1.921	2.950	3.770
596	4.873	1.666	1.590	2.885	3.447
598	4.227	1.272	1.331	2.834	3.129
600	3.644	1.001	1.134	2.797	2.832
602	3.131	0.809	0.974	2.775	2.545
604	2.706	0.670	0.845	2.770	2.281
606	2.362	0.566	0.742	2.783	2.038
608	2.089	0.487	0.657	2.814	1.823
610	1.874		0.587	2.864	1.629
612	1.703		0.528	2.931	1.457
614	1.563		0.474	3.015	1.308

Table 10.1.
(Continued)

λ (nm)	ε_{HHb} $N = 8$	ε_{O_2Hb} $N = 11$	ε_{COHb} $N = 8$	ε_{Hi} $N = 16$	ε_{HiCN} $N = 13$
616	1.456			3.119	1.178
618	1.363			3.239	1.065
620	1.286			3.370	0.967
622	1.223			3.501	0.881
624	1.166			3.627	0.807
626	1.118			3.743	0.740
628	1.076			3.841	0.686
630	1.038			3.916	0.633

λ = 450–630 nm. Rat procedure; HP8450A.
Absorbance measured with $l = 0.010$–0.016 cm and $l = 1.000$ cm (HiCN). N is number of specimens. pH range (Hi) = 6.42–6.90.

wavelengths near the principal absorption maxima and minima in the spectra of the common derivatives are presented in Table 10.3.

Figure 10.1 and Table 10.3 show no differences in the position of the light absorption maxima and minima of the common derivatives of rat and adult human haemoglobin. As these absorption spectra had been measured with a diode array spectrophotometer, the results do not rule out that slight differences in the position of the absorption maxima and minima do exist. Therefore, the maxima and minima in the spectra of these haemoglobin derivatives have also been exactly located with the aid of an Optica CF4 grating spectrophotometer. Measurements were made at intervals of 0.5 nm, with a spectral bandwidth of only 0.06 nm.

The results are presented in Tables 10.4 and 10.5. Part of these data have also been presented in Table 2 of ref. [477]. In this table the mean values of the wavelengths found for the maximum and minimum points in the absorption spectra of the haemoglobin derivatives of each of the specimens are given. This yielded for the α- and β-peaks of O_2Hb 577.0 and 542.2 nm, and for the α- and β-peaks of COHb 569.0 and 539.1 nm. Tables 10.4 and 10.5 show the mean values of the absorptivities measured at the given wavelengths. As to the wavelengths of maximum and minimum absorption, the difference in location is negligible.

Comparing the data of Tables 10.4 and 10.5 with those of Tables 8.5 and 8.6 shows that there is indeed no difference in the position of the principal absorption maxima and minima in the spectra of the common derivatives between rat and human haemoglobin, with a single exception: the maximum of HHb is found in the rat at 553 nm, in the human at 554 nm; this difference is statistically significant ($p < 0.01$) [477]. The much larger difference observed for the maximum of Hi near 500 nm (503.5 nm in the rat, 500 nm in man), which also is discernible in the measurements with the

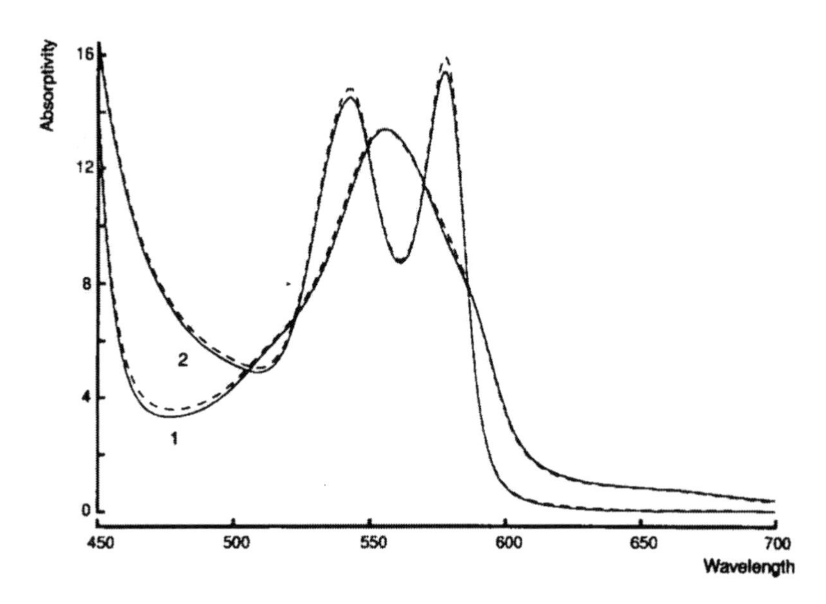

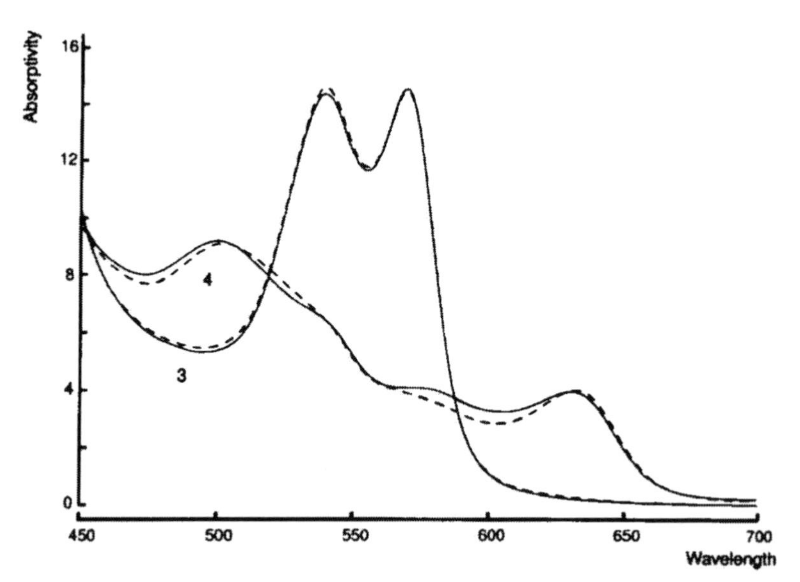

Figure 10.1a. Absorption spectra of the common derivatives of rat haemoglobin in comparison with those of human HbA.

Absorptivity in $L \cdot mmol^{-1} \cdot cm^{-1}$, wavelength in nm. Solid lines: human haemoglobin; dashed lines: rat haemoglobin. Upper panel: HHb (1) and O_2Hb (2); lower panel: COHb (3) and Hi (4). Absorbance measured with l = 0.010–0.016 cm. HP8450A. pH (rat Hi) = 6.42–6.90; pH (human Hi) = 7.14–7.30.

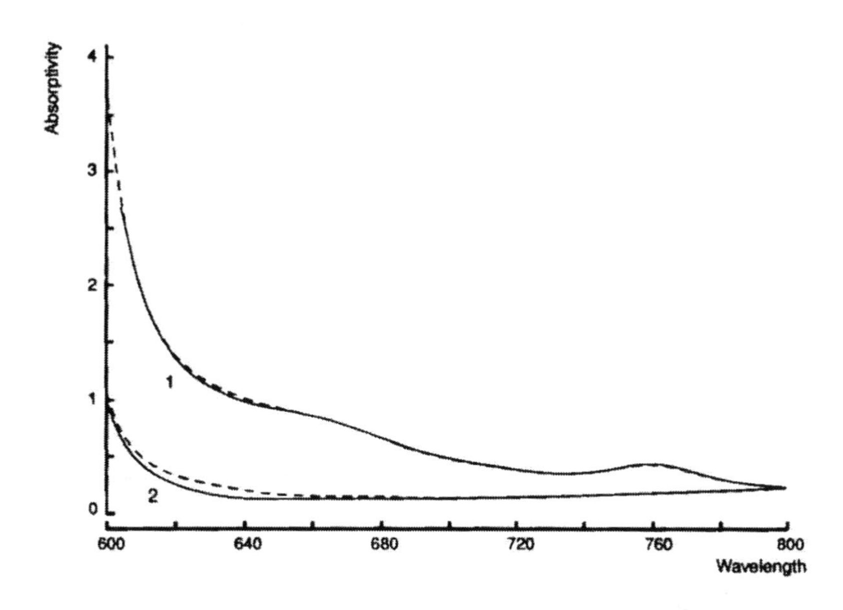

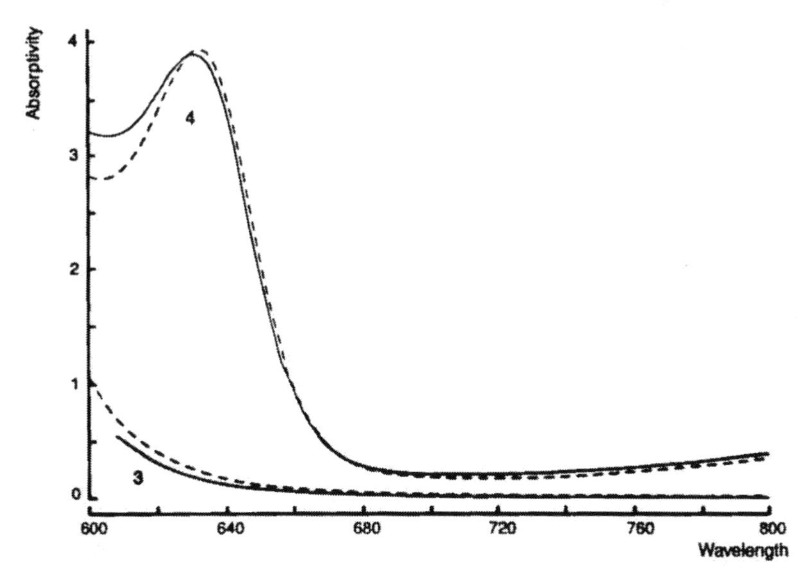

Figure 10.1b. Absorption spectra of the common derivatives of rat haemoglobin in comparison with those of human HbA.

Absorptivity in L · mmol^{-1} · cm^{-1}, wavelength in nm. Solid lines: human haemoglobin; dashed lines: rat haemoglobin. Upper panel: HHb (1) and O_2Hb (2); lower panel: COHb (3) and Hi (4). Absorbance measured with l = 0.200 cm (HHb, O_2Hb and Hi); l = 1.000 cm (COHb). HP8450A. pH (rat Hi) = 6.42–6.90; pH (human Hi) = 7.14–7.30.

Table 10.2.
Absorptivities ($L \cdot mmol^{-1} \cdot cm^{-1}$) of the common derivatives of rat haemoglobin

λ (nm)	ε_{HHb} $N = 3$	ε_{O_2Hb} $N = 3$	ε_{COHb} $N = 2$	ε_{Hi} $N = 4$
600	3.605	1.026	1.070	2.834
602	3.128	0.840	0.967	2.813
604	2.724	0.704	0.864	2.808
606	2.395	0.603	0.776	2.821
608	2.133	0.525	0.699	2.850
610	1.926	0.463	0.633	2.901
612	1.760	0.415	0.576	2.967
614	1.629	0.376	0.526	3.056
616	1.523	0.345	0.483	3.154
618	1.435	0.319	0.444	3.273
620	1.361	0.297	0.408	3.397
622	1.299	0.279	0.376	3.529
624	1.244	0.262	0.346	3.648
626	1.197	0.247	0.319	3.756
628	1.156	0.233	0.293	3.843
630	1.119	0.220	0.269	3.922
632	1.086	0.208	0.248	3.944
634	1.056	0.198	0.227	3.932
636	1.030	0.187	0.208	3.865
638	1.006	0.178	0.191	3.737
640	0.984	0.169	0.175	3.548
642	0.964	0.161	0.162	3.310
644	0.945	0.153	0.150	3.032
646	0.928	0.147	0.139	2.730
648	0.912	0.141	0.130	2.417
650	0.897	0.136	0.123	2.112
652	0.883	0.132	0.116	1.825
654	0.869	0.129	0.110	1.571
656	0.856	0.127	0.105	1.362
658	0.838	0.124	0.099	1.112
660	0.827	0.123	0.094	0.968
662	0.814	0.122	0.090	0.833
664	0.798	0.122	0.086	0.715
666	0.781	0.122	0.082	0.617
668	0.763	0.122	0.078	0.536
670	0.743	0.122	0.075	0.469
672	0.722	0.122	0.072	0.415
674	0.700	0.122	0.069	0.370
676	0.678	0.121	0.066	0.335
678	0.655	0.120	0.063	0.306
680	0.633	0.119	0.060	0.283
682	0.610	0.117	0.058	0.264

Table 10.2.
(Continued)

λ (nm)	ε_{HHb} $N = 3$	ε_{O_2Hb} $N = 3$	ε_{COHb} $N = 2$	ε_{Hi} $N = 4$
684	0.588	0.116	0.056	0.248
686	0.567	0.114	0.054	0.236
688	0.546	0.112	0.052	0.226
690	0.526	0.111	0.050	0.218
692	0.509	0.110	0.048	0.212
694	0.492	0.109	0.047	0.207
696	0.477	0.109	0.045	0.203
698	0.464	0.109	0.044	0.199
700	0.451	0.109	0.043	0.196
702	0.440	0.109	0.042	0.194
704	0.429	0.109	0.041	0.192
706	0.419	0.109	0.040	0.190
708	0.409	0.110	0.040	0.189
710	0.400	0.111	0.039	0.187
712	0.391	0.112	0.038	0.186
714	0.382	0.113	0.038	0.185
716	0.373	0.114	0.037	0.185
718	0.364	0.115	0.037	0.185
720	0.356	0.116	0.036	0.185
722	0.348	0.118	0.036	0.185
724	0.341	0.119	0.035	0.186
726	0.335	0.121	0.035	0.187
728	0.330	0.122	0.034	0.188
730	0.326	0.124	0.034	0.190
732	0.323	0.126	0.034	0.192
734	0.322	0.128	0.033	0.194
736	0.323	0.130	0.033	0.197
738	0.326	0.132	0.033	0.200
740	0.330	0.134	0.033	0.202
742	0.335	0.136	0.032	0.206
744	0.343	0.138	0.032	0.209
746	0.351	0.141	0.032	0.213
748	0.361	0.143	0.032	0.216
750	0.372	0.145	0.031	0.220
752	0.383	0.147	0.031	0.225
754	0.393	0.150	0.031	0.229
756	0.402	0.152	0.030	0.234
758	0.407	0.154	0.030	0.238
760	0.408	0.157	0.030	0.243
762	0.405	0.159	0.030	0.248
764	0.397	0.161	0.029	0.253
766	0.385	0.163	0.029	0.258

Table 10.2.
(Continued)

λ (nm)	ε_{HHb} $N = 3$	ε_{O_2Hb} $N = 3$	ε_{COHb} $N = 2$	ε_{Hi} $N = 4$
768	0.371	0.166	0.029	0.264
770	0.354	0.167	0.028	0.269
772	0.338	0.170	0.028	0.275
774	0.321	0.172	0.028	0.281
776	0.306	0.174	0.028	0.286
778	0.292	0.177	0.027	0.293
780	0.279	0.179	0.027	0.299
782	0.268	0.181	0.027	0.304
784	0.258	0.184	0.027	0.311
786	0.249	0.186	0.026	0.317
788	0.242	0.188	0.026	0.324
790	0.236	0.190	0.026	0.330
792	0.231	0.193	0.026	0.337
794	0.227	0.196	0.025	0.343
796	0.223	0.198	0.025	0.350
798	0.219	0.200	0.025	0.357
800	0.217	0.203	0.025	0.363

Absorbance measured with $l = 0.200$ cm (HHb, O_2Hb, and Hi); $l = 1.000$ cm (COHb). N is number of specimens. pH range (Hi) = 6.42–6.90.
λ = 600–800 nm. Rat procedure; HP8450A.

HP8450A spectrophotometer (Table 10.3) should be considered in relation to the appreciable difference in pH between the two solutions.

Although there is little difference in the position of the principal light absorption maxima and minima, Fig. 10.1 and Table 10.3 show that there are considerable differences in absorptivity between rat and human haemoglobin. The α- and β-peaks of O_2Hb, and the β-peak of COHb are clearly higher in the rat than in man. The differences are statistically significant in spite of the large spread in the data of rat haemoglobin. In the HHb spectrum, the minimum near 476 nm is lower in human than in rat haemoglobin, while in the maximum near 554 nm the absorptivities are equal, although this is the site of the only significant difference in the position of an absorption maximum between rat and human haemoglobin. The lower panel of Fig. 10.1a shows conspicuous differences between the standard spectra of rat and human Hi.

It may be that these differences are partly due to the differences in the preparation of the samples for spectrophotometric analysis. Application of the standard or the SFH procedure (p. 55) to rat blood was not possible. The SFH procedure, which usually results in a clear haemoglobin solution with $ct_{Hb} \approx 8$ mmol/L, resulted, when applied to rat blood, in sticky solutions from which haemoglobin tended to precipitate. Therefore, we

Table 10.3.
Comparison of absorptivities of the common derivatives of rat (Rat procedure) and adult human haemoglobin (Standard procedure)

	Human		Rat		
λ (nm)	ε_{HHb} $N = 6$	*SEM*	ε_{HHb} $N = 8$	*SEM*	*p*
476	3.298	0.013	3.576	0.016	< 0.001
478	3.302	0.012	3.564	0.017	< 0.001
552	13.21	0.044	13.29	0.029	
554	13.36	0.045	13.38	0.029	
556	13.36	0.045	13.35	0.031	
λ (nm)	ε_{O_2Hb} $N = 6$	*SEM*	ε_{O_2Hb} $N = 11$	*SEM*	*p*
508	4.882	0.007	5.037	0.016	< 0.001
540	14.32	0.027	14.65	0.043	< 0.001
542	14.52	0.023	14.81	0.043	< 0.001
544	14.34	0.022	14.62	0.043	< 0.001
560	8.767	0.015	8.868	0.028	< 0.05
576	15.26	0.034	15.72	0.057	< 0.001
578	15.36	0.027	15.87	0.057	< 0.001
λ (nm)	ε_{COHb} $N = 8$	*SEM*	ε_{COHb} $N = 8$	*SEM*	*p*
496	5.225	0.017	5.377	0.033	< 0.002
538	14.30	0.039	14.51	0.055	< 0.01
540	14.27	0.036	14.50	0.061	< 0.01
542	14.04	0.036	14.26	0.058	< 0.01
554	11.63	0.029	11.76	0.062	
568	14.43	0.037	14.34	0.069	
570	14.46	0.039	14.37	0.067	
λ (nm)	ε_{Hi} $N = 15$	*SEM*	ε_{Hi} $N = 16$	*SEM*	*p*
498	9.088	0.028	8.935	0.026	< 0.001
500	9.106	0.029	8.991	0.026	< 0.01
502	9.098	0.028	9.011	0.025	< 0.05
630	3.895	0.021	3.916	0.014	
632	3.896	0.021	3.954	0.014	< 0.05
634	3.857	0.020	3.947	0.014	< 0.001

HP8450A.

Absorptivity in $L \cdot mmol^{-1} \cdot cm^{-1}$; *N* is number of specimens; *SEM* is standard error of the mean. *p* values of Student's *t*-test for unpaired samples given only when $p < 0.05$.

Table 10.4.
Absorption maxima and minima in the spectra of HHb and O_2Hb of rat haemoglobin

λ (nm)	ε_{HHb}	*SEM*	*N*	λ (nm)	ε_{O_2Hb}	*SEM*	*N*
475	3.575	0.046	6	507	5.008	0.037	10
475.5	3.557	0.056	5	507.5	5.001	0.039	10
476	3.549	0.058	5	508	4.997	0.039	10
476.5	3.549	0.060	5	508.5	4.999	0.040	10
477	3.544	0.060	5	509	5.002	0.041	10
477.5	3.552	0.059	5	509.5	5.007	0.042	10
478	3.564	0.058	5	510	5.013	0.043	10
				540	14.90	0.087	8
551	13.32	0.034	6	541	15.01	0.093	8
552	13 39	0.037	6	541.5	15 04	0.094	8
552.5	13.40	0.034	6	542	15.05	0.099	8
553	13.42	0.038	6	542.5	15.05	0.105	8
553.5	13.41	0.036	6	543	15 03	0.110	8
554	13.41	0.037	6	543.5	14.97	0.110	8
554.5	13.40	0.039	6	544	14.91	0.118	8
555	13.38	0.037	6				
555.5	13.32	0.064	6	559	8.865	0.053	9
556	13.30	0.068	6	559.5	8.823	0.049	9
				560	8.801	0.053	9
				560.5	8.796	0.051	9
				561	8.803	0.047	9
				561.5	8.829	0.047	9
				562	8.878	0.047	9
				575	15.60	0.039	9
				576	15.99	0.048	9
				576.5	16.08	0.055	9
				577	16.12	0.059	9
				577.5	16.09	0.060	9
				578	15.98	0.067	9
				578.5	15.79	0.073	9
				579	15.53	0.082	9

Optica CF4.
Absorptivity in $L \cdot mmol^{-1} \cdot cm^{-1}$. *N* is number of specimens. *SEM* is standard error of the mean.

had to resort to stronger dilution and addition of a considerable amount of NaCl and of a phosphate buffer. This resulted in dilute haemoglobin solutions ($ct_{Hb} \approx 2.5$ mmol/L) in hypertonic saline, which contained much haemiglobin, occasionally up to 40%. If F_{Hi} was > 20% the stock solution was rejected.

Table 10.5.
Absorption maxima and minima in the spectra of COHb and Hi of rat haemoglobin

λ (nm)	ε_{COHb}	*SEM*	*N*	λ (nm)	ε_{Hi}	*SEM*	*N*
494	5.396	0.059	7	470	7.575	0.033	10
494.5	5.396	0.060	7	472	7.524	0.033	10
495	5.392	0.059	7	474	7.518	0.035	10
496	5.412	0.057	7	475	7.532	0.042	8
496.5	5.424	0.055	7	476	7.539	0.034	10
497	5.433	0.054	7				
				499	8.945	0.035	10
537	14.69	0.073	7	500	8.972	0.037	10
537.5	14.75	0.075	7	501	8.990	0.035	10
538	14.80	0.079	7	502	9.005	0.036	10
538.5	14.82	0.083	7	502.5	9.011	0.039	10
539	14.84	0.085	7	503	9.014	0.038	10
539.5	14.83	0.091	7	503.5	9.034	0.041	9
540	14.81	0.092	7	504	9.011	0.040	10
541	14 71	0.092	7	504.5	9.005	0.045	8
				505	8.979	0.039	10
552	12.04	0.063	7				
553	11.91	0.057	7	600	2.849	0.035	10
553.5	11.84	0.051	7	601	2.829	0.035	10
554	11.84	0.052	7	602	2.825	0.036	10
554.5	11.82	0.054	7	603	2.822	0.037	10
555	11.84	0.051	7	604	2.810	0.036	9
555.5	11.86	0.054	7	605	2.836	0.036	10
556	11.91	0.053	7				
				630	4.043	0.050	10
567	14.44	0.051	7	630.5	4.054	0.052	10
567.5	14.51	0.061	7	631	4.063	0.052	10
568	14.57	0.066	7	631.5	4.074	0.048	10
568.5	14.61	0.074	7	632	4.069	0.051	10
569	14.62	0.076	7	632.5	4.051	0.051	10
569.5	14.59	0.084	7	633	4.063	0.050	10
570	14.54	0.092	7	633.5	4.052	0.049	10
571	14.36	0.097	7	634	4.035	0.051	10
				640	3.535	0.051	10
				650	1.992	0.044	10

Optica CF4.
Absorptivity in $L \cdot mmol^{-1} \cdot cm^{-1}$. *N* is number of specimens. *SEM* is standard error of the mean.

The stock solutions were far from perfect for preparing solutions of HHb, O_2Hb and COHb by tonometry. Especially for a kind of haemoglobin like that of the rat that is liable to the formation of Hi and probably to further detoriation, tonometry in the protective environment of the erythrocytes would have been preferable. In the applied procedure, a correction of up

Table 10.6.
Hi fractions as determined by MCA for a pure rat haemiglobin solution at various pH values

pH	F_{Hi} (%)
6.42	99.5
6.58	99.2
6.65	100.5
6.69	99.8
6.78	99.3
6.81	101.4
6.90	100.5

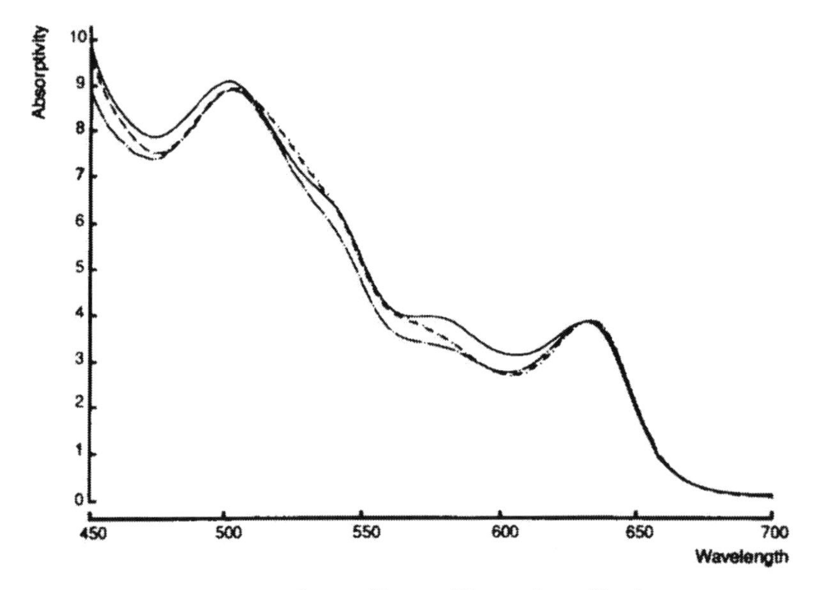

Figure 10.2. Absorption spectra of rat and human Hi at various pH values.
The two rat Hi spectra (· · ·) and (-·-·) with pH = 6.42 and pH ≈ 6.69 coincide. The solid line (——) is the standard human Hi spectrum with pH ≈ 7.21, the other line (——·) represents a human Hi spectrum at pH 6.80. The absorptivity at 630 nm has been normalised to 3.895 $L \cdot mmol^{-1} \cdot cm^{-1}$, the absorptivity of human Hi at pH ≈ 7.21.

to 17% for the presence of Hi in the solutions of the other haemoglobin derivatives (Chapter 7; equation 7.2) was necessary, against a correction of < 2% in human erythrolysates. This large correction and the low total haemoglobin concentration will have diminished the precision of the results. The possibility cannot be ruled out that the generally higher absorptivities of rat haemoglobin in comparison with human haemoglobin are in part due to some interference resulting from the unusual method for preparing the samples.

The most conspicuous differences between rat and human haemoglobin are in the Hi spectra. Figure 10.2 shows a Hi spectrum with pH 6.42, which cannot be distinguished from the standard rat Hi spectrum (pH ≈ 6.69). These spectra are accompanied by two Hi spectra of human haemoglobin. The upper one is the standard spectrum of human Hi based on 15 specimens of human haemoglobin and a pH range of 7.14–7.30. The lower one is a human Hi spectrum at pH = 6.80. The spectra have been normalised to $\varepsilon_{Hi}(630) = 3.895\ L \cdot mmol^{-1} \cdot cm^{-1}$, the absorptivity at 630 nm of the standard human Hi spectrum. There obviously is no pH value at which the Hi spectra of rat and man would coincide. This points to an intrinsic difference between the Hi spectra of rat and human haemoglobin.

It is quite improbable that rat haemiglobin will not bind OH^- ions; it readily forms HiCN. The virtual independence of the Hi spectrum of the pH of the sample, which is also demonstrated by the data of Table 10.6, is obviously caused by the very low OH^- activity due to the low pH. Table 10.6 shows the results of multicomponent analysis of a sample of rat haemoglobin that is completely in the form of Hi. The measurements all give $F_{Hi} \approx 100\%$, in spite of the considerable difference in pH.

CHAPTER 11

ABSORPTION SPECTRA OF BOVINE HAEMOGLOBIN

Bovine blood has frequently been used for *in vitro* experiments concerning the development of optical instruments, for instance oximeters and photometers for other physiological measurements [490]. It is easy to obtain in large quantities, has a good keeping quality and a high suspension stability, because rouleaux formation of the erythrocytes (p. 255) does not occur [44, 198]. These favourable properties and the absence of pathogenic human viruses, *e.g.* human immunodeficiency virus (HIV), have recently promoted its use as a substitute for human blood in the production of reference materials for haemoglobinometry (Chapter 16) and for quality assessment in multicomponent analysis (Chapter 17). However, concern has been expressed as to a potential danger of spongiform encephalitis (BSE; *mad cow* disease).

For the applications mentioned above it was necessary, firstly, to ascertain that the absorptivity at 540 nm of HiCN of bovine haemoglobin did not differ significantly from that of human haemoglobin, and, secondly, to determine the absorption spectra of its common derivatives. In Chapter 5 it has been shown that the value of 10.96 $L \cdot mmol^{-1} \cdot cm^{-1}$ found for $\varepsilon_{HiCN}(540)$ of the cow is within the range of values on which the internationally accepted value is based. In this chapter the absorption spectra of HHb, O_2Hb, COHb and Hi of bovine haemoglobin are presented. Part of these results have been reported earlier [476].

Blood from healthy cows (***Bos tauris***) was obtained at a butcher's and collected in flasks containing 100 USP units of sodium heparin per mL blood. When measurements of $\varepsilon_{HiCN}(540)$ were required, iron free flasks were used. Solutions containing HHb, O_2Hb, COHb and Hi were prepared with the standard procedure described in Chapter 6 and the absorption spectra were determined using an HP8450A diode array spectrophotometer and an Optica CF4 grating spectrophotometer as described in Chapter 7. In the determination of the absorptivities of HHb, measurements with and without sodium dithionite were made. The ε_{HHb} values presented in Tables 11.1, 11.2 and 11.4 have all been obtained with dithionite.

Table 11.1 shows the absorptivities of HHb, O_2Hb, COHb and Hi of bovine haemoglobin for the spectral range of 450–630 nm measured with an HP8450A diode array spectrophotometer in 0.100-cm cuvettes with plan

Table 11.1.
Absorptivities ($L \cdot mmol^{-1} \cdot cm^{-1}$) of the common derivatives of bovine haemoglobin

λ (nm)	ε_{HHb} $N = 10$	ε_{O_2Hb} $N = 8$	ε_{COHb} $N = 8$	ε_{Hi} $N = 8$
450	13.37	16.13	10.13	9.211
452	10.54	14.85	9.429	8.919
454	8.498	13.77	8.868	8.719
456	7.007	12.77	8.364	8.543
458	5.934	11.90	7.959	8.422
460	5.124	11.13	7.585	8.301
462	4.549	10.45	7.281	8.221
464	4.112	9.835	6.991	8.129
466	3.809	9.302	6.758	8.075
468	3.590	8.795	6.523	8.006
470	3.449	8.374	6.342	7.992
472	3.355	7.961	6.149	7.978
474	3.310	7.610	6.011	8.004
476	3.286	7.273	5.869	8.046
478	3.294	6.993	5.765	8.140
480	3.311	6.710	5.665	8.242
482	3.351	6.475	5.597	8.384
484	3.396	6.252	5.514	8.518
486	3.463	6.074	5.457	8.693
488	3.536	5.898	5.390	8.845
490	3.633	5.755	5.347	9.012
492	3.734	5.615	5.309	9.146
494	3.861	5.501	5.293	9.288
496	4.000	5.383	5.281	9.375
498	4.164	5.297	5.306	9.466
500	4.334	5.197	5.337	9.477
502	4.534	5.119	5.407	9.482
504	4.732	5.039	5.487	9.409
506	4.956	4.983	5.619	9.329
508	5.166	4.941	5.774	9.179
510	5.391	4.942	5.998	9.024
512	5.598	4.978	6.267	8.817
514	5.819	5.099	6.644	8.623
516	6.020	5.295	7.102	8.388
518	6.251	5.615	7.676	8.177
520	6.464	6.045	8.328	7.935
522	6.722	6.643	9.096	7.717
524	6.985	7.379	9.879	7.493
526	7.290	8.299	10.71	7.304
528	7.600	9.306	11.51	7.103
530	7.973	10.41	12.30	6.950
532	8.355	11.46	12.99	6.779

Table 11.1.
(Continued)

λ (nm)	ε_{HHb} $N = 10$	ε_{O_2Hb} $N = 8$	ε_{COHb} $N = 8$	ε_{Hi} $N = 8$
534	8.815	12.46	13.62	6.646
536	9.287	13.26	14.03	6.488
538	9.846	13.92	14.27	6.347
540	10.39	14.35	14.26	6.169
542	11.00	14.53	14.02	5.980
544	11.55	14.35	13.59	5.749
546	12.08	13.82	13.09	5.504
548	12.52	12.97	12.54	5.233
550	12.89	11.98	12.10	4.966
552	13.14	10.96	11.77	4.701
554	13.29	10.10	11.60	4.468
556	13.31	9.396	11.60	4.263
558	13.24	8.932	11.82	4.100
560	13.06	8.675	12.18	3.975
562	12.84	8.678	12.71	3.883
564	12.55	8.941	13.33	3.825
566	12.22	9.520	13.91	3.791
568	11.82	10.39	14.29	3.776
570	11.41	11.58	14.31	3.773
572	10.94	12.94	13.87	3.777
574	10.49	14.28	12.99	3.782
576	10.01	15.19	11.74	3.787
578	9.576	15.33	10.26	3.782
580	9.155	14.46	8.691	3.769
582	8.766	12.67	7.149	3.741
584	8.371	10.33	5.779	3.700
586	7.958	7.957	4.599	3.650
588	7.479	5.871	3.660	3.594
590	6.936	4.234	2.935	3.540
592	6.324	3.040	2.369	3.488
594	5.675	2.205	1.939	3.444
596	5.011	1.633	1.609	3.406
598	4.368	1.245	1.352	3.381
600	3.783	0.980	1.150	3.364
602	3.266	0.791	0.989	3.359
604	2.833	0.654	0.857	3.360
606	2.481	0.552	0.751	3.370
608	2.203	0.472	0.662	3.388
610	1.983		0.589	3.413

Table 11.1.
(Continued)

λ (nm)	ε_{HHb} $N = 10$	ε_{O_2Hb} $N = 8$	ε_{COHb} $N = 8$	ε_{Hi} $N = 8$
612	1.808		0.526	3.450
614	1.670		0.474	3.501
616	1.557			3.566
618	1.463			3.641
620	1.383			3.725
622	1.316			3.810
624	1.258			3.889
626	1.207			3.954
628	1.161			4.000
630	1.120			4.025

$\lambda = 450–630$ nm. Standard procedure; HP8450A.
Absorbance measured with $l = 0.010–0.016$ cm. N is number of specimens. $Na_2S_2O_4$ used to prevent reoxygenation of HHb. pH (Hi) = 7.24–7.32.

parallel glass inserts. To obtain reliable results, the measured absorbance should not be less than 0.050 (p. 47). No absorptivity values have been given when the measured absorbance was < 0.050. Table 11.2 shows the absorptivities for the spectral range of 600–800 nm measured with the same instrument; the lightpath was adjusted so that the measured absorbance was > 0.050. Absorptivity values have not been given when the spectrophotometer rejected the measurements because the absorbance of the samples was too high. This happened occasionally in measurements with $l = 0.200$ cm between 600 and 650 nm. The pH range of the Hi samples corresponds with the pH that ensues when a fresh bovine blood sample is haemolysed. The spectra represented by the absorptivities of Tables 11.1 and 11.2 are designated as the **standard spectra** of bovine haemoglobin.

Figure 11.1 shows the absorption spectra of HHb, O_2Hb, COHb and Hi of bovine haemoglobin in comparison with those of human HbA for the spectral ranges of 450–700 nm (Table 11.1) and 600–800 nm (Table 11.2). In Fig. 11.1a, the spectral absorption curves run to 700 nm, but for $\lambda > 600$ nm part of the measurements resulted in $A < 0.050$; these values have not been given in Table 11.1.

Figure 11.1 and Table 11.3 show no differences in the position of the light absorption maxima and minima of the common derivatives of bovine and adult human haemoglobin. As the measurements have been made with a diode array spectrophotometer, which has a limited spectral resolution, this does not rule out that slight differences in the position of the absorption maxima and minima do exist. Of more importance, however, is the presence of any differences between the absorptivities of the common derivatives of human HbA and bovine haemoglobin.

Table 11.2.
Absorptivities ($L \cdot mmol^{-1} \cdot cm^{-1}$) of the common derivatives of bovine haemoglobin

λ (nm)	ε_{HHb} $N = 9$	ε_{O_2Hb} $N = 8$	ε_{COHb} $N = 8$	ε_{Hi} $N = 7$
600		0.990	1.169	3.335
602		0.813	1.013	3.334
604		0.680	0.885	3.331
606		0.579	0.780	3.345
608		0.501	0.692	3.358
610	1.933	0.438	0.619	3.390
612	1.777	0.388	0.556	3.420
614	1.647	0.347	0.503	3.476
616	1.539	0.315	0.456	3.533
618	1.451	0.288	0.415	3.610
620	1.373	0.265	0.368	3.691
622	1.309	0.246	0.340	3.775
624	1.250	0.229	0.314	3.847
626	1.201	0.213	0.290	3.909
628	1.156	0.198	0.268	3.948
630	1.117	0.185	0.248	3.972
632	1.081	0.172	0.230	3.957
634	1.050	0.162	0.213	3.906
636	1.023	0.153	0.199	3.811
638	0.999	0.145	0.186	3.662
640	0.977	0.138	0.174	3.461
642	0.958	0.132	0.163	3.226
644	0.942	0.128	0.154	2.953
646	0.927	0.124	0.146	2.664
648	0.913	0.120	0.138	2.361
650	0.902	0.117	0.132	2.069
652	0.890	0.115	0.125	1.793
654	0.880	0.113	0.119	1.548
656	0.870	0.110	0.114	1.344
658	0.854	0.107	0.108	1.096
660	0.842	0.108	0.103	0.957
662	0.829	0.107	0.098	0.824
664	0.813	0.105	0.094	0.709
666	0.795	0.104	0.089	0.614
668	0.776	0.103	0.085	0.536
670	0.755	0.103	0.082	0.472
672	0.733	0.102	0.078	0.421
674	0.711	0.102	0.075	0.380
676	0.689	0.101	0.072	0.347
678	0.667	0.101	0.069	0.321
680	0.645	0.100	0.067	0.299
682	0.624	0.100	0.064	0.283
684	0.603	0.100	0.062	0.269
686	0.583	0.100	0.060	0.258

Table 11.2.
(Continued)

λ (nm)	ε_{HHb} $N = 9$	ε_{O_2Hb} $N = 8$	ε_{COHb} $N = 8$	ε_{Hi} $N = 7$
688	0.563	0.100	0.058	0.249
690	0.545	0.100	0.057	0.242
692	0.528	0.100	0.055	0.237
694	0.512	0.100	0.054	0.233
696	0.498	0.100	0.052	0.229
698	0.485	0.101	0.051	0.226
700	0.472	0.102	0.050	0.224
702	0.461	0.103	0.049	0.222
704	0.450	0.104	0.049	0.221
706	0.440	0.105	0.048	0.220
708	0.430	0.106	0.047	0.220
710	0.420	0.107	0.046	0.220
712	0.411	0.109	0.046	0.220
714	0.402	0.110	0.045	0.220
716	0.393	0.112	0.045	0.220
718	0.384	0.113	0.044	0.221
720	0.375	0.115	0.044	0.222
722	0.367	0.117	0.043	0.223
724	0.360	0.118	0.043	0.225
726	0.353	0.120	0.043	0.226
728	0.348	0.122	0.042	0.228
730	0.344	0.124	0.042	0.230
732	0.341	0.126	0.042	0.232
734	0.340	0.128	0.041	0.234
736	0.341	0.130	0.041	0.237
738	0.343	0.131	0.041	0.240
740	0.347	0.133	0.040	0.242
742	0.353	0.136	0.040	0.246
744	0.360	0.137	0.040	0.249
746	0.369	0.140	0.039	0.253
748	0.380	0.142	0.039	0.256
750	0.391	0.144	0.039	0.260
752	0.402	0.146	0.039	0.264
754	0.413	0.148	0.038	0.269
756	0.421	0.151	0.038	0.273
758	0.427	0.153	0.038	0.278
760	0.428	0.156	0.037	0.283
762	0.424	0.158	0.037	0.288
764	0.416	0.160	0.037	0.293
766	0.404	0.163	0.037	0.298
768	0.389	0.165	0.036	0.304
770	0.372	0.168	0.036	0.310
772	0.355	0.170	0.036	0.316

Table 11.2.
(Continued)

λ (nm)	ε_{HHb} $N = 9$	ε_{O_2Hb} $N = 8$	ε_{COHb} $N = 8$	ε_{Hi} $N = 7$
774	0.338	0.173	0.036	0.322
776	0.322	0.175	0.035	0.328
778	0.307	0.178	0.035	0.334
780	0.294	0.180	0.035	0.341
782	0.282	0.183	0.035	0.347
784	0.272	0.186	0.034	0.354
786	0.263	0.188	0.034	0.360
788	0.255	0.191	0.034	0.367
790	0.249	0.193	0.034	0.374
792	0.243	0.196	0.034	0.381
794	0.238	0.198	0.033	0.388
796	0.234	0.201	0.033	0.395
798	0.230	0.203	0.033	0.402
800	0.226	0.205	0.033	0.409

$\lambda = 600$–800 nm. Standard procedure; HP8450A.

Absorbance measured with $l = 0.100$ cm (Hi); $l = 0.200$ cm (HHb, O_2Hb, and COHb at $\lambda = 600$–618 nm), $l = 1.000$ cm (COHb at $\lambda = 620$–800 nm). N is number of specimens. $Na_2S_2O_4$ used to prevent reoxygenation of HHb. pH (Hi) $= 7.24$–7.32.

The absorptivities of the derivatives of bovine haemoglobin have been compared with those of adult human haemoglobin by means of Student's t-test for unpaired samples. The results at the wavelengths near the principal absorption maxima and minima in the spectra of the common derivatives are presented in Table 11.3.

The absorption spectra of Fig. 11.1 show that for HHb, O_2Hb, and COHb there are only slight spectral differences between bovine and human haemoglobin, although some are statistically significant (Table 11.3). In the O_2Hb spectrum the minimum at 560 nm is slightly lower for bovine than for human haemoglobin, while at 508 nm the absorptivity slightly exceeds that of human haemoglobin. In the absorption spectrum of COHb the α-peak around 569 nm is somewhat lower for bovine haemoglobin, but in the minimum near 496 nm the spectrum of bovine COHb lies a little above that of human COHb. The maximum in the bovine HHb spectrum at 554 nm seems to be somewhat lower than that in the spectrum of human HHb, but this difference does not reach statistical significance. As shown in Fig. 11.1b, there are, in the red and near-infrared region, no significant spectral differences between bovine and human haemoglobin as far as HHb, O_2Hb and COHb are concerned. The spectra of bovine and human HHb and O_2Hb converge to the isosbestic point near 800 nm.

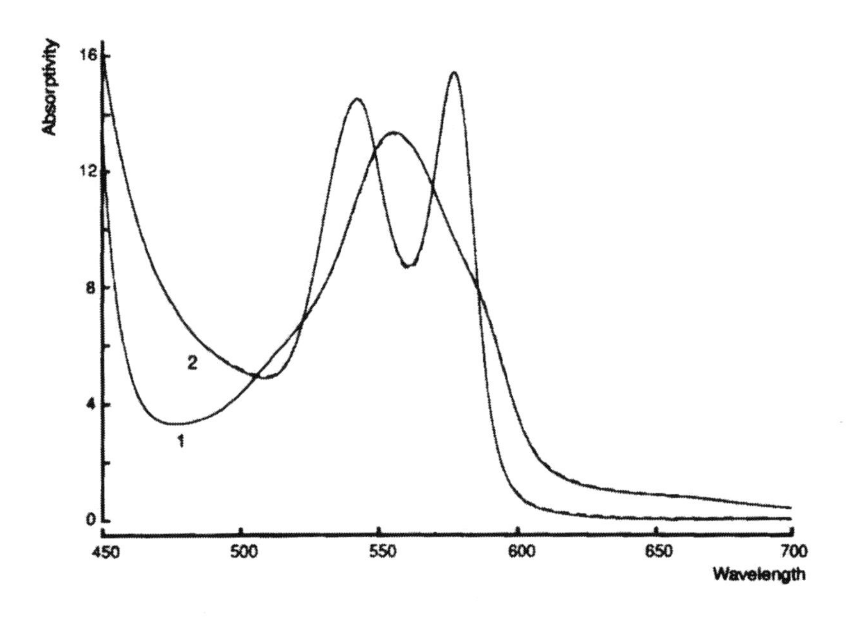

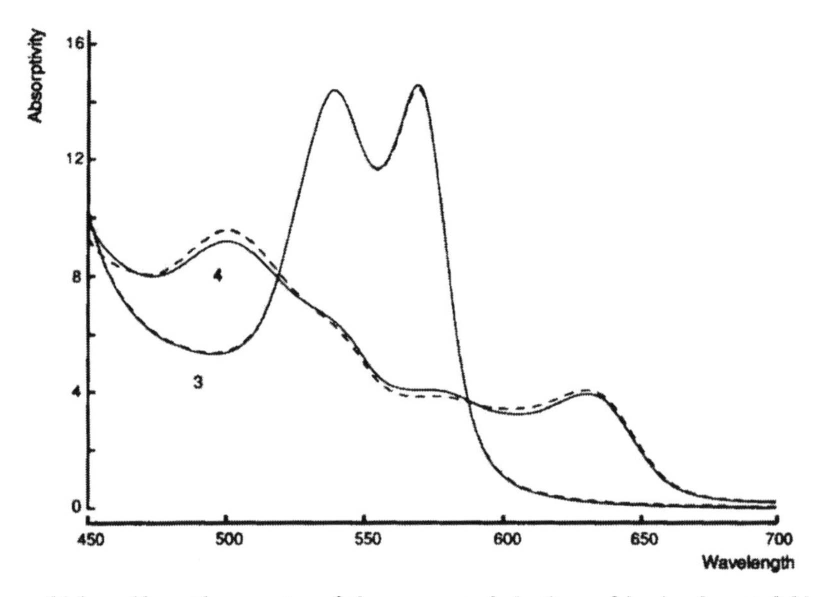

Figure 11.1a. Absorption spectra of the common derivatives of bovine haemoglobin in comparison with those of human HbA.

Absorptivity in $L \cdot mmol^{-1} \cdot cm^{-1}$, wavelength in nm. Solid lines: human haemoglobin; dashed lines: bovine haemoglobin. Upper panel: HHb (1) and O_2Hb (2); lower panel: COHb (3) and Hi (4). Absorbance measured with $l = 0.010$–0.016 cm. HP8450A. pH (bovine Hi) ≈ 7.28; (pH human Hi) ≈ 7.21.

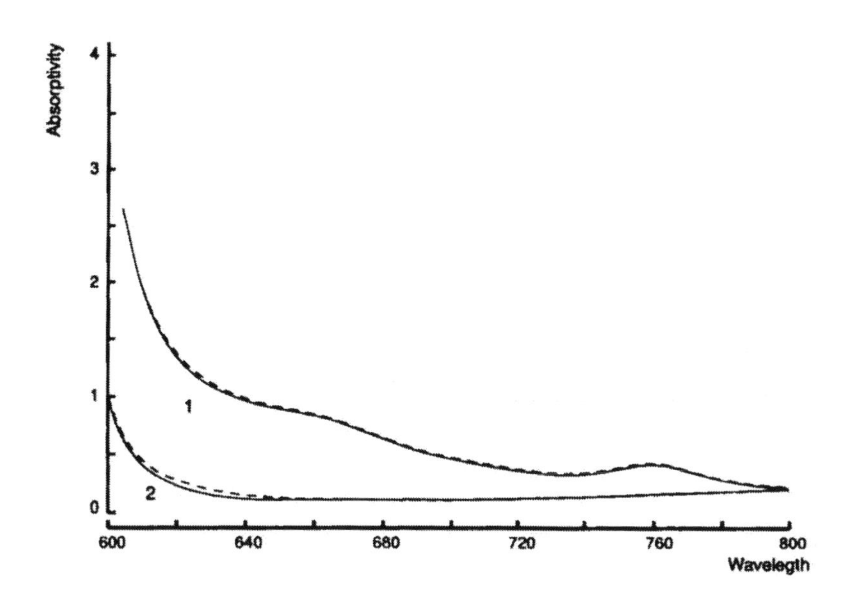

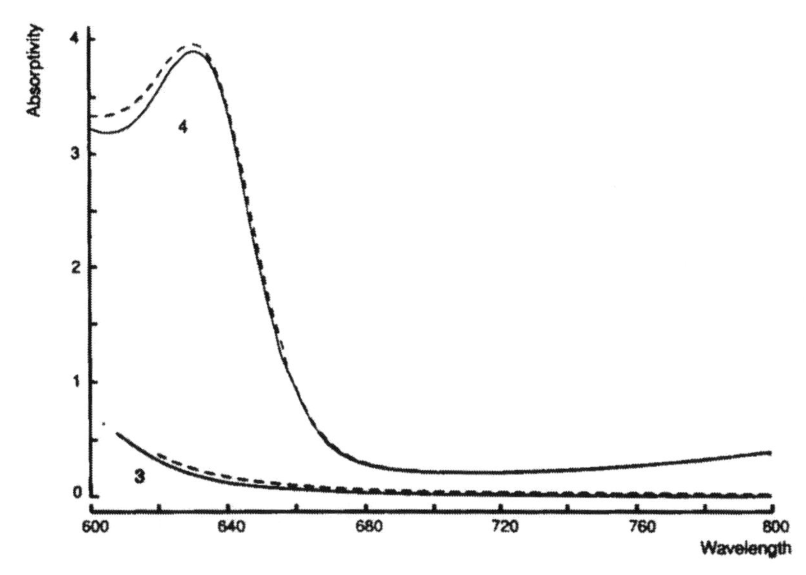

Figure 11.1b. Absorption spectra of the common derivatives of bovine haemoglobin in comparison with those of human HbA.

Absorptivity in $L \cdot mmol^{-1} \cdot cm^{-1}$, wavelength in nm. Solid lines: human haemoglobin; dashed lines: bovine haemoglobin. Upper panel: HHb (1) and O_2Hb (2); lower panel: COHb (3) and Hi (4). Absorbance measured with l = 0.100 cm (Hi); l = 0.200 cm (HHb and O_2Hb); l = 1.000 cm (COHb). HP8450A. pH (bovine Hi) ≈ 7.28; (pH human Hi) ≈ 7.21.

Table 11.3.
Comparison of absorptivities of the common derivatives of bovine and adult human haemoglobin

	Human		Bovine		
λ (nm)	ε_{HHb} $N = 6$	*SEM*	ε_{HHb} $N = 10$	*SEM*	*p*
476	3.298	0.013	3.281	0.012	
478	3.302	0.012	3.289	0.012	
552	13.21	0.044	13.15	0.031	
554	13.36	0.045	13.30	0.032	
556	13.36	0.045	13.32	0.029	
λ (nm)	ε_{O_2Hb} $N = 6$	*SEM*	ε_{O_2Hb} $N = 8$	*SEM*	*p*
508	4.882	0.007	4.941	0.009	< 0.001
540	14.32	0.027	14.35	0.027	
542	14.52	0.023	14.53	0.031	
544	14.34	0.022	14.35	0.027	
560	8.767	0.015	8.675	0.013	< 0.001
576	15.26	0.034	15.19	0.034	
578	15.36	0.027	15.33	0.034	
λ (nm)	ε_{COHb} $N = 8$	*SEM*	ε_{COHb} $N = 8$	*SEM*	*p*
496	5.225	0.017	5.281	0.010	< 0.01
538	14.30	0.039	14.27	0.025	
540	14.27	0.036	14.26	0.031	
542	14.04	0.036	14.02	0.026	
554	11.63	0.029	11.60	0.022	
568	14.43	0.037	14.29	0.027	< 0.01
570	14.46	0.039	14.31	0.022	< 0.01
λ (nm)	ε_{Hi} $N = 15$	*SEM*	ε_{Hi} $N = 8$	*SEM*	*p*
498	9.088	0.028	9.466	0.026	< 0.001
500	9.106	0.029	9.477	0.024	< 0.001
502	9.098	0.028	9.482	0.026	< 0.001
630	3.895	0.021	4.025	0.016	< 0.001
632	3.896	0.021	4.020	0.016	< 0.001
634	3.857	0.020	3.976	0.016	< 0.001

Standard procedure; HP8450A.

Absorptivity in $L \cdot mmol^{-1} \cdot cm^{-1}$; *N* is number of specimens, *SEM* is standard error of the mean. *p* values of Student's *t*-test for unpaired samples given only when $p < 0.05$.

Table 11.4.
Absorptivities ($L \cdot mmol^{-1} \cdot cm^{-1}$) of the common derivatives of bovine haemoglobin at selected wavelengths between 600 and 1000 nm

λ (nm)	ε_{HHb} $N = 8$	ε_{O_2Hb} $N = 7$	ε_{COHb} $N = 8$	ε_{Hi} $N = 7$
600	3.439	0.918	1.085	3.288
630	1.105	0.167	0.218	3.900
660	0.845	0.099	0.087	0.871
680	0.637	0.094	0.051	0.257
700	0.467	0.097	0.037	0.198
750	0.406	0.142	0.028	0.247
775	0.310	0.173	0.026	0.317
800	0.221	0.206	0.024	0.403
805	0.215	0.211	0.024	0.420
840	0.198	0.255	0.021	0.532
845	0.197	0.260	0.021	0.546
880	0.207	0.293	0.018	0.627
904	0.216	0.305	0.017	0.683
920	0.211	0.308	0.016	0.718
940	0.184	0.307	0.014	0.754
960	0.138	0.298	0.011	0.777
1000	0.062	0.275	0.010	0.796

Standard procedure; Optica CF4.
Absorbance measured with $l = 0.100$ cm (Hi at $\lambda = 600$ and 630 nm); $l = 0.200$ cm (HHb and O_2Hb, COHb at $\lambda = 600$ nm, Hi at $\lambda = 660-1000$ nm); $l = 1.000$ cm (COHb at $\lambda = 630-1000$ nm). N is number of specimens. $Na_2S_2O_4$ used to prevent reoxygenation of HHb. pH (Hi) = 7.24–7.32. In the wavelength range 630–1000 nm a red sensitive photomultiplier tube with a red filter has been used.

Table 11.4 presents the absorptivities of the common derivatives of bovine haemoglobin for selected wavelengths in the spectral range of 600–1000 nm, measured with an Optica CF4 grating spectrophotometer. Wavelength selection was necessary to complete this series of manual measurements within a reasonable time. Wavelengths of importance for biomedical applications have preferentially been chosen.

Comparing the values of $\varepsilon_{HHb}(800/805)$ and $\varepsilon_{O_2Hb}(800/805)$ in Table 11.4 (0.221/0.215 and 0.206/0.211) gives the impression that, for bovine haemoglobin, the isosbestic point of HHb and O_2Hb, lies a little above 800 nm, the established human value [272]. This, however, appears to be an effect of the addition of $Na_2S_2O_4$. In the corresponding determinations of ε_{HHb} without use of $Na_2S_2O_4$ the values at 800 and 805 were 0.201 and 0.197 $L \cdot mmol^{-1} \cdot cm^{-1}$, clearly demonstrating that there actually is no difference in the position of the isosbestic point between bovine and human haemoglobin (compare p. 145).

For a further discussion of these measurements and an explanation of the slight differences between the data of this table and those of Table 11.2, see Chapter 15.

In the lower panel of Fig. 11.1a and in Table 11.3 the standard bovine Hi spectrum (pH ≈ 7.28) is compared with the standard human Hi spectrum (pH ≈ 7.21). These pH values are the mid-point values of the pH range chosen for the standard Hi spectra of the two species because these ranges correspond with the pH values that ensue when fresh bovine or human blood is haemolysed. The absorption spectra of Fig. 11.2 demonstrate that the differences found between bovine and human Hi as shown in Fig. 11.1 and in Table 11.3 are definitely not caused solely by the difference in pH. The upper panel of Fig. 11.2 shows the influence of pH on the bovine Hi absorption spectrum. As in human Hi (Fig. 8.2), a deviation from the chosen standard pH of 7.28 in the alkaline direction is accompanied by a stronger spectral change than is seen with a deviation in the acid direction. In the lower panel of Fig. 11.2 the same bovine Hi spectra are shown in comparison with the standard Hi spectrum of human HbA. For easy comparison the absorptivities at 630 nm have been normalised to 4.025 $L \cdot mmol^{-1} \cdot cm^{-1}$, the value of $\varepsilon_{Hi}(630)$ in the standard bovine Hi spectrum. The spectra also show that it is improbable that the only difference between bovine and human Hi would be a difference in OH^- ion affinity. In that case there would have been a pH value at which the spectra could be made to coincide. Hence, it appears that there is an intrinsic difference in absorption between bovine and human Hi.

Proper selection of the pH value of the standard Hi spectrum for the haemoglobin of each animal species is especially important in view of the use of these spectra in multicomponent analysis of haemoglobin derivatives as described in Chapter 8 for human haemoglobin. Therefore, the pH of the

Table 11.5.
Hi fractions as determined by MCA for a pure bovine haemiglobin solution at various pH values

pH	F_{Hi} (%)
6.83	102.5
6.95	101.4
7.18	100.7
7.28	100.0
7.38	97.4
7.51	94.9
7.70	90.4

MCA = multicomponent analysis with an HP8450A spectrophotometer in the system $HHb/O_2Hb/COHb/Hi$ (Chapter 17). The standard Hi spectrum in the calculation was recorded at pH ≈ 7.28.

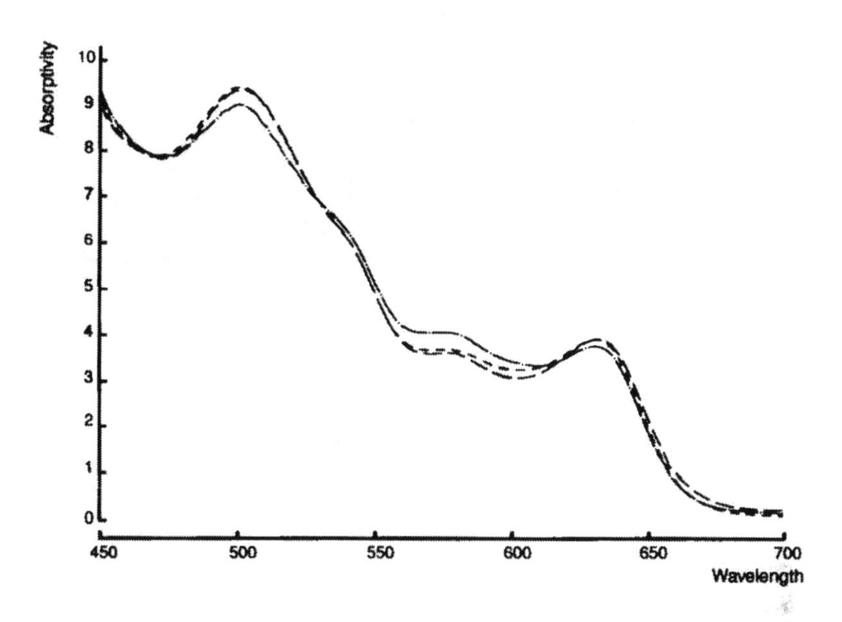

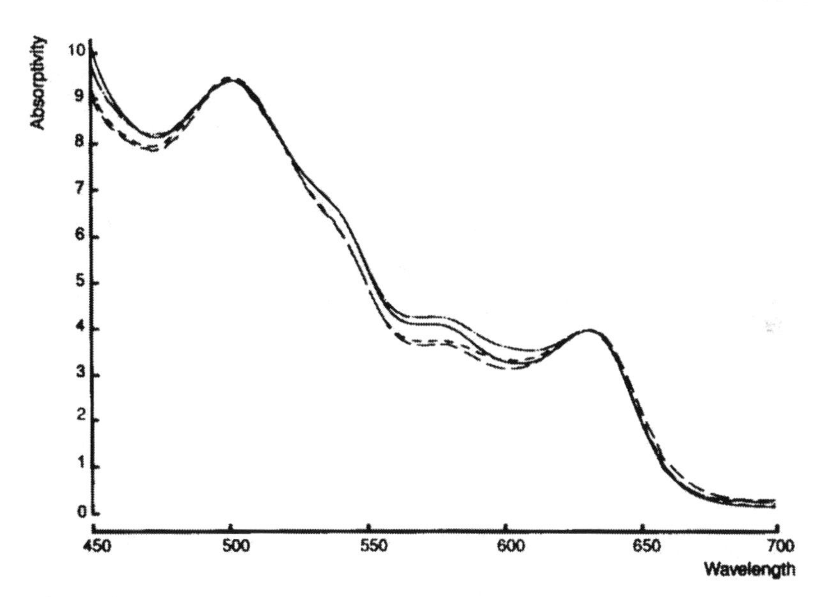

Figure 11.2. Absorption spectra of bovine and human Hi at various pH values.

The upper panel shows the absorption spectra of bovine Hi at pH = 6.95 (— — —), 7.28 (- - - -) and 7.51 (·——·). The lower panel shows the Hi spectra with an absorptivity curve of human Hi at pH ≈ 7.21 added (——). The absorptivity at 630 nm has been normalised to 4.025 $L \cdot mmol^{-1} \cdot cm^{-1}$, the absorptivity of bovine Hi at pH ≈ 7.28.

standard Hi spectrum should be within the range of values which may be expected to ensue when fresh whole blood is haemolysed.

The experiment of Table 11.5 gives support to the adequacy of pH $\approx$ 7.28 for the bovine standard Hi spectrum. Multicomponent analysis in the system HHb/O_2Hb/COHb/Hi was repeatedly performed on a single specimen of bovine haemoglobin that contained only Hi, while the pH of the samples was increased stepwise. The increase in pH was accompanied by a rise in F_{Hi}, reaching 100% at pH = 7.28. The change in F_{Hi} increases towards the alkaline end of the series, showing that more HiOH is formed in each step when the steep part of the titration curve is approached.

CHAPTER 12

ABSORPTION SPECTRA OF PIG HAEMOGLOBIN

The pig is increasingly used as an experimental animal in cardiovascular research. In these experiments it is often mandatory to keep the respiratory and circulatory state of the animal stable in the presence of severe interventions during long lasting general anaesthesia. This requires repeated and often continuous measurements of the oxygen saturation and other blood gas parameters, for which optical methods are the most suitable. Therefore it seemed worthwhile to investigate whether the absorption spectra of the common derivatives of pig haemoglobin deviated to such an extent from those of the corresponding derivatives of human HbA that appreciable errors in the application of oximetry and multicomponent analysis of haemoglobin derivatives could be expected to occur. The results of this investigation have not been published elsewhere.

Pig blood has frequently been used as a substitute for human blood in laboratory experiments for testing materials and methods for extracorporeal circulation and optical methods for the determination of oxygen saturation and other physiologically important quantities [490]. It is easily available and in many respects, *e.g.* rouleaux formation and suspension stability [44, 198], it is similar to human blood.

The investigation of the absorption spectra of pig haemoglobin started, as usual, with the determination of the absorptivity of HiCN at 540 nm, $\varepsilon_{HiCN}(540)$, the anchor value for the whole of haemoglobin spectrophotometry, as described in Chapter 5. This yielded for $\varepsilon_{HiCN}(540)$ a value of 10.92 $L \cdot mmol^{-1} \cdot cm^{-1}$ (Tables 5.1 and 5.2), which is within the range of values underlying the internationally accepted value of 11.0 $L \cdot mmol^{-1} \cdot cm^{-1}$. Hence, the latter value was used in the calculation of the absorptivities of the haemoglobin derivatives in the pig.

Blood from healthy pigs (*Sus scrofa vera*) was obtained at a butcher's and collected in flasks containing 100 USP units of sodium heparin per mL blood. When measurements of $\varepsilon_{HiCN}(540)$ were required, iron free flasks were used. Solutions containing HHb, O_2Hb, COHb and Hi were prepared with the standard procedure described in Chapter 6 and the absorption spectra were determined using HP8450A and Optica CF4 spectrophotometers as described in Chapter 7. In the determination of the absorptivities of HHb, measurements with and without sodium dithionite were made. The ε_{HHb} values presented in Tables 12.1, 12.2, and 12.4 have all been obtained with the addition of a trace of $Na_2S_2O_4$ to the HHb solutions after deoxygenation by tonometry.

Table 12.1 shows the absorptivities of HHb, O_2Hb, COHb and Hi of pig haemoglobin for the spectral range of 450–630 nm measured with an HP8450A diode array spectrophotometer in 0.100-cm cuvettes with plan parallel glass inserts. To obtain reliable results, the measured absorbance should not be less than 0.050 (p. 47). No absorptivity values have been given when the measured absorbance was < 0.050. Table 12.2 shows the absorptivities for the spectral range of 600–800 nm measured with the same instrument; the lightpath was adjusted so that the measured absorbance was > 0.050. Absorptivity values have not been given when the spectrophotometer rejected the measurements because the absorbance of the samples was too high. This happened occasionally in measurements with $l = 0.200$ cm between 600 and 650 nm. The pH range of the Hi samples corresponds with the pH that ensues when a fresh pig blood sample is haemolysed. The spectra represented by the absorptivities of Tables 12.1 and 12.2 are designated as the **standard spectra** of pig haemoglobin.

Figure 12.1 shows the absorption spectra of HHb, O_2Hb, COHb and Hi of porcine haemoglobin in comparison with those of human HbA for the spectral ranges of 450–700 nm (Table 12.1) and 600–800 nm (Table 12.2). In Fig. 12.1a, the spectral absorption curves run to 700 nm, but for $\lambda > 600$ nm part of the measurements have been made with $A < 0.050$; these values have not been given in Table 12.1.

Figure 12.1 and Table 12.3 show no differences in the position of the light absorption maxima and minima of the common derivatives between pig and adult human haemoglobin. As the measurements have been made with a diode array spectrophotometer, which has a limited spectral resolution, this does not rule out that slight differences in the position of the maxima and minima do exist. This has not been further investigated by spectrophotometry with high spectral resolution as was the case in the dog and the rat.

Table 12.3 shows the results of a comparison of the absorptivities of the common derivatives between pig and human haemoglobin, *i.e.* a comparison of the data of Table 12.1 with those of Table 8.1. Student's t-test for unpaired samples was used. The table gives the results at the wavelengths near the principal maxima and minima in the absorption spectra of the common derivatives.

The absorption spectra of Fig. 12.1 show that there is little difference in the absorptivities of HHb, O_2Hb and COHb between porcine and human adult haemoglobin. Most distinct, and statistically significant, are the lower minimum at 560 nm of O_2Hb, and the lower α-peak and the lower minimum at 554 nm of COHb in the spectral absorption curves of pig haemoglobin (Fig. 12.1a). The slightly lower maximum in the porcine HHb spectrum

Table 12.1.
Absorptivities ($L \cdot mmol^{-1} \cdot cm^{-1}$) of the common derivatives of pig haemoglobin

λ (nm)	ε_{HHb} $N = 8$	ε_{O_2Hb} $N = 10$	ε_{COHb} $N = 8$	ε_{Hi} $N = 7$
450	13.11	15.88	10.11	9.000
452	10.31	14.62	9.399	8.733
454	8.326	13.53	8.834	8.542
456	6.857	12.55	8.324	8.380
458	5.808	11.73	7.917	8.265
460	5.018	10.95	7.539	8.153
462	4.454	10.29	7.239	8.063
464	4.029	9.668	6.949	7.975
466	3.734	9.154	6.709	7.916
468	3.516	8.650	6.479	7.851
470	3.379	8.237	6.295	7.824
472	3.287	7.835	6.106	7.803
474	3.243	7.490	5.963	7.826
476	3.217	7.156	5.818	7.875
478	3.229	6.873	5.713	7.965
480	3.241	6.596	5.608	8.065
482	3.285	6.372	5.533	8.210
484	3.327	6.146	5.449	8.354
486	3.398	5.973	5.386	8.523
488	3.464	5.801	5.315	8.674
490	3.562	5.663	5.270	8.856
492	3.663	5.525	5.230	9.000
494	3.797	5.417	5.219	9.142
496	3.934	5.304	5.204	9.246
498	4.102	5.216	5.234	9.332
500	4.272	5.122	5.266	9.363
502	4.470	5.047	5.342	9.376
504	4.671	4.971	5.425	9.311
506	4.900	4.918	5.561	9.230
508	5.109	4.876	5.719	9.078
510	5.337	4.888	5.951	8.931
512	5.546	4.932	6.224	8.726
514	5.765	5.062	6.611	8.532
516	5.973	5.266	7.069	8.300
518	6.201	5.595	7.671	8.085
520	6.420	6.038	8.328	7.844
522	6.680	6.646	9.104	7.631
524	6.940	7.393	9.888	7.400
526	7.252	8.318	10.75	7.208
528	7.563	9.331	11.54	7.022
530	7.939	10.44	12.36	6.861
532	8.329	11.48	13.04	6.704
534	8.787	12.48	13.69	6.572

Table 12.1.
(Continued)

λ (nm)	ε_{HHb} $N = 8$	ε_{O_2Hb} $N = 10$	ε_{COHb} $N = 8$	ε_{Hi} $N = 7$
536	9.269	13.25	14.10	6.426
538	9.824	13.92	14.36	6.283
540	10.37	14.31	14.28	6.107
542	10.99	14.49	14.06	5.911
544	11.52	14.26	13.57	5.676
546	12.07	13.73	13.05	5.433
548	12.49	12.85	12.47	5.143
550	12.88	11.88	12.04	4.878
552	13.12	10.87	11.67	4.601
554	13.28	10.01	11.53	4.369
556	13.29	9.321	11.52	4.162
558	13.22	8.868	11.74	4.000
560	13.03	8.622	12.11	3.873
562	12.82	8.646	12.66	3.782
564	12.51	8.927	13.26	3.722
566	12.19	9.522	13.87	3.690
568	11.78	10.42	14.22	3.672
570	11.37	11.63	14.26	3.666
572	10.90	12.97	13.80	3.668
574	10.45	14.30	12.92	3.669
576	9.973	15.18	11.64	3.666
578	9.542	15.32	10.13	3.654
580	9.120	14.43	8.523	3.627
582	8.737	12.62	6.979	3.582
584	8.347	10.29	5.607	3.531
586	7.943	7.886	4.469	3.454
588	7.472	5.812	3.564	3.380
590	6.941	4.213	2.865	3.314
592	6.347	3.029	2.333	3.243
594	5.717	2.210	1.927	3.181
596	5.073	1.655	1.617	3.134
598	4.446	1.276	1.380	3.099
600	3.866	1.014	1.198	3.076
602	3.351	0.832	1.055	3.068
604	2.919	0.699	0.945	3.074
606	2.568	0.602	0.857	3.094
608	2.292	0.528	0.788	3.125
610	2.076	0.471	0.730	3.170
612	1.906		0.678	3.230
614	1.770		0.626	3.306

Table 12.1.
(Continued)

λ (nm)	ε_{HHb} $N = 8$	ε_{O_2Hb} $N = 10$	ε_{COHb} $N = 8$	ε_{Hi} $N = 7$
616	1.657		0.572	3.397
618	1.569		0.519	3.505
620	1.491		0.465	3.623
622	1.417			3.734
624	1.345			3.843
626	1.280			3.934
628	1.219			4.010
630	1.163			4.060

λ = 450–630 nm. Standard procedure; HP8450A.

Absorbance measured with l = 0.010–0.016 cm. N is number of specimens. $Na_2S_2O_4$ used to prevent reoxygenation of HHb. pH (Hi) = 7.16–7.29.

Table 12.2.
Absorptivities ($L \cdot mmol^{-1} \cdot cm^{-1}$) of the common derivatives of pig haemoglobin

λ (nm)	ε_{HHb} $N = 10$	ε_{O_2Hb} $N = 9$	ε_{COHb} $N = 8$	ε_{Hi} $N = 7$
600		0.988		3.078
602		0.818		3.073
604		0.692		3.077
606		0.600		3.099
608		0.528		3.127
610	2.000	0.474		3.172
612	1.843	0.432		3.232
614	1.717	0.399		3.308
616	1.615	0.373		3.395
618	1.526	0.350		3.501
620	1.449	0.329		3.612
622	1.379	0.307		3.716
624	1.315	0.284	0.378	3.830
626	1.255	0.260	0.342	3.900
628	1.199	0.236	0.310	3.985
630	1.150	0.212	0.281	4.020
632	1.105	0.191	0.255	4.020
634	1.067	0.172	0.233	3.980
636	1.033	0.157	0.214	3.880
638	1.005	0.144	0.198	3.745
640	0.980	0.135	0.184	3.533
642	0.959	0.128	0.172	3.259

Table 12.2.
(Continued)

λ (nm)	ε_{HHb} $N = 10$	ε_{O_2Hb} $N = 9$	ε_{COHb} $N = 8$	ε_{Hi} $N = 7$
644	0.941	0.123	0.162	2.949
646	0.926	0.120	0.154	2.620
648	0.913	0.117	0.146	2.328
650	0.901	0.115	0.140	2.064
652	0.889	0.114	0.134	1.790
654	0.879	0.113	0.128	1.551
656	0.870	0.114	0.123	1.299
658	0.853	0.112	0.116	1.174
660	0.841	0.111	0.112	1.014
662	0.828	0.110	0.108	0.866
664	0.812	0.110	0.103	0.738
666	0.794	0.109	0.099	0.636
668	0.775	0.109	0.095	0.558
670	0.754	0.108	0.091	0.489
672	0.732	0.108	0.088	0.435
674	0.710	0.108	0.085	0.394
676	0.688	0.108	0.082	0.360
678	0.666	0.107	0.079	0.332
680	0.645	0.107	0.077	0.310
682	0.624	0.107	0.074	0.293
684	0.603	0.107	0.072	0.279
686	0.583	0.107	0.070	0.267
688	0.564	0.107	0.069	0.259
690	0.546	0.107	0.067	0.252
692	0.529	0.107	0.066	0.246
694	0.514	0.107	0.064	0.241
696	0.500	0.108	0.063	0.238
698	0.487	0.108	0.062	0.236
700	0.475	0.109	0.061	0.233
702	0.464	0.110	0.060	0.232
704	0.453	0.111	0.059	0.231
706	0.443	0.112	0.058	0.230
708	0.433	0.113	0.057	0.230
710	0.423	0.114	0.057	0.229
712	0.414	0.115	0.056	0.230
714	0.405	0.117	0.055	0.230
716	0.396	0.118	0.055	0.231
718	0.388	0.119	0.054	0.231
720	0.379	0.121	0.054	0.232
722	0.371	0.123	0.053	0.234
724	0.364	0.124	0.053	0.235
726	0.358	0.126	0.052	0.236
728	0.353	0.128	0.052	0.237

Table 12.2.
(Continued)

λ (nm)	ε_{HHb} $N = 10$	ε_{O_2Hb} $N = 9$	ε_{COHb} $N = 8$	ε_{Hi} $N = 7$
730	0.349	0.130	0.051	0.239
732	0.347	0.131	0.051	0.241
734	0.346	0.133	0.050	0.242
736	0.346	0.135	0.050	0.245
738	0.349	0.137	0.050	0.247
740	0.353	0.139	0.049	0.251
742	0.359	0.141	0.049	0.254
744	0.366	0.143	0.048	0.257
746	0.376	0.145	0.048	0.260
748	0.386	0.147	0.048	0.264
750	0.397	0.149	0.047	0.267
752	0.408	0.151	0.047	0.271
754	0.419	0.153	0.047	0.275
756	0.428	0.155	0.046	0.280
758	0.433	0.157	0.046	0.284
760	0.435	0.159	0.046	0.289
762	0.431	0.162	0.045	0.294
764	0.423	0.164	0.045	0.299
766	0.410	0.166	0.045	0.304
768	0.395	0.168	0.044	0.309
770	0.378	0.170	0.044	0.316
772	0.361	0.172	0.044	0.321
774	0.344	0.174	0.043	0.326
776	0.328	0.177	0.043	0.332
778	0.313	0.179	0.043	0.339
780	0.299	0.181	0.042	0.346
782	0.287	0.183	0.042	0.352
784	0.277	0.186	0.042	0.358
786	0.268	0.188	0.041	0.364
788	0.260	0.190	0.041	0.371
790	0.254	0.192	0.041	0.379
792	0.248	0.195	0.040	0.385
794	0.243	0.197	0.040	0.392
796	0.238	0.199	0.040	0.399
798	0.234	0.201	0.039	0.405
800	0.230	0.204	0.039	0.413

λ = 600–800 nm. Standard procedure; HP8450A.

Absorbance measured with l = 0.100 cm (Hi); l = 0.200 cm (HHb and O_2Hb), l = 1.000 cm (COHb). N is number of specimens. $Na_2S_2O_4$ used to prevent reoxygenation of HHb. pH (Hi) = 7.16–7.29.

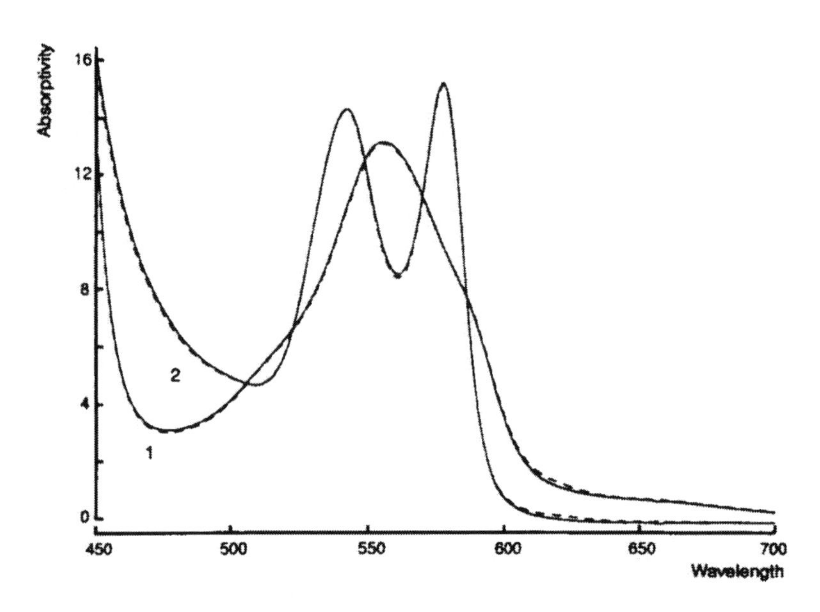

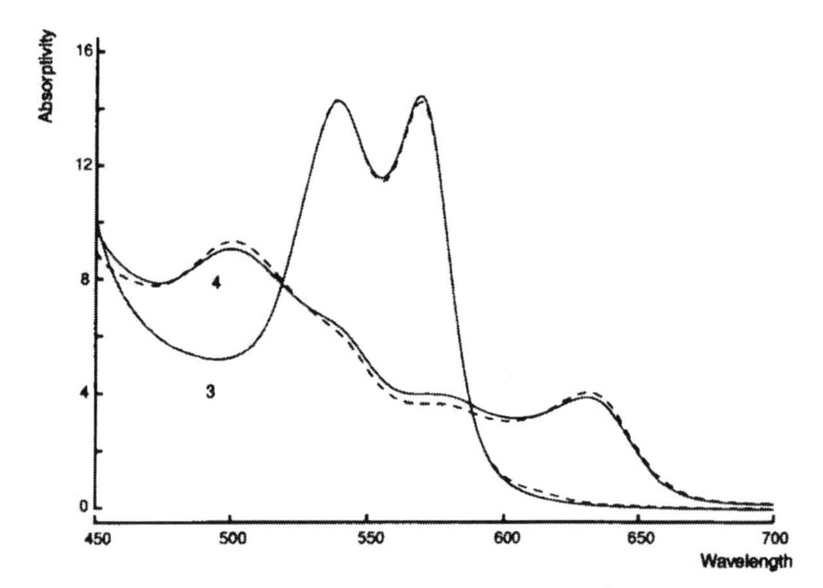

Figure 12.1a. Absorption spectra of the common derivatives of pig haemoglobin in comparison with those of human HbA.

Absorptivity in $L \cdot mmol^{-1} \cdot cm^{-1}$; wavelength in nm. Solid lines: human haemoglobin; dashed lines: pig haemoglobin. Upper panel: HHb (1) and O_2Hb (2); lower panel: COHb (3) and Hi (4). Absorbance measured with $l = 0.010–0.016$ cm. HP8450A. pH (pig Hi) ≈ 7.23; pH (human Hi) ≈ 7.21.

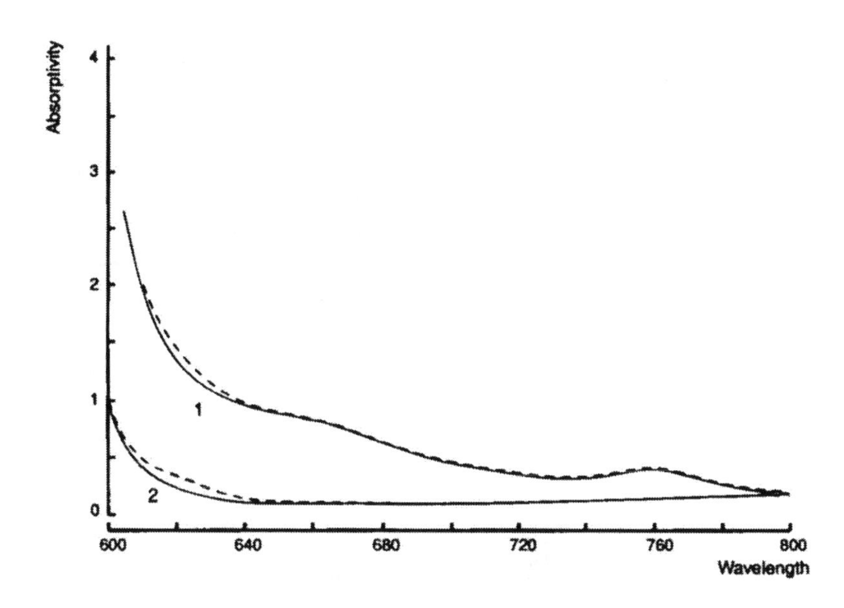

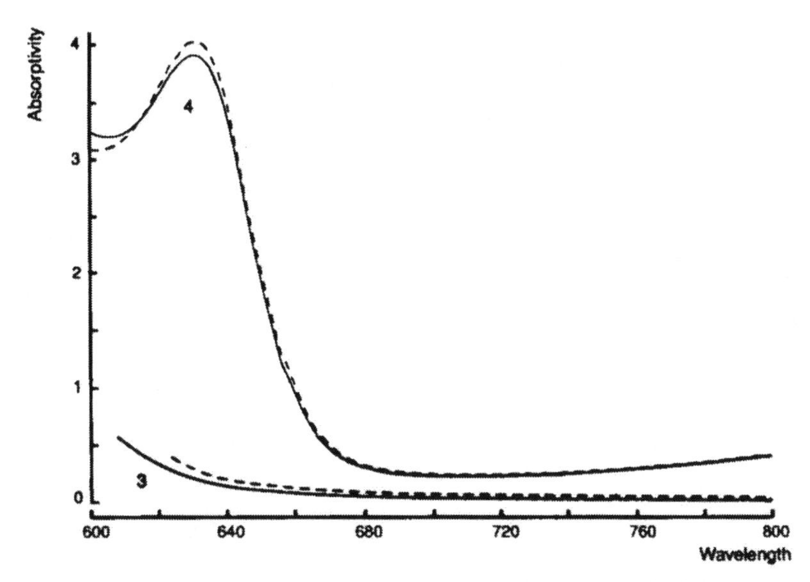

Figure 12.1b. Absorption spectra of the common derivatives of pig haemoglobin in comparison with those of human HbA.

Absorptivity in $L \cdot mmol^{-1} \cdot cm^{-1}$; wavelength in nm. Solid lines: human haemoglobin; dashed lines: pig haemoglobin. Upper panel: HHb (1) and O_2Hb (2); lower panel: COHb (3) and Hi (4). Absorbance measured with $l = 0.100$ cm (Hi); $l = 0.200$ cm (HHb and O_2Hb); $l = 1.000$ cm (COHb). HP8450A. pH (pig Hi) ≈ 7.23; pH (human Hi) ≈ 7.21.

Table 12.3.
Comparison of absorptivities of the common derivatives of pig and adult human haemoglobin

	Human		Pig		
λ (nm)	ε_{HHb} $N = 6$	*SEM*	ε_{HHb} $N = 8$	*SEM*	*p*
476	3.298	0.013	3.217	0.016	< 0.01
478	3.302	0.012	3.228	0.016	< 0.01
552	13.21	0.044	13.12	0.025	
554	13.36	0.045	13.28	0.021	
556	13.36	0.045	13.29	0.026	
λ (nm)	ε_{O_2Hb} $N = 6$	*SEM*	ε_{O_2Hb} $N = 10$	*SEM*	*p*
508	4.882	0.007	4.876	0.010	
540	14.32	0.027	14.31	0.029	
542	14.52	0.023	14.49	0.028	
544	14.34	0.022	14.26	0.025	< 0.05
560	8.767	0.015	8.622	0.017	< 0.001
576	15.26	0.034	15.18	0.036	
578	15.36	0.027	15.32	0.039	
λ (nm)	ε_{COHb} $N = 8$	*SEM*	ε_{COHb} $N = 8$	*SEM*	*p*
496	5.225	0.017	5.204	0.010	
538	14.30	0.039	14.36	0.021	
540	14.27	0.036	14.28	0.017	
542	14.04	0.036	14.06	0.015	
554	11.63	0.029	11.53	0.016	< 0.01
568	14.43	0.037	14.22	0.022	< 0.001
570	14.46	0.039	14.26	0.022	< 0.001
λ (nm)	ε_{Hi} $N = 15$	*SEM*	ε_{Hi} $N = 7$	*SEM*	*p*
498	9.088	0.028	9.332	0.032	< 0.001
500	9.106	0.029	9.363	0.032	< 0.001
502	9.098	0.028	9.376	0.033	< 0.001
630	3.895	0.021	4.060	0.021	< 0.001
632	3.896	0.021	4.072	0.021	< 0.001
634	3.857	0.020	4.044	0.022	< 0.001

Standard procedure; HP8450A.

Absorptivity in $L \cdot mmol^{-1} \cdot cm^{-1}$; *N* is number of specimens, *SEM* is standard error of the mean. *p* values of Student's *t*-test for unpaired samples given only when $p < 0.05$.

does not reach statistical significance. As shown in Fig. 12.1b, in the region > 640 nm, there are no spectral differences between porcine and human haemoglobin as far as HHb, O_2Hb and COHb are concerned. The spectra

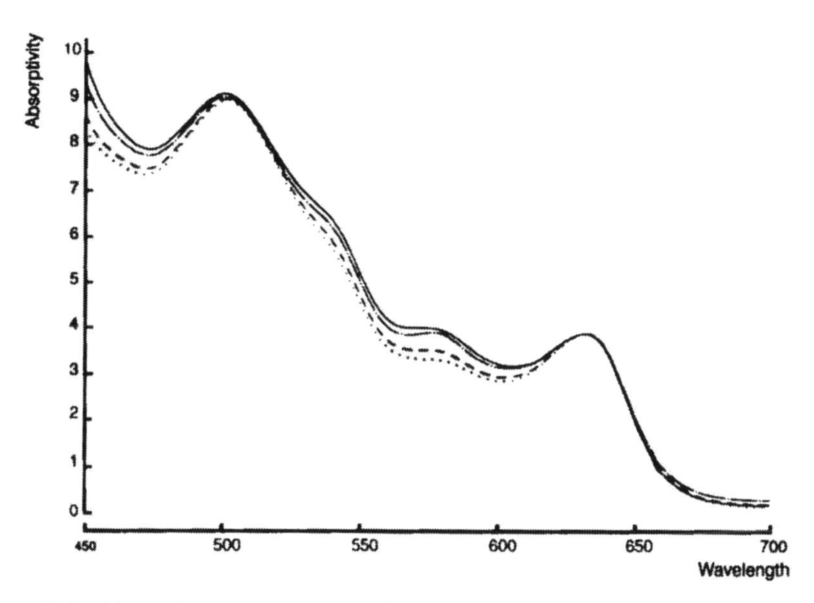

Figure 12.2. Absorption spectra of pig and human Hi at various pH values.
Shown are two pig Hi spectra, obtained at pH = 6.99 (· · ·), and pH = 7.36 (·——·——·), together with the standard pig Hi spectrum (pH ≈ 7.23) (- - - -) and the standard Hi spectrum of human HbA (pH ≈ 7.21) (——). The absorptivity at 630 nm has been normalised to 3.895 $L \cdot mmol^{-1} \cdot cm^{-1}$, the absorptivity of human Hi at pH ≈ 7.21.

of porcine and human HHb and O_2Hb converge to the isosbestic point near 800 nm.

Most conspicuous are the differences in absorptivity between the Hi spectra of pig and man. The lower panel of Fig. 12.1a shows considerable differences between the standard porcine Hi spectrum (pH = 7.16–7.29) and the standard human Hi spectrum (pH = 7.14–7.30). These spectral differences, however, should be interpreted taking any differences in pH into account. Figure 12.2 shows the standard Hi spectra of pig and man, and two individual Hi spectra of the pig. The data given for ε_{Hi} in Table 12.3 relate to the two standard Hi spectra measured in samples with pH = 7.16–7.29 (pig Hi) and pH = 7.14–7.30 (human Hi). These ranges correspond with the pH values which ensue when fresh porcine, *c.q.* human whole blood is haemolysed. The individual porcine Hi spectra have been measured at pH = 6.99 and pH = 7.36. For easy comparison, the absorptivity at 630 nm of all four spectra has been normalised to 3.895 $L \cdot mmol^{-1} \cdot cm^{-1}$, the absorptivity at this wavelength in the standard Hi spectrum of human HbA. The similarity between the porcine Hi spectrum at pH = 7.36 and the standard human Hi spectrum suggests that the observed spectral difference between the two kinds of Hi mainly results from a difference in OH^- ion affinity.

Table 12.4 presents the absorptivities of the common derivatives of pig haemoglobin for selected wavelengths in the spectral range of 600–1000 nm,

Table 12.4.
Absorptivities ($L \cdot mmol^{-1} \cdot cm^{-1}$) of the common derivatives of pig haemoglobin at selected wavelengths between 600 and 1000 nm

λ (nm)	ε_{HHb} $N = 7$	ε_{O_2Hb} $N = 7$	ε_{COHb} $N = 8$	ε_{Hi} $N = 7$
600		0.936	1.166	
630	1.141	0.186	0.258	3.921
660	0.849	0.088	0.095	0.883
680	0.644	0.088	0.059	0.258
700	0.472	0.091	0.045	0.197
750	0.422	0.137	0.035	0.245
775	0.320	0.166	0.032	0.314
800	0.227	0.194	0.029	0.402
805	0.221	0.200	0.028	0.419
840	0.205	0.241	0.024	0.540
845	0.204	0.246	0.023	0.556
880	0.213	0.276	0.020	0.652
904	0.220	0.290	0.018	0.711
920	0.213	0.291	0.016	0.749
940	0.184	0.290	0.014	0.786
960	0.138	0.282	0.011	0.808
1000	0.062	0.252	0.009	0.827

Standard procedure; Optica CF4.

Absorbance measured with $l = 0.100$ cm (Hi at $\lambda = 600$ and 630 nm); $l = 0.200$ cm (HHb and O_2Hb, COHb at $\lambda = 600$ nm, Hi at $\lambda = 660-1000$ nm); $l = 1.000$ cm (COHb at $\lambda = 630-1000$ nm). N is number of specimens. $Na_2S_2O_4$ used to prevent reoxygenation of HHb. pH (Hi) = 7.16–7.29. In the wavelength range 630–1000 nm a red sensitive photomultiplier tube and a red filter have been used.

measured with an Optica CF4 grating spectrophotometer. Wavelength selection was necessary to complete this series of manual measurements within a reasonable time. Wavelengths of importance for biomedical applications have been chosen preferentially.

As mentioned in Chapter 6, the addition of $Na_2S_2O_4$, which binds molecular oxygen, causes side-reactions that influence the HHb spectrum. In Chapter 8 it has been shown that when HHb samples are prepared by tonometry and subsequently measured, some reoxygenation easily occurs, even when the spectrophotometer cuvette is filled directly from the tonometer. Therefore, the use of $Na_2S_2O_4$ has been reintroduced, but only a tiny amount is added after deoxygenation by tonometry. This procedure has also been used in the study of the spectral properties of pig haemoglobin.

The uncertainty that is introduced even by the use of only a little $Na_2S_2O_4$ especially comes to light in the interpretation of the data of Table 12.4. The absorptivity values of HHb and O_2Hb given in this table suggest that the isosbestic point in the spectra of these two derivatives of pig haemoglobin is

Table 12.5.
ε_{HHb} of pig haemoglobin for λ = 630–1000 nm, measured with and without a trace of sodium dithionite to prevent reoxygenation

	With $Na_2S_2O_4$		Without $Na_2S_2O_4$	
λ (nm)	ε_{HHb} $N = 7$	*SEM*	ε_{HHb} $N = 7$	*SEM*
630	1.141	0.013	1.116	0.019
660	0.849	0.009	0.820	0.010
680	0.644	0.013	0.604	0.011
700	0.472	0.010	0.437	0.010
750	0.422	0.009	0.393	0.008
775	0.320	0.008	0.295	0.007
800	0.227	0.007	0.208	0.005
805	0.221	0.007	0.203	0.005
840	0.205	0.006	0.194	0.004
845	0.204	0.006	0.194	0.004
880	0.213	0.005	0.207	0.003
904	0.220	0.005	0.215	0.003
920	0.213	0.005	0.209	0.003
940	0.184	0.005	0.181	0.003
960	0.138	0.005	0.137	0.003
1000	0.062	0.004	0.063	0.002

Standard procedure. Optica CF4.
Absorptivity in $L \cdot mmol^{-1} \cdot cm^{-1}$. Absorbance measured with l = 0.200 cm. *SEM* is standard error of the mean.

situated beyond 800 nm, the value found for human HbA. This may be either a real property of pig haemoglobin or a side-effect of $Na_2S_2O_4$, or the two possible causes may each be responsible for part of the effect.

Table 12.5, showing ε_{HHb} for the wavelength range of 630–1000 nm, measured with and without addition of $Na_2S_2O_4$, has been added to solve this problem. The values with $Na_2S_2O_4$ correspond with those of Table 12.4. Plotting of the two spectral absorption curves of pig HHb, the one with and the one without $Na_2S_2O_4$, together with the O_2Hb curve from Table 12.4, yields a crossover point at ~820 nm for the HHb curve with, and at ~810 nm for the HHb curve without $Na_2S_2O_4$. This confirms that $Na_2S_2O_4$ may cause a shift of the isosbestic point to a longer wavelength; it also shows that in pig haemoglobin the isosbestic point of HHb and O_2Hb indeed lies on the infrared side of that of human haemoglobin.

For a further discussion of the measurements underlying the absorptivities presented in Table 12.4 and an explanation of the slight differences between these data and those of Table 12.2, see Chapter 15.

CHAPTER 13

ABSORPTION SPECTRA OF HORSE HAEMOGLOBIN

Horse haemoglobin was selected for determination of the absorptivity of the common haemoglobin derivatives because of the need of multicomponent analysis of horse haemoglobin in veterinary medicine. There was little experience with handling horse blood in our laboratory, except that it was known from earlier investigations that horse blood has a low suspension stability due to a strong tendency of the erythrocytes to form rouleaux [44, 198] (p. 255). This proved to be no serious drawback in working up the samples of horse haemoglobin for spectrophotometry. The results given in this chapter have not yet been published elsewhere.

Similar to the procedure applied in the other animal species, the investigation of the absorption spectra of horse haemoglobin began with the determination of the absorptivity of HiCN at 540 nm, $\varepsilon_{HiCN}(540)$, the anchor value for the whole of haemoglobin spectrophotometry, as described in Chapter 5. This yielded a value of 10.98 $L \cdot mmol^{-1} \cdot cm^{-1}$ for $\varepsilon_{HiCN}(540)$ (Tables 5.1 and 5.2), which is within the range of values underlying the internationally accepted value of 11.0 $L \cdot mmol^{-1} \cdot cm^{-1}$. Hence, the latter value was also used in the calculation of the absorptivities of the haemoglobin derivatives in the horse.

Blood from healthy horses (*Equus caballus*) was obtained from the Department of Animal Physiology of the University of Utrecht. The blood was collected in flasks containing 100 USP units of sodium heparin per mL blood. When measurements of $\varepsilon_{HiCN}(540)$ were required, iron free flasks were used. The flasks were transported in a cooled container and the blood was processed as soon as possible after arrival in the laboratory. Solutions containing HHb, O_2Hb, COHb and Hi were prepared with the standard procedure described in Chapter 6 and the absorption spectra were determined using HP8450A and Optica CF4 spectrophotometers as described in Chapter 7. In the determination of the absorptivities of HHb, measurements with and without sodium dithionite were made. The ε_{HHb} values presented in Tables 13.1, 13.2, and 13.4 have all been obtained with the addition of a trace of $Na_2S_2O_4$ to the HHb solutions after deoxygenation by tonometry.

Table 13.1 shows the absorptivities of HHb, O_2Hb, COHb and Hi of horse haemoglobin for the spectral range of 450–630 nm measured with

Table 13.1.
Absorptivities ($L \cdot mmol^{-1} \cdot cm^{-1}$) of the common derivatives of horse haemoglobin

λ (nm)	ε_{HHb} $N = 11$	ε_{O_2Hb} $N = 10$	ε_{COHb} $N = 10$	ε_{Hi} $N = 7$
450	13.30	16.30	10.22	8.669
452	10.44	15.01	9.519	8.427
454	8.403	13.92	8.948	8.253
456	6.920	12.91	8.443	8.108
458	5.854	12.05	8.028	8.009
460	5.060	11.26	7.649	7.912
462	4.491	10.58	7.338	7.844
464	4.064	9.950	7.044	7.765
466	3.767	9.409	6.808	7.716
468	3.550	8.899	6.575	7.656
470	3.412	8.464	6.391	7.646
472	3.316	8.051	6.204	7.630
474	3.276	7.697	6.061	7.657
476	3.253	7.348	5.923	7.708
478	3.260	7.061	5.820	7.800
480	3.277	6.764	5.721	7.903
482	3.317	6.524	5.651	8.050
484	3.359	6.290	5.572	8.189
486	3.418	6.102	5.513	8.370
488	3.491	5.920	5.448	8.529
490	3.578	5.773	5.405	8.704
492	3.681	5.625	5.362	8.856
494	3.802	5.504	5.348	8.998
496	3.939	5.383	5.336	9.105
498	4.099	5.291	5.356	9.199
500	4.270	5.190	5.383	9.225
502	4.460	5.112	5.451	9.238
504	4.657	5.024	5.530	9.170
506	4.877	4.967	5.661	9.086
508	5.086	4.913	5.809	8.935
510	5.313	4.911	6.026	8.775
512	5.518	4.941	6.292	8.560
514	5.739	5.050	6.657	8.365
516	5.938	5.226	7.096	8.119
518	6.167	5.529	7.659	7.886
520	6.390	5.938	8.299	7.635
522	6.649	6.512	9.041	7.407
524	6.906	7.221	9.810	7.177
526	7.219	8.112	10.63	6.969
528	7.528	9.099	11.41	6.773
530	7.900	10.19	12.19	6.609
532	8.286	11.23	12.88	6.447
534	8.747	12.23	13.49	6.312

Table 13.1.
(Continued)

λ (nm)	ε_{HHb} $N = 11$	ε_{O_2Hb} $N = 10$	ε_{COHb} $N = 10$	ε_{Hi} $N = 7$
536	9.216	13.05	13.94	6.163
538	9.762	13.73	14.18	6.024
540	10.31	14.18	14.19	5.846
542	10.91	14.41	13.98	5.661
544	11.45	14.28	13.56	5.430
546	12.00	13.82	13.08	5.191
548	12.43	13.02	12.57	4.915
550	12.80	12.06	12.14	4.649
552	13.06	11.07	11.80	4.386
554	13.21	10.21	11.64	4.154
556	13.24	9.483	11.63	3.953
558	13.16	8.991	11.82	3.792
560	12.98	8.688	12.17	3.664
562	12.77	8.653	12.69	3.575
564	12.48	8.866	13.27	3.517
566	12.16	9.390	13.87	3.481
568	11.76	10.20	14.25	3.468
570	11.36	11.33	14.29	3.461
572	10.90	12.63	13.87	3.464
574	10.44	13.95	13.02	3.466
576	9.973	14.88	11.79	3.466
578	9.535	15.14	10.35	3.460
580	9.110	14.39	8.807	3.444
582	8.721	12.77	7.300	3.413
584	8.326	10.58	5.927	3.372
586	7.912	8.297	4.758	3.317
588	7.437	6.229	3.808	3.257
590	6.899	4.552	3.058	3.196
592	6.295	3.289	2.479	3.137
594	5.655	2.384	2.032	3.083
596	4.997	1.755	1.686	3.040
598	4.359	1.324	1.416	3.007
600	3.774	1.030	1.206	2.988
602	3.255	0.821	1.037	2.984
604	2.818	0.672	0.902	2.992
606	2.463	0.561	0.791	3.013
608	2.180	0.477	0.700	3.047
610	1.957		0.625	3.097
612	1.781		0.560	3.160

Table 13.1.
(Continued)

λ (nm)	ε_{HHb} $N = 11$	ε_{O_2Hb} $N = 10$	ε_{COHb} $N = 10$	ε_{Hi} $N = 7$
614	1.641		0.505	3.241
616	1.529		0.458	3.339
618	1.435			3.447
620	1.354			3.564
622	1.288			3.681
624	1.231			3.790
626	1.181			3.885
628	1.136			3.960
630	1.096			4.009

λ = 450–630 nm. Standard procedure; HP8450A.

Absorbance measured with l = 0.010–0.016 cm. N is number of specimens. $Na_2S_2O_4$ used to prevent reoxygenation of HHb. pH (Hi) = 7.10–7.25.

Table 13.2.
Absorptivities ($L \cdot mmol^{-1} \cdot cm^{-1}$) of the common derivatives of horse haemoglobin

λ (nm)	ε_{HHb} $N = 10$	ε_{O_2Hb} $N = 10$	ε_{COHb} $N = 10$	ε_{Hi} $N = 7$
600		1.014		2.989
602		0.821		2.986
604		0.679	0.870	2.991
606		0.573		3.016
608		0.492	0.688	3.042
610	1.907	0.428	0.617	3.092
612	1.748	0.378	0.556	3.150
614	1.617	0.337	0.503	3.231
616	1.506	0.304	0.457	3.320
618	1.416	0.277	0.418	3.429
620	1.339	0.253	0.370	3.536
622	1.275	0.234	0.344	3.655
624	1.219	0.216	0.319	3.758
626	1.171	0.200	0.295	3.850
628	1.128	0.186	0.273	3.918
630	1.090	0.173	0.255	3.961
632	1.056	0.162	0.237	3.971
634	1.027	0.152	0.222	3.939
636	1.000	0.144	0.208	3.853
638	0.979	0.138	0.196	3.704
640	0.958	0.133	0.185	3.503
642	0.941	0.129	0.176	3.260
644	0.926	0.126	0.168	2.980

Table 13.2.
(Continued)

λ (nm)	ε_{HHb} $N = 10$	ε_{O_2Hb} $N = 10$	ε_{COHb} $N = 10$	ε_{Hi} $N = 7$
646	0.913	0.125	0.161	2.681
648	0.901	0.124	0.155	2.370
650	0.891	0.123	0.149	2.072
652	0.881	0.123	0.143	1.791
654	0.873	0.122	0.138	1.540
656	0.860	0.118	0.134	1.332
658	0.843	0.115	0.127	1.075
660	0.837	0.119	0.122	0.933
662	0.824	0.118	0.117	0.796
664	0.809	0.117	0.112	0.677
666	0.791	0.116	0.108	0.579
668	0.771	0.115	0.103	0.498
670	0.751	0.114	0.099	0.433
672	0.729	0.112	0.095	0.379
674	0.706	0.112	0.091	0.337
676	0.684	0.111	0.088	0.302
678	0.661	0.110	0.085	0.275
680	0.639	0.109	0.082	0.253
682	0.618	0.109	0.080	0.326
684	0.597	0.109	0.078	0.222
686	0.576	0.108	0.076	0.211
688	0.557	0.108	0.074	0.203
690	0.539	0.108	0.072	0.196
692	0.522	0.108	0.070	0.191
694	0.506	0.108	0.069	0.187
696	0.491	0.109	0.068	0.184
698	0.478	0.109	0.067	0.181
700	0.466	0.110	0.066	0.180
702	0.455	0.111	0.065	0.179
704	0.444	0.112	0.064	0.178
706	0.434	0.113	0.063	0.178
708	0.425	0.114	0.062	0.178
710	0.415	0.115	0.062	0.178
712	0.406	0.117	0.061	0.179
714	0.397	0.118	0.060	0.180
716	0.388	0.120	0.060	0.181
718	0.380	0.121	0.059	0.182
720	0.371	0.123	0.059	0.183
722	0.364	0.125	0.059	0.185
724	0.356	0.126	0.058	0.187
726	0.350	0.128	0.058	0.189
728	0.345	0.130	0.057	0.191
730	0.341	0.131	0.057	0.193
732	0.338	0.133	0.057	0.195

Table 13.2.
(Continued)

λ (nm)	ε_{HHb} $N = 10$	ε_{O_2Hb} $N = 10$	ε_{COHb} $N = 10$	ε_{Hi} $N = 7$
734	0.337	0.135	0.056	0.198
736	0.338	0.137	0.056	0.201
738	0.340	0.139	0.056	0.204
740	0.344	0.141	0.056	0.207
742	0.349	0.142	0.055	0.210
744	0.356	0.144	0.055	0.214
746	0.365	0.146	0.055	0.218
748	0.375	0.149	0.054	0.222
750	0.386	0.151	0.054	0.226
752	0.397	0.153	0.054	0.231
754	0.407	0.155	0.054	0.235
756	0.415	0.157	0.053	0.240
758	0.420	0.159	0.053	0.245
760	0.420	0.161	0.053	0.250
762	0.417	0.164	0.052	0.256
764	0.409	0.166	0.052	0.261
766	0.397	0.168	0.052	0.267
768	0.383	0.171	0.052	0.273
770	0.367	0.173	0.051	0.279
772	0.351	0.175	0.051	0.285
774	0.335	0.178	0.051	0.291
776	0.320	0.180	0.051	0.298
778	0.306	0.182	0.050	0.304
780	0.293	0.185	0.050	0.311
782	0.282	0.187	0.050	0.318
784	0.272	0.189	0.050	0.325
786	0.263	0.192	0.049	0.332
788	0.256	0.194	0.049	0.338
790	0.249	0.197	0.049	0.345
792	0.244	0.199	0.049	0.352
794	0.239	0.201	0.048	0.359
796	0.234	0.204	0.048	0.367
798	0.231	0.206	0.048	0.374
800	0.227	0.208	0.048	0.381

λ = 600–800 nm. Standard procedure; HP8450A.

Absorbance measured with l = 0.100 cm (Hi); l = 0.200 cm (HHb and O_2Hb), l = 1.000 cm (COHb). N is number of specimens. $Na_2S_2O_4$ used to prevent reoxygenation of HHb. pH (Hi) = 7.10–7.25.

an HP8450A diode array spectrophotometer in 0.100-cm cuvettes with plan parallel glass inserts. To obtain reliable results, the measured absorbance should not be less than 0.050 (p. 47). No absorptivity values have been given when the measured absorbance was < 0.050. Table 13.2 shows the absorptivities for the spectral range of 600–800 nm measured with the same instrument; the lightpath was adjusted so that the measured absorbance was > 0.050. Absorptivity values have also not been given when the spectrophotometer rejected the measurements because the absorbance of the samples was too high. This happened occasionally in measurements with l = 0.200 cm between 600 and 650 nm. The pH range of the Hi samples corresponds with the pH that ensues when a fresh horse blood sample is haemolysed. The spectra represented by the absorptivities of Tables 13.1 and 13.2 are designated as the **standard spectra** of horse haemoglobin.

Figure 13.1 shows the absorption spectra of HHb, O_2Hb, COHb and Hi of equine haemoglobin in comparison with those of human HbA for the spectral ranges of 450–700 nm (Table 13.1) and 600–800 nm (Table 13.2). In Fig. 13.1a the spectral absorption curves run to 700 nm, but for $\lambda > 600$ nm part of the measurements resulted in $A < 0.050$; these values have not been given in Table 13.1.

Figure 13.1 and Table 13.3 show no clear differences in the position of the light absorption maxima and minima of the common derivatives between horse and adult human haemoglobin. However, it would seem that the α-peaks of O_2Hb and COHb of the horse are slightly displaced to the right in comparison with the corresponding peaks of human haemoglobin. The α-peak of horse O_2Hb also seems to be a little narrower. This, and the exact positions of the other absorption maxima and minima in the spectra of horse haemoglobin, have not been further investigated by high resolution spectrophotometry as in the case of the dog and the rat.

Table 13.3 shows the results of a comparison of the absorptivities of the common derivatives between horse and human haemoglobin, *i.e.* a comparison of the data of Table 13.1 with those of Table 8.1. Student's t-test for unpaired samples was used. The table gives the results at the wavelengths near the principal absorption maxima and minima in the spectra of the common derivatives.

In Fig. 13.1a, some differences may be noted between the absorptivities of equine and human HHb, O_2Hb and COHb. Although the differences are small, the p-values of Table 13.3 indicate that they are statistically significant. Moreover, they are definitely of such a magnitude that they must be taken into account in multicomponent analysis of haemoglobin derivatives (Chapter 17). If there is a difference, the absorptivity of the equine haemoglobin derivative is generally lower than that of the corresponding derivative of human haemoglobin. This holds for the maximum of HHb near 555 nm, and for the α- and β-peaks of O_2Hb and COHb, as well as for the minimum in the O_2Hb spectrum near 560 nm. As is shown in Fig. 13.1b, there are, in the red and near infrared region, no significant spectral differences between equine and human haemoglobin as far as HHb, O_2Hb and COHb are concerned. The spectra of equine and human HHb and O_2Hb converge to the isosbestic point near 800 nm.

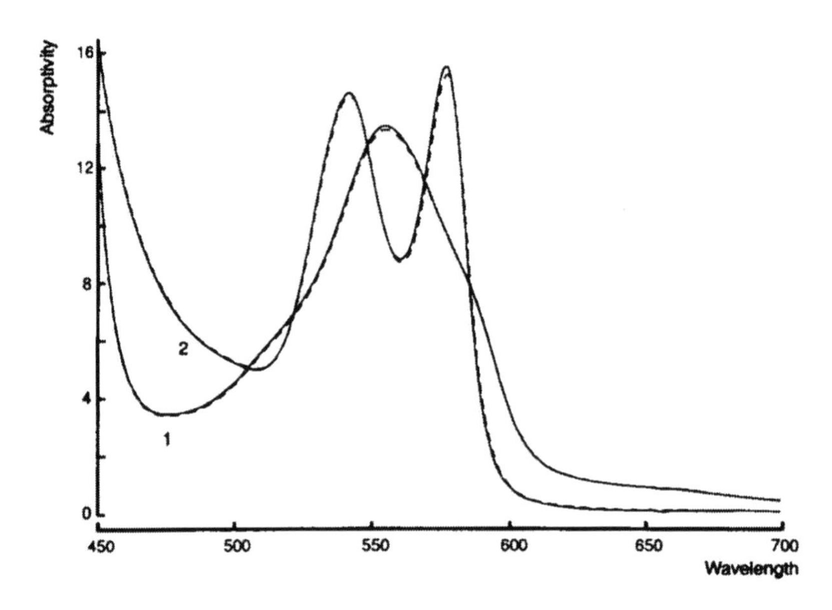

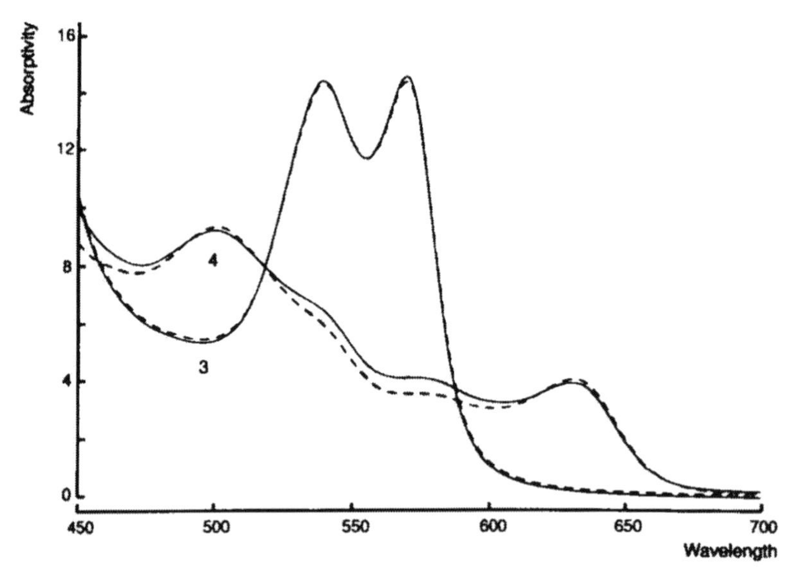

Figure 13.1a. Absorption spectra of the common derivatives of horse haemoglobin in comparison with those of human HbA.

Absorptivity in $L \cdot mmol^{-1} \cdot cm^{-1}$; wavelength in nm. Solid lines: human haemoglobin; dashed lines: horse haemoglobin. Upper panel: HHb (1) and O_2Hb (2); lower panel: COHb (3) and Hi (4). Absorbance measured with $l = 0.010$–0.016 cm. HP8450A. pH (horse Hi) ≈ 7.17; pH (human Hi) ≈ 7.21.

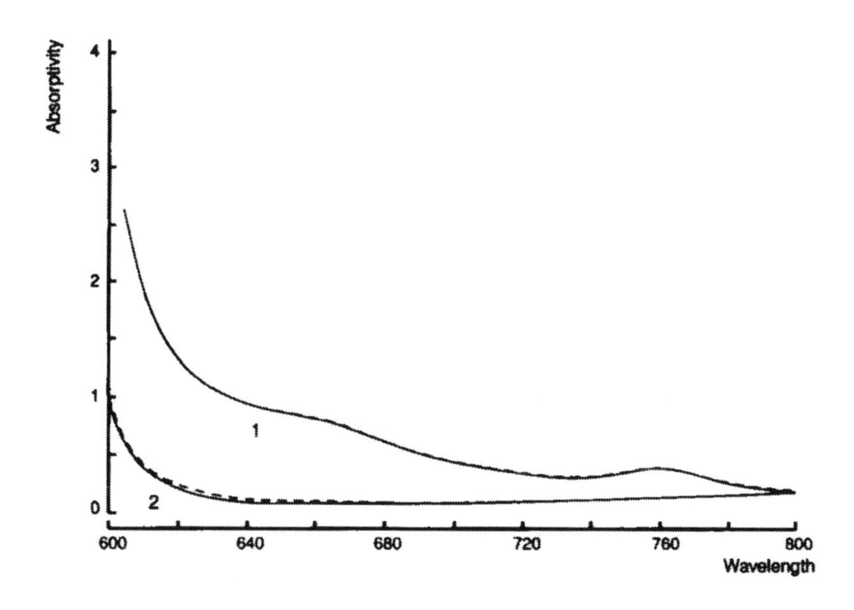

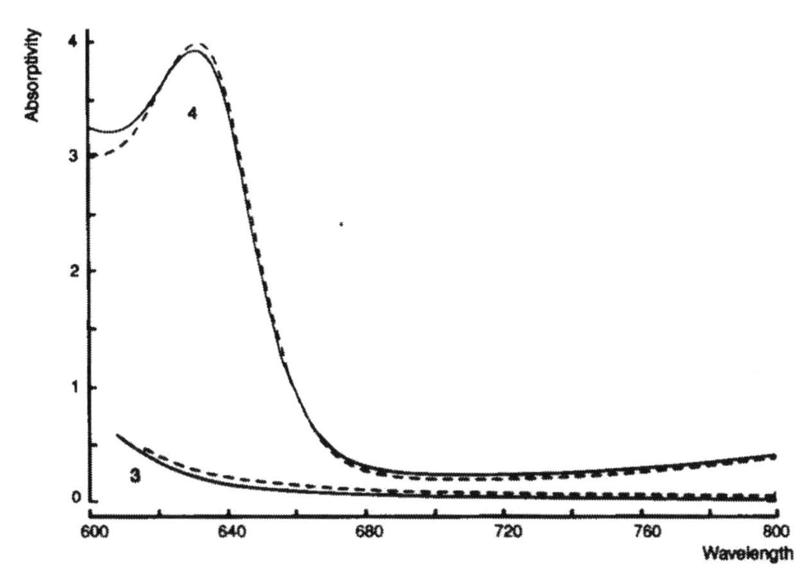

Figure 13.1b. Absorption spectra of the common derivatives of horse haemoglobin in comparison with those of human HbA.

Absorptivity in $L \cdot mmol^{-1} \cdot cm^{-1}$; wavelength in nm. Solid lines: human haemoglobin; dashed lines: horse haemoglobin. Upper panel: HHb (1) and O_2Hb (2); lower panel: COHb (3) and Hi (4). Absorbance measured with $l = 0.100$ cm (Hi); $l = 0.200$ cm (HHb and O_2Hb); $l = 1.000$ cm (COHb). HP8450A. pH (horse Hi) ≈ 7.17; pH (human Hi) ≈ 7.21.

Table 13.3.
Comparison of absorptivities of the common derivatives of horse and adult human haemoglobin

	Human		Horse		
λ (nm)	ε_{HHb} $N = 6$	*SEM*	ε_{HHb} $N = 11$	*SEM*	*p*
476	3.298	0.013	3.254	0.013	< 0.05
478	3.302	0.012	3.260	0.012	< 0.05
552	13.21	0.044	13.06	0.013	< 0.001
554	13.36	0.045	13.20	0.016	< 0.002
556	13.36	0.045	13.24	0.013	< 0.01
λ (nm)	ε_{O_2Hb} $N = 6$	*SEM*	ε_{O_2Hb} $N = 10$	*SEM*	*p*
508	4.882	0.007	4.913	0.006	< 0.01
540	14.32	0.027	14.18	0.026	< 0.01
542	14.52	0.023	14.41	0.025	
544	14.34	0.022	14.28	0.026	
560	8.767	0.015	8.688	0.014	< 0.005
576	15.26	0.034	14.88	0.031	< 0.001
578	15.36	0.027	15.14	0.027	< 0.001
λ (nm)	ε_{COHb} $N = 8$	*SEM*	ε_{COHb} $N = 10$	*SEM*	*p*
496	5.225	0.017	5.336	0.007	< 0.001
538	14.30	0.039	14.18	0.017	< 0.01
540	14.27	0.036	14.19	0.015	< 0.05
542	14.04	0.036	13.98	0.017	
554	11.63	0.029	11.64	0.015	
568	14.43	0.037	14.25	0.017	< 0.001
570	14.46	0.039	14.29	0.016	< 0.001
λ (nm)	ε_{Hi} $N = 15$	*SEM*	ε_{Hi} $N = 7$	*SEM*	*p*
498	9.088	0.028	9.199	0.018	< 0.05
500	9.106	0.029	9.225	0.018	< 0.02
502	9.098	0.028	9.238	0.021	< 0.01
630	3.895	0.021	4.009	0.016	< 0.005
632	3.896	0.021	4.023	0.015	< 0.001
634	3.857	0.020	3.994	0.014	< 0.001

Standard procedure; HP8450A.
Absorptivity in $L \cdot mmol^{-1} \cdot cm^{-1}$; *N* is number of specimens, *SEM* is standard error of the mean. *p* values of Student's *t*-test for unpaired samples given only when $p < 0.05$.

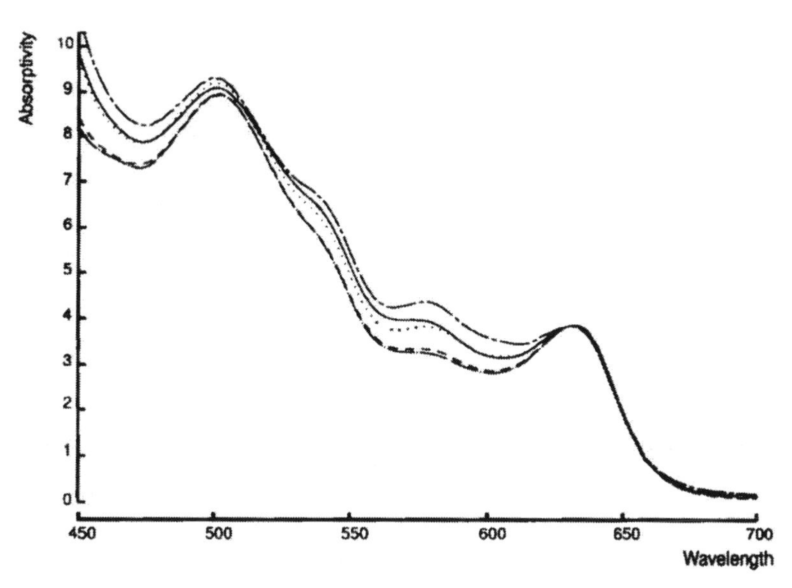

Figure 13.2. Absorption spectra of horse and human Hi at various pH values.
Shown are three horse Hi spectra, obtained at pH = 7.00 (·—·—·), pH = 7.50 (· · ·), and pH = 7.65 (— - — -), together with the standard horse Hi spectrum (pH ≈ 7.17) (- - - -) and the standard Hi spectrum of human HbA (pH ≈ 7.21) (——). The absorptivity at 630 nm has been normalised to 3.895 $L \cdot mmol^{-1} \cdot cm^{-1}$, the absorptivity of human Hi at pH ≈ 7.21.

As in the other animal species, the most distinct differences in relation to human haemoglobin are found in the equine Hi spectrum. In the lower panel of Fig. 13.1a the two maxima in the Hi spectrum, near 500 and 630 nm are clearly higher in the horse, while between 520 and 620 nm the Hi spectrum of horse haemoglobin definitely lies below that of human haemoglobin. As always, these differences should be interpreted in light of the prevailing differences in pH. To this end, Fig. 13.2 shows three equine Hi spectra at different pH in combination with the standard equine Hi spectrum (pH = 7.10–7.25) and the standard human Hi spectrum (pH = 7.14–7.30), with the absorptivity at 630 nm normalised to 3.895 $L \cdot mmol^{-1} \cdot cm^{-1}$, the value of $\varepsilon_{Hi}(630)$ in the standard human Hi spectrum. The figure shows the expected increase of the influence of a change in pH at higher pH values. The similarity of the standard human Hi spectrum and the equine Hi spectrum at pH = 7.50 suggests that the differences between the standard equine Hi spectrum (pH ≈ 7.17) and the standard human Hi spectrum (pH ≈ 7.21) are to a considerable extent the result of a difference in OH^- ion affinity. On the other hand, it seems improbable that there is a pH value at which the Hi spectra of horse and man will coincide exactly. Hence, there also is an intrinsic spectral difference between the two kinds of haemoglobin.

Table 13.4 presents the absorptivities of the common derivatives of horse haemoglobin for selected wavelengths in the spectral range of 600–1000 nm, measured with an Optica CF4 grating spectrophotometer. Wavelength selec-

Table 13.4.
Absorptivities ($L \cdot mmol^{-1} \cdot cm^{-1}$) of the common derivatives of horse haemoglobin at selected wavelengths between 600 and 1000 nm

λ (nm)	ε_{HHb} $N = 10$	ε_{O_2Hb} $N = 7$	ε_{COHb} $N = 10$	ε_{Hi} $N = 7$
600	3.341	0.956	1.127	2.972
630	1.066	0.147	0.221	3.954
660	0.830	0.099	0.102	0.855
680	0.623	0.091	0.064	0.219
700	0.456	0.093	0.050	0.162
750	0.403	0.137	0.041	0.218
775	0.304	0.165	0.039	0.299
800	0.218	0.196	0.037	0.378
805	0.212	0.201	0.036	0.395
840	0.195	0.239	0.033	0.522
845	0.194	0.245	0.033	0.537
880	0.193	0.275	0.030	0.633
904	0.213	0.288	0.029	0.701
920	0.207	0.292	0.028	0.738
940	0.178	0.289	0.025	0.776
960	0.131	0.282	0.022	0.803
1000	0.055	0.255	0.021	0.818

Standard procedure. Optica CF4.

Absorbance measured with $l = 0.100$ cm (Hi at $\lambda = 600$ and 630 nm); $l = 0.200$ cm (HHb and O_2Hb, COHb at $\lambda = 600$ nm ($N = 6$), Hi at $\lambda = 660-1000$ nm); $l = 1.000$ cm (COHb at $\lambda = 630-1000$ nm). N is number of specimens. $Na_2S_2O_4$ used to prevent reoxygenation of HHb. pH (Hi) = 7.10–7.25. In the wavelength range 630–1000 nm a red sensitive photomultiplier tube with a red filter has been used.

tion was necessary to complete this series of manual measurements within a reasonable time. Wavelengths of importance for biomedical applications have been chosen preferentially.

For a discussion of the measurements underlying the absorptivities presented in Table 13.4 and an explanation of the slight differences between these data and those of Table 13.2 see Chapter 15.

CHAPTER 14

ABSORPTION SPECTRA OF SHEEP HAEMOGLOBIN

The sheep and the (foetal) lamb are frequently used as experimental animals in perinatal pathophysiology and paediatric cardiology. In our experience [80, 81, 144, 145, 388, 389], these experiments often require repeated and continuous measurements of the oxygen saturation and other blood gas parameters, for which optical methods are most suitable. Therefore, it seemed worthwhile to investigate, also in the sheep, whether the absorption spectra of the common haemoglobin derivatives deviated to such an extent from those of the corresponding derivatives of human HbA that appreciable errors in the application of oximetry and multicomponent analysis of haemoglobin derivatives could be expected to occur when instruments calibrated for application in humans are used. The results of this investigation have not been published elsewhere.

The fact that experiments on lambs are often performed in the perinatal period [1, 2, 364, 365] made it desirable to also investigate if any differences existed between the spectral properties of foetal and adult sheep haemoglobin. To this end we completed the study of the absorption spectra of sheep haemoglobin with a few similar series of measurements of haemoglobin derivatives prepared from the blood of neonatal lambs. The procedures described for human HbF in Chapters 6 and 7 were followed. In fact, this is a rather crude approach because of the diversity of sheep haemoglobin. In the adult sheep there are at least three genetically determined types of haemoglobin, each with a different oxygen affinity [79]. In addition there is foetal haemoglobin that can be distinguished from the adult haemoglobins [244]. Simply comparing 'adult' and 'foetal' sheep haemoglobin as if they were homogeneous substances thus does not take the complexity of the system properly into account. Lacking the means for a more thorough approach, the pilot study described in this chapter was considered to shed at least some light on the size of the differences, if any, which may occur.

The investigation of the absorption spectra of sheep haemoglobin started, similar to that in the other animal species, with the determination of the absorptivity of HiCN at 540 nm, $\varepsilon_{HiCN}(540)$, the value on which all other absorptivity values are based (Chapter 5). This yielded a value of $10.98\ L \cdot mmol^{-1} \cdot cm^{-1}$ for $\varepsilon_{HiCN}(540)$ (Tables 5.1 and 5.2), which is within

Table 14.1.
Absorptivities ($L \cdot mmol^{-1} \cdot cm^{-1}$) of the common derivatives of adult sheep haemoglobin

λ (nm)	ε_{HHb} $N = 7$	ε_{O_2Hb} $N = 7$	ε_{COHb} $N = 6$	ε_{Hi} $N = 7$
450	13.65	15.91	10.17	8.824
452	10.74	14.62	9.459	8.568
454	8.677	13.54	8.899	8.390
456	7.149	12.54	8.393	8.247
458	6.049	11.71	7.987	8.148
460	5.222	10.97	7.605	8.057
462	4.632	10.29	7.303	7.988
464	4.183	9.679	7.010	7.915
466	3.876	9.180	6.771	7.875
468	3.644	8.679	6.535	7.823
470	3.496	8.262	6.352	7.819
472	3.400	7.848	6.159	7.813
474	3.352	7.514	6.013	7.854
476	3.325	7.186	5.869	7.907
478	3.327	6.899	5.765	8.014
480	3.344	6.628	5.656	8.127
482	3.378	6.401	5.590	8.279
484	3.422	6.184	5.494	8.435
486	3.486	6.004	5.442	8.615
488	3.559	5.871	5.366	8.775
490	3.650	5.711	5.331	8.965
492	3.755	5.585	5.279	9.120
494	3.880	5.473	5.264	9.272
496	4.016	5.359	5.251	9.388
498	4.186	5.268	5.274	9.487
500	4.353	5.181	5.307	9.519
502	4.547	5.104	5.375	9.542
504	4.752	5.031	5.460	9.479
506	4.971	4.975	5.589	9.396
508	5.191	4.937	5.749	9.255
510	5.418	4.945	5.982	9.094
512	5.621	4.991	6.258	8.894
514	5.847	5.115	6.648	8.685
516	6.048	5.318	7.114	8.445
518	6.278	5.641	7.713	8.207
520	6.497	6.075	8.388	7.951
522	6.761	6.677	9.163	7.721
524	7.017	7.429	9.968	7.481
526	7.328	8.360	10.82	7.261
528	7.635	9.362	11.64	7.056
530	8.015	10.47	12.44	6.881
532	8.401	11.54	13.16	6.706
534	8.864	12.56	13.79	6.555

Table 14.1.
(Continued)

λ (nm)	ε_{HHb} $N=7$	ε_{O_2Hb} $N=7$	ε_{COHb} $N=6$	ε_{Hi} $N=7$
536	9.335	13.32	14.22	6.395
538	9.889	14.00	14.44	6.243
540	10.44	14.42	14.40	6.060
542	11.05	14.57	14.13	5.858
544	11.61	14.36	13.66	5.623
546	12.15	13.81	13.12	5.373
548	12.59	12.88	12.56	5.085
550	12.97	11.95	12.09	4.822
552	13.22	10.89	11.74	4.542
554	13.37	10.05	11.57	4.303
556	13.39	9.350	11.58	4.088
558	13.31	8.904	11.81	3.919
560	13.14	8.661	12.19	3.785
562	12.92	8.690	12.77	3.689
564	12.62	8.989	13.40	3.624
566	12.30	9.600	14.03	3.586
568	11.91	10.57	14.39	3.566
570	11.50	11.84	14.42	3.561
572	11.04	13.18	13.95	3.569
574	10.58	14.56	13.07	3.574
576	10.10	15.44	11.79	3.578
578	9.667	15.59	10.29	3.576
580	9.237	·14.68	8.681	3.565
582	8.845	12.77	7.126	3.533
584	8.448	10.64	5.729	3.500
586	8.027	8.040	4.561	3.443
588	7.551	5.891	3.625	3.386
590	7.009	4.399	2.900	3.334
592	6.403	3.100	2.348	3.277
594	5.759	2.250	1.925	3.225
596	5.096	1.706	1.600	3.186
598	4.452	1.327	1.349	3.158
600	3.865	1.054	1.153	3.139
602	3.346	0.865	0.997	3.135
604	2.910	0.729	0.870	3.142
606	2.556	0.632	0.766	3.161
608	2.276	0.549	0.681	3.190
610	2.057	0.485	0.611	3.230
612	1.883		0.550	3.284
614	1.747		0.499	3.356

Table 14.1.
(Continued)

λ (nm)	ε_{HHb} $N = 7$	ε_{O_2Hb} $N = 7$	ε_{COHb} $N = 6$	ε_{Hi} $N = 7$
616	1.635		0.453	3.442
618	1.542			3.538
620	1.460			3.644
622	1.392			3.747
624	1.329			3.850
626	1.272			3.934
628	1.222			4.005
630	1.175			4.049

λ = 450–630 nm. Standard procedure; HP8450A.
Absorbance measured with l = 0.010–0.016 cm. N is number of specimens. $Na_2S_2O_4$ used to prevent reoxygenation of HHb. pH range (Hi) = 7.13–7.23.

the range of values underlying the internationally accepted value of 11.0 $L \cdot mmol^{-1} \cdot cm^{-1}$. Hence, the latter value was also used in the calculation of the absorptivities of the haemoglobin derivatives in the sheep.

ABSORPTIVITY OF ADULT HAEMOGLOBIN

Blood from healthy sheep (*Ovis aries*) was obtained from the Central Animal Laboratory of the University of Groningen and collected in flasks containing 100 USP units of sodium heparin per mL blood. When measurements of $\varepsilon_{HiCN}(540)$ were required, iron free flasks were used. Solutions containing HHb, O_2Hb, COHb and Hi were prepared with the standard procedure described in Chapter 6 and the absorption spectra were determined using HP8450A and Optica CF4 spectrophotometers as described in Chapter 7. In the determination of the absorptivities of HHb, measurements with and without sodium dithionite were made. The ε_{HHb} values presented in Tables 14.1, 14.2, and 14.4 have all been obtained with the addition of a trace of $Na_2S_2O_4$ to the HHb solutions after deoxygenation by tonometry.

Table 14.1 shows the absorptivities of HHb, O_2Hb, COHb and Hi of adult sheep haemoglobin for the spectral range of 450–630 nm, measured with an HP8450A diode array spectrophotometer in 0.100-cm cuvettes with plan parallel glass inserts. To obtain reliable results, the measured absorbance should not be less than 0.050 (p. 47). No absorptivity values have been given when the measured absorbance was < 0.050. Table 14.2 shows the absorptivities for the spectral range of 600–800 nm measured with the same instrument; the lightpath was adjusted so that the measured absorbance was > 0.050. Absorptivity values have also not been given when the spectrophotometer rejected the measurements because the absorbance of the

Table 14.2.
Absorptivities ($L \cdot mmol^{-1} \cdot cm^{-1}$) of the common derivatives of adult sheep haemoglobin

λ (nm)	ε_{HHb} $N = 7$	ε_{O_2Hb} $N = 7$	ε_{COHb} $N = 6$	ε_{Hi} $N = 7$
600		1.044	1.157	3.156
602		0.866	1.008	3.153
604		0.735	0.886	3.158
606		0.639	0.786	3.178
608	2.226	0.560	0.703	3.202
610	2.026	0.497	0.634	3.244
612	1.859	0.448	0.575	3.295
614	1.730	0.408	0.524	3.365
616	1.618	0.375	0.478	3.441
618	1.528	0.348	0.419	3.538
620	1.448	0.323	0.393	3.639
622	1.380	0.302	0.364	3.737
624	1.319	0.282	0.336	3.836
626	1.264	0.265	0.311	3.913
628	1.215	0.248	0.288	3.979
630	1.171	0.233	0.267	4.012
632	1.132	0.220	0.249	4.016
634	1.097	0.208	0.232	3.989
636	1.066	0.198	0.217	3.896
638	1.040	0.190	0.204	3.758
640	1.016	0.184	0.192	3.560
642	0.997	0.178	0.181	3.311
644	0.980	0.174	0.172	3.002
646	0.966	0.172	0.164	2.696
648	0.953	0.169	0.156	2.397
650	0.941	0.168	0.149	2.118
652	0.930	0.167	0.143	1.838
654	0.920	0.166	0.137	1.589
656	0.908	0.160	0.132	1.334
658	0.890	0.162	0.125	1.174
660	0.882	0.164	0.121	1.021
662	0.869	0.163	0.117	0.875
664	0.853	0.163	0.113	0.748
666	0.835	0.162	0.109	0.646
668	0.815	0.161	0.105	0.568
670	0.793	0.161	0.102	0.499
672	0.771	0.160	0.098	0.445
674	0.749	0.159	0.095	0.404
676	0.726	0.158	0.093	0.370
678	0.704	0.159	0.090	0.343
680	0.682	0.157	0.088	0.321
682	0.660	0.157	0.086	0.303
684	0.639	0.157	0.084	0.290
686	0.619	0.156	0.082	0.278

Table 14.2.
(Continued)

λ (nm)	ε_{HHb} $N = 7$	ε_{O_2Hb} $N = 7$	ε_{COHb} $N = 6$	ε_{Hi} $N = 7$
688	0.600	0.156	0.080	0.269
690	0.581	0.156	0.079	0.262
692	0.564	0.155	0.077	0.257
694	0.548	0.155	0.076	0.252
696	0.534	0.156	0.075	0.249
698	0.520	0.156	0.073	0.246
700	0.508	0.156	0.072	0.244
702	0.496	0.157	0.071	0.242
704	0.485	0.157	0.071	0.241
706	0.474	0.158	0.070	0.241
708	0.463	0.159	0.069	0.241
710	0.453	0.159	0.068	0.240
712	0.443	0.160	0.068	0.240
714	0.433	0.161	0.067	0.241
716	0.423	0.162	0.066	0.241
718	0.414	0.163	0.066	0.241
720	0.405	0.165	0.065	0.243
722	0.396	0.165	0.065	0.244
724	0.388	0.167	0.064	0.245
726	0.381	0.168	0.064	0.247
728	0.375	0.170	0.063	0.248
730	0.371	0.171	0.063	0.250
732	0.368	0.172	0.062	0.251
734	0.367	0.174	0.062	0.253
736	0.367	0.176	0.061	0.255
738	0.370	0.177	0.061	0.258
740	0.374	0.179	0.061	0.262
742	0.379	0.180	0.060	0.264
744	0.387	0.182	0.060	0.266
746	0.396	0.184	0.059	0.270
748	0.407	0.187	0.059	0.274
750	0.418	0.188	0.059	0.277
752	0.430	0.190	0.058	0.281
754	0.441	0.192	0.058	0.285
756	0.450	0.194	0.057	0.289
758	0.455	0.196	0.057	0.295
760	0.456	0.198	0.057	0.299
762	0.453	0.200	0.056	0.305
764	0.444	0.202	0.056	0.309
766	0.431	0.204	0.056	0.315
768	0.415	0.206	0.055	0.320
770	0.397	0.209	0.055	0.328

Table 14.2.
(Continued)

λ (nm)	ε_{HHb} $N = 7$	ε_{O_2Hb} $N = 7$	ε_{COHb} $N = 6$	ε_{Hi} $N = 7$
772	0.379	0.210	0.055	0.332
774	0.362	0.212	0.054	0.338
776	0.345	0.214	0.054	0.344
778	0.329	0.216	0.053	0.351
780	0.315	0.218	0.053	0.357
782	0.303	0.220	0.053	0.364
784	0.292	0.222	0.053	0.369
786	0.283	0.224	0.052	0.376
788	0.275	0.226	0.052	0.383
790	0.268	0.228	0.052	0.390
792	0.262	0.230	0.051	0.395
794	0.257	0.232	0.051	0.402
796	0.252	0.235	0.051	0.409
798	0.248	0.236	0.050	0.414
800	0.245	0.239	0.050	0.422

$\lambda = 600-800$ nm. Standard procedure; HP8450A.

Absorbance measured with $l = 0.100$ cm (Hi); $l = 0.200$ cm (HHb and O_2Hb, COHb at $\lambda = 600-616$ nm); $l = 1.000$ cm (COHb at $\lambda = 618-800$ nm). N is number of specimens. $Na_2S_2O_4$ used to prevent reoxygenation of HHb. pH range (Hi) = 7.13–7.23.

samples was too high. This happened occasionally in measurements with $l = 0.200$ cm between 600 and 650 nm. The pH range of the Hi samples corresponds with the pH that ensues when a fresh sheep blood sample is haemolysed. The spectra represented by the absorptivities of Tables 14.1 and 14.2 are designated as the **standard spectra** of adult sheep haemoglobin.

Figure 14.1 shows the absorption spectra of HHb, O_2Hb, COHb and Hi of adult ovine haemoglobin in comparison with those of human HbA for the spectral ranges of 450–700 nm (Table 14.1) and 600–800 nm (Table 14.2). In Fig. 14.1a the spectral absorption curves run to 700 nm, but for $\lambda >$ 600 nm part of the measurements resulted in $A < 0.050$; these values have not been given in Table 14.1.

Figure 14.1 and Table 14.3 show no clear differences in the position of the light absorption maxima and minima of the common derivatives between adult sheep and adult human haemoglobin, although it seems that the maximum near 500 nm in the ovine Hi spectrum is situated at a slightly longer wavelength than that in the human Hi spectrum. As the measurements have been made with a diode array spectrophotometer, which has a limited spectral resolution, this does not rule out that other slight differences in the position of the maxima and minima do exist. This has not been further investigated by spectrophotometry with high spectral resolution as was the case in the dog and the rat.

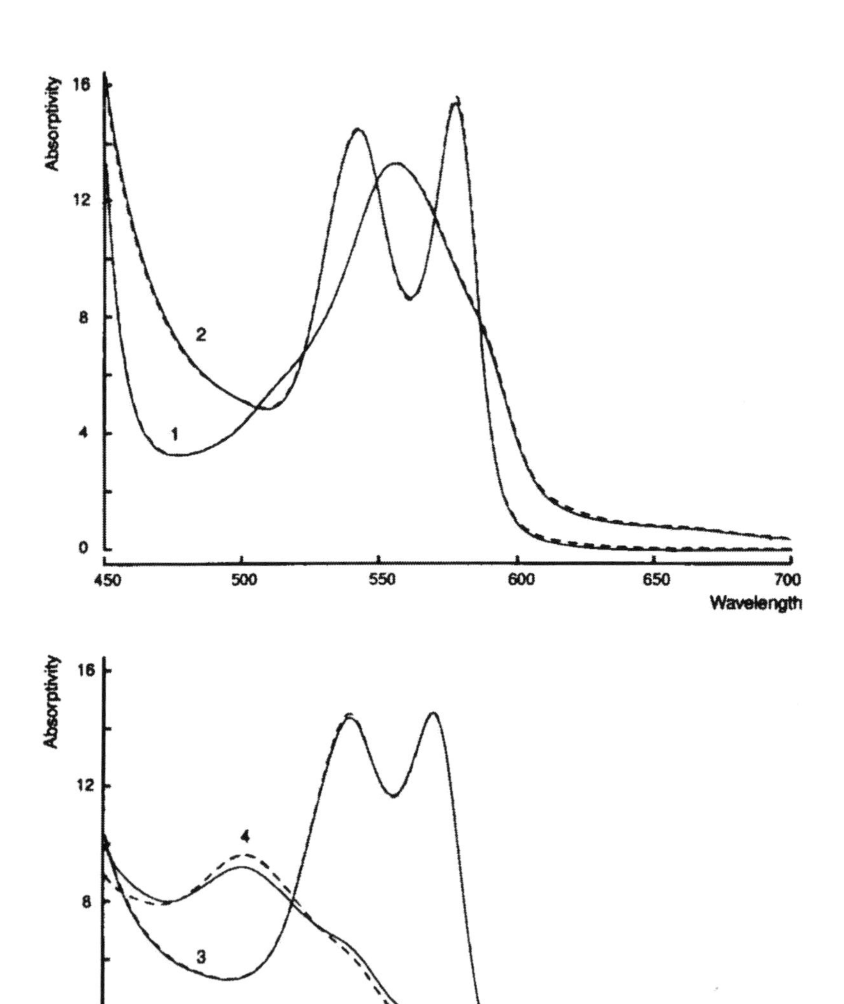

Figure 14.1a. Absorption spectra of the common derivatives of adult sheep haemoglobin in comparison with those of human HbA.

Absorptivity in $L \cdot mmol^{-1} \cdot cm^{-1}$; wavelength in nm. Solid lines: human haemoglobin; dashed lines: adult sheep haemoglobin. Upper panel: HHb (1) and O_2Hb (2); lower panel: COHb (3) and Hi (4). Absorbance measured with $l = 0.010–0.016$ cm. HP8450A. pH (sheep Hi) ≈ 7.17; pH (human Hi) ≈ 7.21.

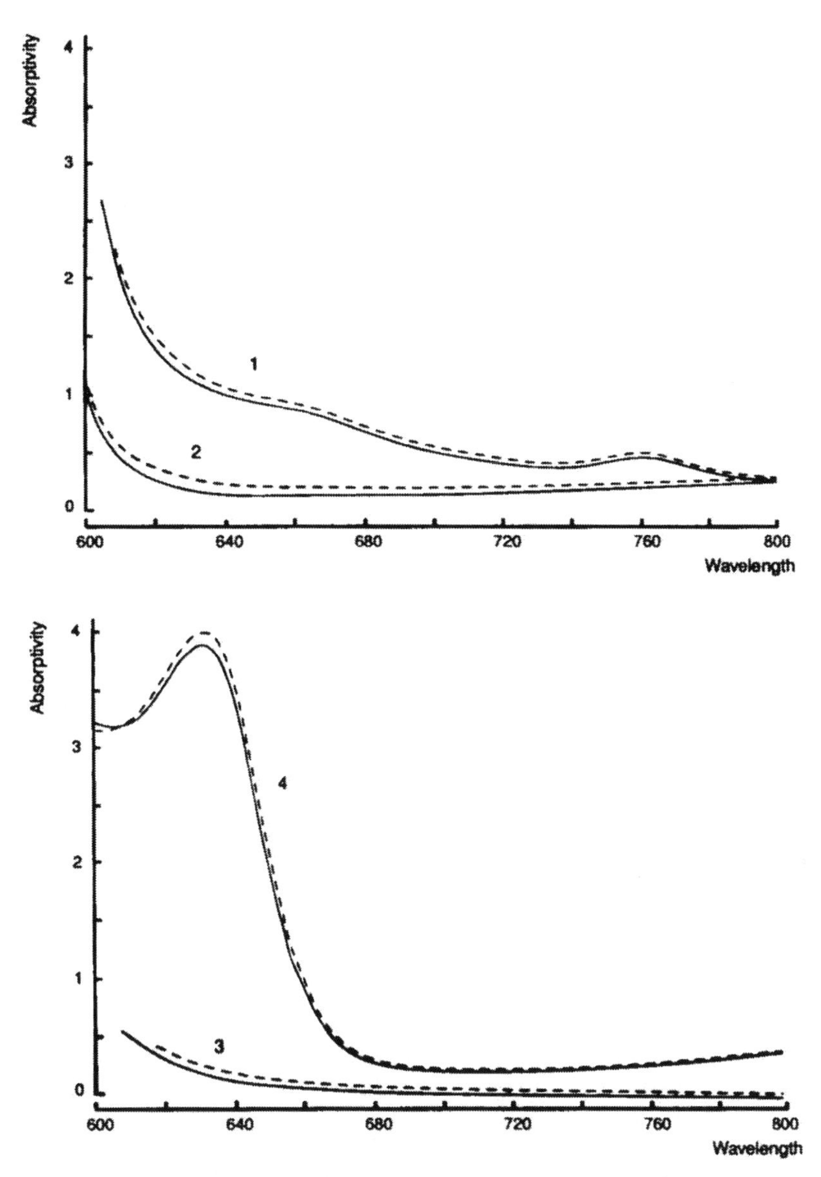

Figure 14.1b. Absorption spectra of the common derivatives of adult sheep haemoglobin in comparison with those of human HbA.

Absorptivity in $L \cdot mmol^{-1} \cdot cm^{-1}$; wavelength in nm. Solid lines: human haemoglobin; dashed lines: adult sheep haemoglobin. Upper panel: HHb (1) and O_2Hb (2); lower panel: COHb (3) and Hi (4). Absorbance measured with $l = 0.100$ cm (Hi); $l = 0.200$ cm (HHb and O_2Hb); $l = 1.000$ cm (COHb). HP8450A. pH (sheep Hi) ≈ 7.17; pH (human Hi) ≈ 7.21.

Table 14.3.
Comparison of absorptivities of the common derivatives of sheep and adult human haemoglobin

	Human		Sheep		
λ (nm)	ε_{HHb} $N = 6$	*SEM*	ε_{HHb} $N = 7$	*SEM*	*p*
476	3.298	0.013	3.325	0.014	
478	3.302	0.012	3.327	0.011	
552	13.21	0.044	13.22	0.035	
554	13.36	0.045	13.37	0.037	
556	13.36	0.045	13.38	0.035	
λ (nm)	ε_{O_2Hb} $N = 6$	*SEM*	ε_{O_2Hb} $N = 7$	*SEM*	*p*
508	4.882	0.007	4.937	0.012	< 0.01
540	14.32	0.027	14.42	0.032	
542	14.52	0.023	14.57	0.033	
544	14.34	0.022	14.36	0.035	
560	8.767	0.015	8.661	0.021	< 0.005
576	15.26	0.034	15.44	0.039	< 0.01
578	15.36	0.027	15.59	0.041	< 0.001
λ (nm)	ε_{COHb} $N = 8$	*SEM*	ε_{COHb} $N = 6$	*SEM*	*p*
496	5.225	0.017	5.251	0.010	
538	14.30	0.039	14.44	0.023	< 0.02
540	14.27	0.036	14.40	0.021	< 0.02
542	14.04	0.036	14.13	0.016	
554	11.63	0.029	11.57	0.010	
568	14.43	0.037	14.39	0.022	
570	14.46	0.039	14.42	0.016	
λ (nm)	ε_{Hi} $N = 15$	*SEM*	ε_{Hi} $N = 7$	*SEM*	*p*
498	9.088	0.028	9.487	0.050	< 0.001
500	9.106	0.029	9.519	0.049	< 0.001
502	9.098	0.028	9.542	0.054	< 0.001
630	3.895	0.021	4.049	0.019	< 0.001
632	3.896	0.021	4.062	0.021	< 0.001
634	3.857	0.020	4.031	0.022	< 0.001

Standard procedure. HP8450A.

Absorptivity in $L \cdot mmol^{-1} \cdot cm^{-1}$; *N* is number of specimens, *SEM* is standard error of the mean. *p* values of Student's *t*-test for unpaired samples given only when $p < 0.05$.

Table 14.4.
Comparison of absorptivities of HHb, O_2Hb and COHb of sheep and human haemoglobin at various wavelengths between 620 and 750 nm

	Human		Sheep		
λ (nm)	ε_{HHb} $N = 6$	*SEM*	ε_{HHb} $N = 7$	*SEM*	*p*
620	1.338	0.010	1.448	0.015	< 0.001
630	1.085	0.008	1.171	0.011	< 0.001
640	0.953	0.006	1.016	0.008	< 0.001
700	0.459	0.005	0.508	0.007	< 0.001
750	0.377	0.005	0.418	0.005	< 0.001
λ (nm)	ε_{O_2Hb} $N = 6$	*SEM*	ε_{O_2Hb} $N = 7$	*SEM*	*p*
620	0.218	0.004	0.323	0.007	< 0.001
630	0.139	0.004	0.233	0.007	< 0.001
640	0.104	0.004	0.184	0.007	< 0.001
700	0.105	0.004	0.156	0.008	< 0.001
750	0.148	0.004	0.188	0.007	< 0.001
λ (nm)	ε_{COHb} $N = 4$	*SEM*	ε_{COHb} $N = 6$	*SEM*	*p*
620	0.316	0.006	0.393	0.006	< 0.001
630	0.196	0.007	0.267	0.006	< 0.001
640	0.125	0.006	0.192	0.006	< 0.001
700	0.028	0.001	0.072	0.007	< 0.001
750	0.019	0.001	0.059	0.006	< 0.001

Standard procedure. HP8450A.
Absorptivity in $L \cdot mmol^{-1} \cdot cm^{-1}$; *N* is number of specimens; *SEM* is standard error of the mean. *p* values of Student's *t*-test for unpaired samples.

Table 14.3 shows the results of a comparison of the absorptivities of the common derivatives between sheep and human haemoglobin, *i.e.* a comparison of the data of Table 14.1 with those of Table 8.1. Student's *t*-test for unpaired samples was used. The table gives the results at wavelengths near the principal absorption maxima and minima in the spectra of the common derivatives.

The upper panel of Fig. 14.1a shows no difference between the adult ovine and adult human HHb spectra. In the O_2Hb and COHb spectra, however, there are some discernible differences. The α-peak of ovine O_2Hb is higher while the absorption minimum of O_2Hb at 560 nm is lower; in the COHb spectrum the β-peak is higher for sheep haemoglobin. These differences are statistically significant (Table 14.3). Figure 14.1b shows that above 610 nm the ovine HHb, O_2Hb and COHb spectra lie above the human ones. Table 14.4 demonstrates that these differences are statistically significant.

Table 14.5.
Absorptivities ($L \cdot mmol^{-1} \cdot cm^{-1}$) of the common derivatives of sheep haemoglobin at selected wavelengths between 600 and 1000 nm

λ (nm)	ε_{HHb} $N = 6$	ε_{O_2Hb} $N = 6$	ε_{COHb} $N = 6$	ε_{Hi} $N = 7$
600	3.485	0.955	1.067	3.146
630	1.146	0.201	0.226	4.015
660	0.873	0.136	0.092	0.889
680	0.674	0.131	0.060	0.269
700	0.494	0.131	0.047	0.210
750	0.423	0.172	0.038	0.257
775	0.318	0.199	0.035	0.325
800	0.230	0.226	0.033	0.411
805	0.224	0.232	0.029	0.426
840	0.206	0.271	0.028	0.544
845	0.205	0.276	0.026	0.558
880	0.213	0.305	0.025	0.650
904	0.221	0.316	0.025	0.710
920	0.215	0.318	0.023	0.746
940	0.187	0.316	0.021	0.785
960	0.141	0.306	0.018	0.809
1000	0.063	0.275	0.016	0.830

Standard procedure. Optica CF4.
Absorbance measured with $l = 0.100$ cm (Hi at $\lambda = 600$ and 630 nm); $l = 0.200$ cm (HHb and O_2Hb, COHb at $\lambda = 600$ nm, Hi at $\lambda = 660$–1000 nm); $l = 1.000$ cm (COHb at $\lambda = 630$–1000 nm). N is number of specimens. $Na_2S_2O_4$ used to prevent reoxygenation of HHb. pH (Hi) = 7.13–7.23. In the wavelength range 630–1000 nm a red sensitive photomultiplier tube and a red filter have been used.

Both the human and ovine HHb and O_2Hb spectra converge to the isosbestic point at 800 nm.

The clear and statistically significant differences between the standard adult ovine and adult human Hi spectra as shown in the lower panel of Fig. 14.1a should of course be considered while taking any differences in pH into account. Because the difference between the pH ranges of the two standard Hi spectra is rather small (pH = 7.14–7.30 for human Hi; pH = 7.13–7.23 for ovine Hi), the spectral differences are probably not due to a difference in pH, but may result from a difference in OH^--ion affinity. If that were the case the ratio HiOH/Hi would be different between the two kinds of Hi at the same pH, but, on the other hand, there would be two different pH values at which the HiOH/Hi ratios would be equal. In the absence of intrinsic differences between the two kinds of Hi, the spectral absorption curves then would coincide. This nearly seemed to be the case in the comparison of pig and human Hi (Fig. 12.2).

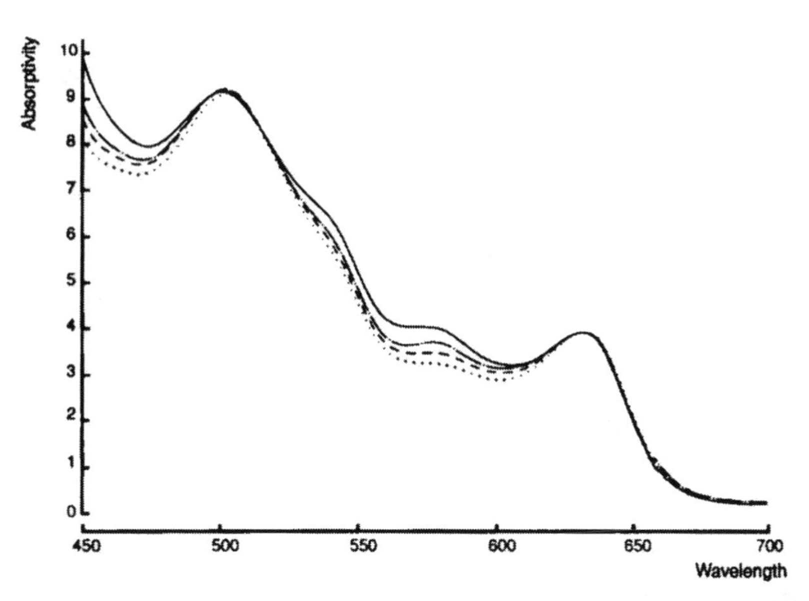

Figure 14.2. Absorption spectra of adult sheep and adult human Hi at various pH values. Shown are two ovine Hi spectra, obtained at pH = 6.81 (· · ·), and pH = 7.23 (·——·), together with the standard sheep Hi spectrum (pH ≈ 7.17) (- - - -) and the standard Hi spectrum of human HbA (pH ≈ 7.21) (——). The absorptivity at 630 nm has been normalised to 3.895 $L \cdot mmol^{-1} \cdot cm^{-1}$, the absorptivity of human Hi at pH ≈ 7.21.

Figure 14.2 shows the standard Hi spectra of sheep and man, and two individual Hi spectra of the sheep. The individual ovine Hi spectra have been measured at pH = 6.81 and pH = 7.23. For easy comparison the absorptivity at 630 nm of all four spectra has been normalised to 3.895 $L \cdot mmol^{-1} \cdot cm^{-1}$, the absorptivity at this wavelength in the standard Hi spectrum of human HbA. The ovine Hi spectra are all quite dissimilar from the standard human Hi spectrum, suggesting an intrinsic difference between the two kinds of Hi, but lacking an ovine Hi spectrum measured at a considerably higher pH it is difficult to decide whether a difference in OH^--ion affinity contributes appreciably to the observed difference between the two standard Hi spectra.

Table 14.5 presents the absorptivities of the common derivatives of adult sheep haemoglobin for selected wavelengths in the spectral range of 600–1000 nm, measured with an Optica CF4 grating spectrophotometer. Wavelength selection was necessary to complete this series of manual measurements within a reasonable time. Wavelengths of importance for biomedical applications have preferably been chosen.

Figure 14.1b and Table 14.4 have shown that, in the spectral region > 610 nm, the O_2Hb and COHb spectra of sheep haemoglobin lie above those of human haemoglobin. This is corroborated by a comparison of the data of Tables 14.5 and 8.4. The data of Table 14.5 suggest that the isosbestic wavelength of adult sheep HHb and O_2Hb in the near infrared is

Table 14.6.
ε_{HHb} of adult ovine haemoglobin for $l = 600-1000$ nm, measured with and without a trace of sodium dithionite to prevent reoxygenation

	With $Na_2S_2O_4$		Without $Na_2S_2O_4$	
λ (nm)	ε_{HHb} $N = 6$	*SEM*	ε_{HHb} $N = 6$	*SEM*
600	3.485	0.025	3.438	0.050
630	1.146	0.018	1.044	0.011
660	0.873	0.009	0.793	0.007
680	0.674	0.013	0.586	0.007
700	0.494	0.006	0.426	0.005
750	0.423	0.004	0.384	0.004
775	0.318	0.004	0.288	0.002
800	0.230	0.003	0.208	0.001
805	0.224	0.003	0.203	0.001
840	0.206	0.002	0.197	0.001
845	0.205	0.002	0.198	0.001
880	0.213	0.001	0.211	0.001
904	0.221	0.001	0.221	0.002
920	0.215	0.001	0.216	0.002
940	0.187	0.001	0.190	0.002
960	0.141	0.001	0.147	0.003
1000	0.063	0.001	0.075	0.003

Standard procedure. Optica CF4.
Absorptivity in $L \cdot mmol^{-1} \cdot cm^{-1}$. Absorbance measured with $l = 0.200$ cm. *SEM* is standard error of the mean.

not significantly different from 800 nm. However, for exactly locating the isosbestic point, the effect of the added trace of sodium dithionite should be taken into account.

To this end, Table 14.6 has been added, showing ε_{HHb} for the wavelength range of 630–1000 nm, when measured with and without addition of $Na_2S_2O_4$. The values with $Na_2S_2O_4$ correspond with those of Table 14.5. Plotting of the two spectral absorption curves of ovine HHb, the one with and the one without $Na_2S_2O_4$, together with the curve of O_2Hb from Table 14.5, yields a crossover point at ~ 800 nm for the HHb curve with, and at ~ 794 nm without $Na_2S_2O_4$. This demonstrates that, in adult sheep haemoglobin, the isosbestic point of HHb and O_2Hb is found below 800 nm, the established value of human haemoglobin.

For a further discussion of the measurements underlying the absorptivities presented in Table 14.5 and an explanation of the slight differences between these data and those of Table 14.2 see Chapter 15.

ABSORPTIVITY OF FOETAL HAEMOGLOBIN

Blood from neonatal lambs was obtained from the Central Animal Laboratory of the University of Groningen. It was collected, immediately after delivery, into a flask containing 100 USP units of sodium heparin per mL blood. Samples containing HHb, O_2Hb, COHb and Hi were prepared with the standard procedure described in Chapter 6 and the absorption spectra were determined using an HP8450A diode array spectrophotometer and an Optica CF4 grating spectrophotometer as described in Chapter 7. The fraction of foetal haemoglobin was determined as described for human HbF in Chapter 7. This gave for the three specimens that yielded acceptable spectral absorption curves foetal haemoglobin fractions of 87.5, 86.5 and 92.5%. The absorptivities were calculated using equation 7.3. In the determination of the absorptivities of HHb, measurements with and without sodium dithionite were made. The ε_{HHb} values presented in Tables 14.7, 14.8, and 14.10 have all been obtained with the addition of a trace of $Na_2S_2O_4$ to the HHb solutions after deoxygenation by tonometry.

Table 14.7 shows the absorptivities of HHb, O_2Hb, COHb and Hi of foetal sheep haemoglobin for the spectral range of 450–630 nm, measured with an HP8450A diode array spectrophotometer in 0.100-cm cuvettes with plan parallel glass inserts. No absorptivity values have been given when the measured absorbance was < 0.050. Table 14.8 shows the absorptivities for the spectral range of 600–800 nm measured with the same instrument; the lightpath was adjusted so that the measured absorbance was > 0.050. Absorptivity values have also not been given when the spectrophotometer rejected the measurements because the absorbance of the samples was too high. This happened occasionally in measurements with $l = 0.200$ cm between 600 and 650 nm.

Figure 14.3 and Table 14.9 show no clear differences in the position of the light absorption maxima and minima of the common derivatives between foetal and adult sheep haemoglobin. As the measurements have been made with a diode array spectrophotometer, which has a limited spectral resolution, this does not rule out that slight differences in the position of the maxima and minima do exist.

Table 14.9 shows the results of a comparison of the absorptivities of the common derivatives between foetal and adult sheep haemoglobin, *i.e.* a comparison of the data of Table 14.7 with those of Table 14.1. Student's t-test for unpaired samples was used. The table gives the results at wavelengths near the principal absorption maxima and minima in the spectra of the common derivatives.

The upper panel of Fig. 14.3a shows no differences between the O_2Hb and HHb spectra of foetal and adult sheep haemoglobin. In the lower panel of Fig. 14.3a it seems that the α- and β-peaks of COHb of foetal haemoglobin are slightly lower than those of adult sheep haemoglobin, but

Table 14.7.
Absorptivities ($L \cdot mmol^{-1} \cdot cm^{-1}$) of the common derivatives of foetal sheep haemoglobin

λ (nm)	ε_{HHb} $N = 2$	ε_{O_2Hb} $N = 3$	ε_{COHb} $N = 3$	ε_{Hi} $N = 2$
450	13.33	15.89	10.12	10.02
452	10.47	14.64	9.433	9.647
454	8.480	13.58	8.876	9.353
456	7.002	12.60	8.376	9.070
458	5.948	11.78	7.982	8.862
460	5.179	10.99	7.600	8.653
462	4.610	10.37	7.299	8.472
464	4.182	9.745	7.016	8.299
466	3.896	9.223	6.785	8.176
468	3.679	8.714	6.555	8.062
470	3.542	8.311	6.364	7.998
472	3.446	7.904	6.183	7.935
474	3.402	7.571	6.037	7.952
476	3.386	7.241	5.896	7.948
478	3.391	6.956	5.793	8.022
480	3.406	6.676	5.698	8.092
482	3.448	6.452	5.614	8.218
484	3.481	6.242	5.536	8.325
486	3.549	6.058	5.462	8.486
488	3.614	5.875	5.401	8.604
490	3.700	5.753	5.359	8.740
492	3.795	5.615	5.317	8.870
494	3.928	5.507	5.291	8.980
496	4.063	5.391	5.292	9.068
498	4.228	5.302	5.317	9.140
500	4.390	5.201	5.353	9.170
502	4.577	5.124	5.421	9.190
504	4.784	5.049	5.508	9.133
506	5.009	4.995	5.630	9.080
508	5.206	4.952	5.800	8.960
510	5.435	4.969	6.007	8.850
512	5.642	5.010	6.295	8.709
514	5.852	5.142	6.672	8.552
516	6.061	5.346	7.138	8.381
518	6.289	5.690	7.722	8.239
520	6.508	6.146	8.380	8.053
522	6.770	6.767	9.154	7.902
524	7.034	7.527	9.950	7.721
526	7.341	8.478	10.79	7.583
528	7.669	9.522	11.57	7.418
530	8.045	10.64	12.37	7.300
532	8.439	11.68	13.07	7.161
534	8.895	12.67	13.70	7.040

Table 14.7.
(Continued)

λ (nm)	ε_{HHb} $N = 2$	ε_{O_2Hb} $N = 3$	ε_{COHb} $N = 3$	ε_{Hi} $N = 2$
536	9.375	13.44	14.11	6.911
538	9.928	14.12	14.32	6.759
540	10.46	14.51	14.28	6.578
542	11.07	14.64	14.01	6.379
544	11.60	14.38	13.52	6.138
546	12.13	13.79	13.01	5.890
548	12.55	12.88	12.44	5.607
550	12.93	11.89	12.00	5.344
552	13.13	10.87	11.65	5.083
554	13.30	10.02	11.51	4.867
556	13.30	9.340	11.51	4.677
558	13.21	8.905	11.74	4.535
560	13.03	8.682	12.13	4.425
562	12.82	8.731	12.70	4.343
564	12.51	9.049	13.31	4.284
566	12.19	9.683	13.92	4.235
568	11.80	10.62	14.29	4.191
570	11.39	11.87	14.30	4.139
572	10.93	13.29	13.79	4.082
574	10.47	14.63	12.87	4.021
576	10.00	15.47	11.57	3.956
578	9.557	15.51	10.08	3.889
580	9.129	14.44	8.497	3.810
582	8.733	12.52	6.976	3.722
584	8.329	10.05	5.625	3.624
586	7.905	7.674	4.495	3.523
588	7.421	5.628	3.593	3.417
590	6.873	4.024	2.891	3.312
592	6.261	2.897	2.351	3.215
594	5.609	2.108	1.933	3.126
596	4.944	1.563	1.612	3.049
598	4.302	1.196	1.362	2.987
600	3.717	0.946	1.170	2.945
602	3.205	0.769	1.013	2.921
604	2.773	0.635	0.885	2.911
606	2.429	0.537	0.783	2.919
608	2.154	0.460	0.698	2.944
610	1.939		0.629	2.985
612	1.766		0.572	3.045
614	1.631		0.526	3.122

Table 14.7.
(Continued)

λ (nm)	ε_{HHb} $N = 2$	ε_{O_2Hb} $N = 3$	ε_{COHb} $N = 3$	ε_{Hi} $N = 2$
616	1.523		0.484	3.215
618	1.432			3.316
620	1.356			3.425
622	1.294			3.536
624	1.240			3.636
626	1.191			3.726
628	1.149			3.796
630	1.112			3.839

λ = 450–630 nm. Standard procedure. HP8450A.
Absorbance measured with l = 0.010–0.016 cm. N is number of specimens. $Na_2S_2O_4$ used to prevent reoxygenation of HHb. pH (Hi) ≈ 6.97.

Table 14.8.
Absorptivities (L · mmol^{-1} · cm^{-1}) of the common derivatives of foetal sheep haemoglobin

λ (nm)	ε_{HHb} $N = 2$	ε_{O_2Hb} $N = 3$	ε_{COHb} $N = 3$	ε_{Hi} $N = 2$
600		0.931		2.871
602		0.763		2.849
604		0.639		2.834
606		0.545		2.845
608	2.102	0.473		2.865
610	1.913	0.415		2.909
612	1.741	0.369		2.961
614	1.611	0.332		3.037
616	1.500	0.301		3.118
618	1.413	0.276	0.439	3.220
620	1.339	0.255	0.406	3.323
622	1.275	0.236	0.376	3.434
624	1.220	0.221	0.348	3.522
626	1.174	0.206	0.324	3.613
628	1.132	0.193	0.302	3.670
630	1.096	0.181	0.282	3.717
632	1.063	0.171	0.265	3.719
634	1.035	0.161	0.248	3.682
636	1.008	0.153	0.233	3.591
638	0.985	0.146	0.219	3.447
640	0.965	0.140	0.206	3.251
642	0.946	0.135	0.195	3.021

Table 14.8.
(Continued)

λ (nm)	ε_{HHb} $N = 2$	ε_{O_2Hb} $N = 3$	ε_{COHb} $N = 3$	ε_{Hi} $N = 2$
644	0.929	0.130	0.184	2.758
646	0.915	0.127	0.175	2.476
648	0.901	0.124	0.166	2.183
650	0.889	0.121	0.158	1.903
652	0.877	0.119	0.151	1.643
654	0.867	0.117	0.144	1.417
656	0.859	0.114	0.139	1.235
658	0.843	0.110	0.132	0.999
660	0.829	0.113	0.127	0.875
662	0.816	0.112	0.123	0.758
664	0.800	0.112	0.118	0.656
666	0.783	0.111	0.115	0.572
668	0.764	0.111	0.111	0.503
670	0.744	0.111	0.108	0.447
672	0.722	0.111	0.105	0.402
674	0.701	0.110	0.102	0.365
676	0.679	0.110	0.099	0.335
678	0.658	0.110	0.096	0.311
680	0.636	0.110	0.094	0.291
682	0.616	0.109	0.092	0.276
684	0.595	0.109	0.090	0.263
686	0.575	0.109	0.088	0.252
688	0.556	0.108	0.086	0.243
690	0.538	0.108	0.084	0.236
692	0.521	0.108	0.083	0.230
694	0.506	0.108	0.081	0.225
696	0.491	0.108	0.080	0.221
698	0.478	0.109	0.078	0.217
700	0.466	0.109	0.077	0.214
702	0.455	0.110	0.076	0.212
704	0.444	0.111	0.075	0.209
706	0.434	0.112	0.074	0.207
708	0.424	0.113	0.073	0.206
710	0.414	0.114	0.073	0.205
712	0.405	0.115	0.072	0.204
714	0.396	0.117	0.071	0.203
716	0.387	0.118	0.070	0.202
718	0.379	0.119	0.070	0.202
720	0.370	0.121	0.069	0.202
722	0.363	0.122	0.069	0.203
724	0.355	0.124	0.068	0.203
726	0.349	0.125	0.067	0.203
728	0.344	0.127	0.067	0.205
730	0.340	0.128	0.066	0.206

Table 14.8.
(Continued)

λ (nm)	ε_{HHb} $N = 2$	ε_{O_2Hb} $N = 3$	ε_{COHb} $N = 3$	ε_{Hi} $N = 2$
732	0.338	0.130	0.066	0.208
734	0.337	0.132	0.065	0.209
736	0.338	0.133	0.065	0.211
738	0.340	0.135	0.065	0.214
740	0.344	0.137	0.064	0.216
742	0.350	0.139	0.064	0.219
744	0.358	0.141	0.063	0.221
746	0.367	0.143	0.063	0.224
748	0.377	0.145	0.062	0.228
750	0.388	0.147	0.062	0.231
752	0.399	0.149	0.062	0.234
754	0.410	0.151	0.061	0.237
756	0.419	0.153	0.061	0.241
758	0.424	0.155	0.060	0.245
760	0.425	0.157	0.060	0.249
762	0.421	0.160	0.060	0.253
764	0.413	0.162	0.059	0.258
766	0.401	0.165	0.059	0.262
768	0.386	0.167	0.058	0.267
770	0.368	0.169	0.058	0.271
772	0.351	0.172	0.058	0.276
774	0.334	0.174	0.057	0.282
776	0.319	0.177	0.057	0.287
778	0.304	0.179	0.057	0.292
780	0.291	0.181	0.056	0.298
782	0.279	0.184	0.056	0.303
784	0.269	0.187	0.056	0.309
786	0.261	0.189	0.055	0.315
788	0.254	0.192	0.055	0.321
790	0.247	0.194	0.055	0.327
792	0.242	0.196	0.054	0.333
794	0.237	0.199	0.054	0.339
796	0.233	0.201	0.054	0.345
798	0.229	0.203	0.053	0.352
800	0.226	0.206	0.053	0.359

λ = 600–800 nm. Standard procedure; HP8450A.

Absorbance measured with l = 0.100 cm (Hi); l = 0.200 cm (HHb and O_2Hb); l = 1.000 cm (COHb). N is number of specimens. $Na_2S_2O_4$ used to prevent reoxygenation of HHb. pH (Hi) ≈ 6.97.

Table 14.9.
Comparison of absorptivities of the common derivatives of adult and foetal sheep haemoglobin

	Adult sheep		Foetal sheep		
λ (nm)	ε_{HHb} $N = 7$	*SEM*	ε_{HHb} $N = 2$	*SEM*	*p*
476	3.325	0.014	3.387	0.009	
478	3.327	0.011	3.393	0.010	< 0.05
552	13.22	0.035	13.14	0.059	
554	13.37	0.037	13.30	0.059	
556	13.38	0.035	13.30	0.057	
λ (nm)	ε_{O_2Hb} $N = 7$	*SEM*	ε_{O_2Hb} $N = 3$	*SEM*	*p*
508	4.937	0.012	4.952	0.003	
540	14.42	0.032	14.51	0.044	
542	14.57	0.033	14.64	0.037	
544	14.36	0.035	14.39	0.037	
560	8.661	0.021	8.681	0.023	
576	15.44	0.039	15.47	0.052	
578	15.59	0.041	15.50	0.043	
λ (nm)	ε_{COHb} $N = 6$	*SEM*	ε_{COHb} $N = 3$	*SEM*	*p*
496	5.251	0.010	5.292	0.011	< 0.05
538	14.44	0.023	14.32	0.088	
540	14.40	0.021	14.28	0.070	
542	14.13	0.016	14.01	0.072	
554	11.57	0.010	11.51	0.042	
568	14.39	0.022	14.29	0.081	
570	14.42	0.016	14.30	0.078	
λ (nm)	ε_{Hi} $N = 7$	*SEM*	ε_{Hi} $N = 2$	*SEM*	*p*
498	9.487	0.050	9.145	0.095	< 0.02
500	9.519	0.049	9.170	0.099	< 0.02
502	9.542	0.054	9.185	0.103	< 0.02
630	4.049	0.019	3.839	0.021	< 0.002
632	4.062	0.021	3.848	0.031	< 0.002
634	4.031	0.022	3.815	0.039	< 0.01

Standard procedure. HP8450A.

Absorptivity in $L \cdot mmol^{-1} \cdot cm^{-1}$; *N* is number of specimens; *SEM* is standard error of the mean. *p* values of Student's *t*-test for unpaired samples given only when $p < 0.05$.

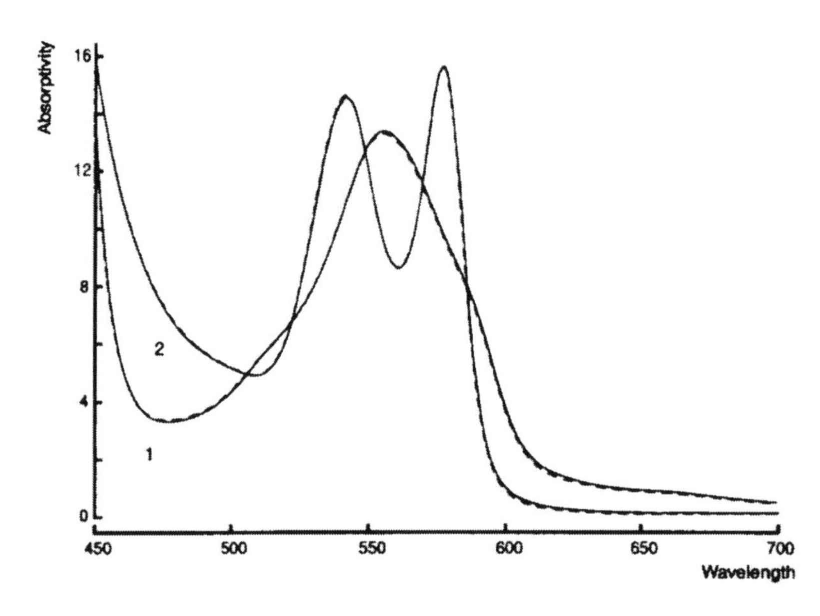

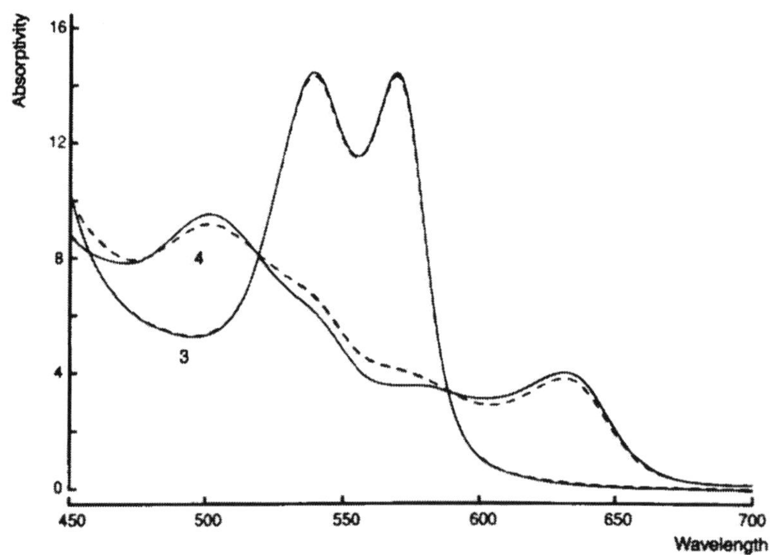

Figure 14.3a. Absorption spectra of the common derivatives of foetal sheep haemoglobin in comparison with those of adult sheep haemoglobin.

Absorptivity in $L \cdot mmol^{-1} \cdot cm^{-1}$, wavelength in nm. Solid lines: adult sheep haemoglobin; dashed lines: foetal sheep haemoglobin. Upper panel: HHb (1) and O_2Hb (2); lower panel: COHb (3) and Hi (4). Absorbance measured with l = 0.010–0.016 cm. HP8450A. pH (adult sheep Hi) ≈ 7.17; pH (foetal sheep Hi) ≈ 6.97.

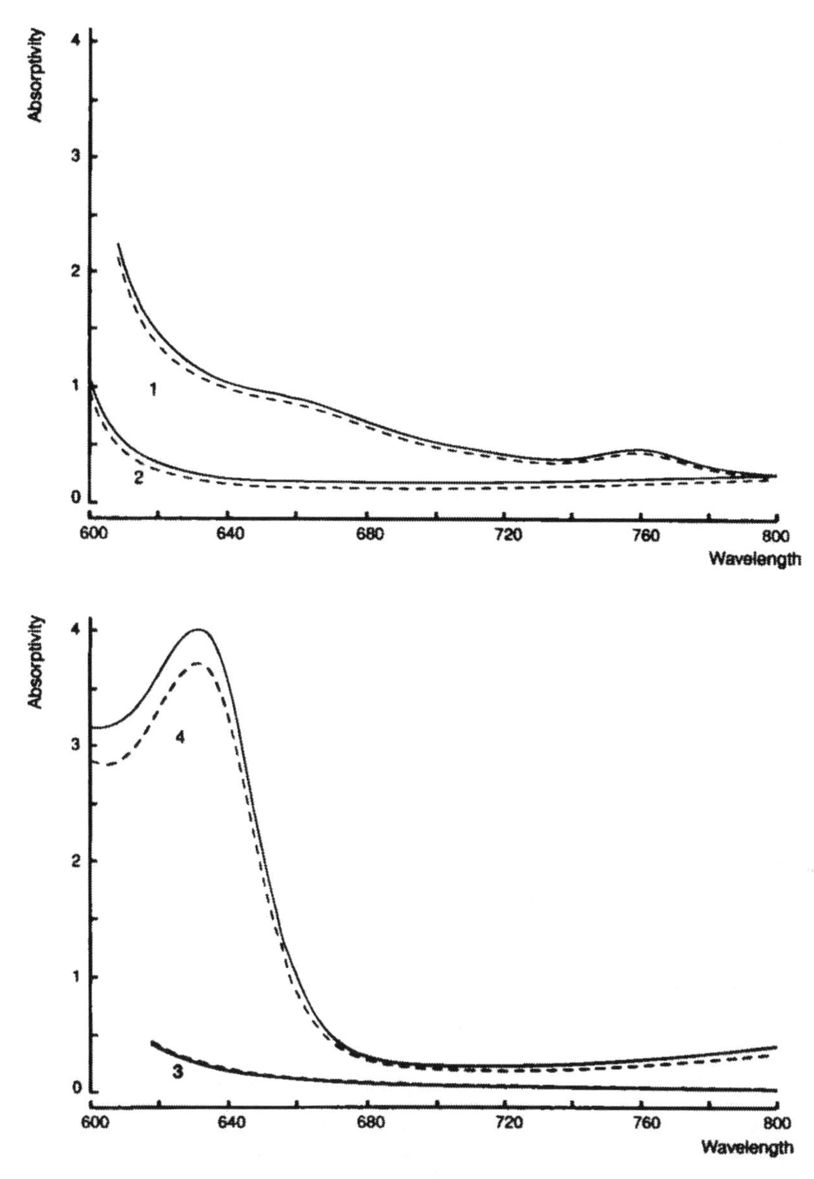

Figure 14.3b. Absorption spectra of the common derivatives of foetal sheep haemoglobin in comparison with those of adult sheep haemoglobin.

Absorptivity in $L \cdot mmol^{-1} \cdot cm^{-1}$, wavelength in nm. Solid lines: adult sheep haemoglobin; dashed lines: foetal sheep haemoglobin. Upper panel: HHb (1) and O_2Hb (2); lower panel: COHb (3) and Hi (4). Absorbance measured with l = 0.100 cm (Hi); l = 0.200 cm (HHb and O_2Hb); l = 1.000 cm (COHb). HP8450A. pH (adult sheep Hi) $\approx$ 7.17; pH (foetal sheep Hi) $\approx$ 6.97.

Table 14.10.
Comparison of absorptivities of HHb, O_2Hb and COHb of adult and foetal sheep haemoglobin at various wavelengths between 620 and 750 nm

	Adult sheep		Foetal sheep		
λ (nm)	ε_{HHb} $N = 7$	*SEM*	ε_{HHb} $N = 2$	*SEM*	*p*
620	1.448	0.015	1.339	0.010	< 0.01
630	1.171	0.011	1.096	0.005	< 0.02
640	1.016	0.008	0.965	0.005	< 0.02
700	0.508	0.007	0,466	0.001	< 0.02
750	0.418	0.005	0.388	0.001	< 0.02
λ (nm)	ε_{O_2Hb} $N = 7$	*SEM*	ε_{O_2Hb} $N = 3$	*SEM*	*p*
620	0.323	0.007	0.255	0.012	< 0.001
630	0.233	0.007	0.181	0.013	< 0.01
640	0.184	0.007	0.140	0.013	< 0.02
700	0.156	0.008	0.109	0.004	< 0.01
750	0.188	0.007	0.147	0.002	< 0.01
λ (nm)	ε_{COHb} $N = 6$	*SEM*	ε_{COHb} $N = 3$	*SEM*	*p*
620	0.393	0.006	0.406	0.014	
630	0.267	0.006	0.282	0.013	
640	0.192	0.006	0.206	0.012	
700	0.072	0.007	0.077	0.007	
750	0.059	0.006	0.062	0.006	

Standard procedure; HP8450A.

Absorptivity in $L \cdot mmol^{-1} \cdot cm^{-1}$; *N* is number of specimens; *SEM* is standard error of the mean. *p* values of Student's *t*-test for unpaired samples given only when $p < 0.05$.

Table 14.9 demonstrates that this difference is not statistically significant. In the minimum near 496 nm, however, ε_{COHb} is significantly higher in foetal than in adult sheep haemoglobin. In Fig. 14.3b it can be seen that the foetal HHb and O_2Hb spectra run below those of the adult sheep, while there is no difference between the spectral absorption curves of foetal and adult COHb. The statistical significance of the difference between the foetal and adult HHb and O_2Hb spectra is demonstrated by the data of Table 14.10. The difference is corroborated by a comparison of the data of Tables 14.5 and 14.11.

Figure 14.3 shows clear differences between the foetal and adult ovine Hi spectra; Table 14.9 demonstrates that these differences are statistically significant in spite of the fact that the standard adult ovine Hi spectrum is compared with the mean of only two Hi spectra of foetal ovine haemoglobin. To analyse the possible influence of the pH of the solutions on these differences Figs 14.4 and 14.5 have been added.

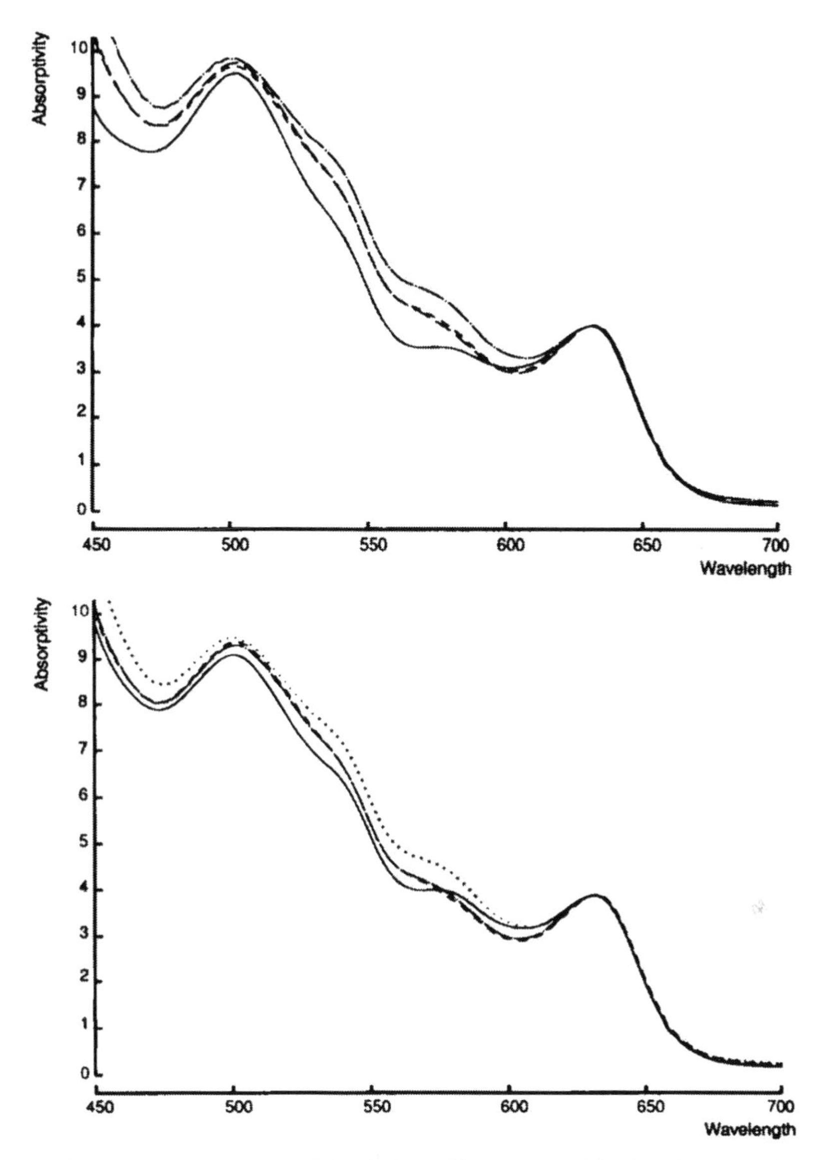

Figure 14.4. Absorption spectra of foetal sheep Hi at various pH values in comparison with the standard adult sheep and the standard adult human Hi spectra.

Upper panel: foetal ovine Hi spectra obtained at pH = 6.83 (— — —), and pH = 7.32 (·——·), standard sheep Hi spectrum (pH ≈ 7.17) (——) and the mean of the two foetal Hi spectra (pH ≈ 6.97) (- - - -) that underlie the data given in Table 14.7 and Fig. 14.3. The absorptivity at 630 nm has been normalised to 4.049 $L \cdot mmol^{-1} \cdot cm^{-1}$, the absorptivity at 630 nm of the standard sheep Hi spectrum. Lower panel: foetal Hi spectra at pH = 6.83 (- - - -) and 7.32 (· · ·), mean foetal Hi spectrum (pH ≈ 6.97) (·——·), and standard adult human Hi spectrum (pH ≈ 7.21) (——). The absorptivity at 630 nm has been normalised to 3.895 $L \cdot mmol^{-1} \cdot cm^{-1}$, the absorptivity at 630 nm of the standard human Hi spectrum.

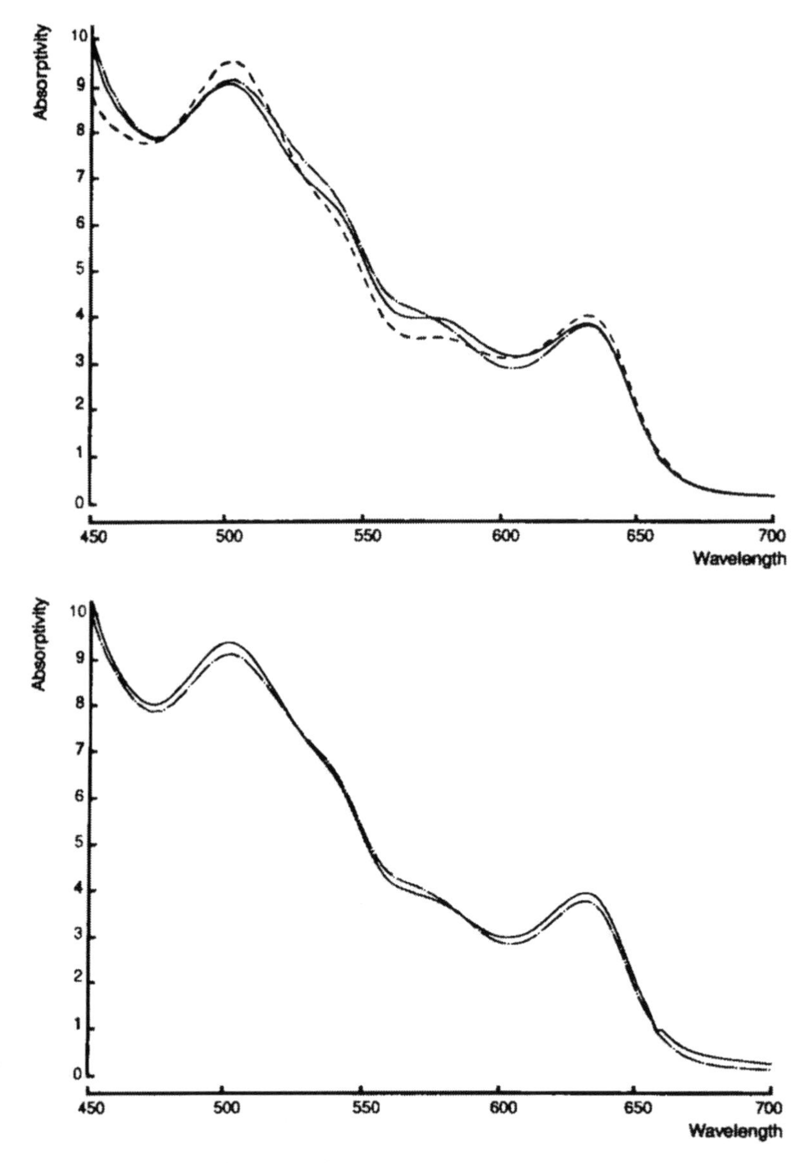

Figure 14.5. Upper panel: mean absorption spectrum of foetal sheep Hi (pH ≈ 6.97) (·——·), standard adult sheep Hi spectrum (pH ≈ 7.17) (- - - -), and standard adult human Hi spectrum (pH ≈ 7.21). (——). Lower panel: mean absorption spectrum of foetal sheep Hi (pH ≈ 6.97) (·——·) and standard foetal human Hi spectrum (pH ≈ 7.08) (——).

Figure 14.4 shows in the upper panel the mean foetal sheep Hi spectrum (pH ≈ 6.93) based on the two specimens of foetal sheep Hi at pH = 6.83 and 7.10 (Table 14.7; Fig. 14.3), together with two individual foetal sheep Hi spectra with pH = 6.83 and 7.32, and the standard sheep Hi spectrum (pH ≈ 7.17). For easy comparison the absorptivity at 630 nm of all four spectra has been normalised to 4.049 $L \cdot mmol^{-1} \cdot cm^{-1}$, the absorptivity at this wavelength of the standard Hi spectrum of adult sheep haemoglobin. The individual foetal Hi spectrum at pH = 6.83 can hardly be distinguished from the mean foetal Hi spectrum with pH ≈ 6.97, again showing that in this pH range, where the OH^--ion activity is still low, the influence of changes in pH on the shape and position of the spectral absorption curve of Hi is but slight. The foetal Hi spectra are all quite dissimilar from the standard adult sheep Hi spectrum, suggesting an intrinsic difference between the two kinds of Hi. The lower panel of Fig. 14.4 gives evidence that the spectral difference between foetal sheep Hi and adult human Hi is also, for the most part, caused by an intrinsic difference in molecular structure because it seems quite improbable that by a further change in pH the foetal ovine and adult human spectral absorption curves of Hi can be made to coincide.

The upper panel of Fig. 14.5 clearly shows the considerable differences between the Hi absorption spectra of foetal ovine, adult ovine and adult human haemoglobin. The lower panel, on the other hand, demonstrates a striking

Table 14.11.
Absorptivities ($L \cdot mmol^{-1} \cdot cm^{-1}$) of the common derivatives of foetal sheep haemoglobin at selected wavelengths between 600 and 1000 nm

λ (nm)	ε_{HHb} $N = 2$	ε_{O_2Hb} $N = 3$	ε_{COHb} $N = 3$	ε_{Hi} $N = 2$
600	3.605	0.795	1.063	2.807
630	1.172	0.169	0.254	3.585
660	0.840	0.099	0.108	0.793
680	0.626	0.096	0.076	0.259
700	0.463	0.096	0.068	0.194
750	0.400	0.140	0.055	0.218
775	0.297	0.171	0.047	0.275
800	0.215	0.204	0.044	0.346
805	0.209	0.209	0.042	0.361
840	0.197	0.252	0.040	0.467
845	0.197	0.257	0.038	0.481
880	0.209	0.290	0.036	0.567
904	0.218	0.302	0.034	0.624
920	0.215	0.308	0.032	0.654
940	0.190	0.305	0.030	0.689
960	0.149	0.297	0.027	0.710
1000	0.078	0.268	0.024	0.725

Standard procedure; Optica CF4.

Absorbance measured with l = 0.100 cm (HHb, O_2Hb and COHb at λ = 600 nm, Hi at λ = 600 and 630 nm); l = 0.200 cm (HHb and O_2Hb at λ = 630–1000 nm, Hi at λ = 660–1000 nm); l = 1.000 cm (COHb at λ = 630–1000 nm). N is number of specimens. $Na_2S_2O_4$ used to prevent reoxygenation of HHb. pH (Hi) = 6.83 and 7.10. In the wavelength range 630–1000 nm a red sensitive photomultiplier tube and a red filter have been used.

similarity between the Hi spectra of ovine and human foetal haemoglobin. This is the more surprising because of the functional differences between the two kinds of haemoglobin and the different relationship of their oxygen affinity to the presence of 2,3-diphosphoglycerate [22].

Table 14.11 presents the absorptivities of the common derivatives of foetal sheep haemoglobin for selected wavelengths in the spectral range of 600–1000 nm, measured with an Optica CF4 grating spectrophotometer. Wavelength selection was necessary to complete this series of manual measurements within a reasonable time. Wavelengths of importance for biomedical applications have been chosen preferentially. For a discussion of the measurements underlying the absorptivities presented in Table 14.11 and an explanation of the slight differences between these data and those of Table 14.8 see Chapter 15.

COMMENTS

The most striking observation concerning the absorption spectra of sheep haemoglobin is the high absorptivity of HHb and O_2Hb at wavelengths > 610 nm. Table 14.12 compares the absorptivities of O_2Hb at a few wavelengths in the red/infrared region of several animal species. O_2Hb has been chosen for the comparison instead of HHb to avoid any interfering effect of the use of sodium dithionite. The table demonstrates that the relatively high absorptivity at wavelengths > 610 nm is unique for the sheep within the group of animal species of which haemoglobin absorption spectra have been measured. The table also shows that the effect extends beyond the range shown in Fig. 14.3.

It should be noted that the high red/infrared absorptivity is not present in the HHb, O_2Hb and COHb spectra of foetal sheep haemoglobin; in

Table 14.12.
Absorptivity of O_2Hb at some wavelengths in the red/infrared region for different kinds of haemoglobin

Haemoglobin λ (nm):	680	700	750	800	880	940
Human A (8.4)	0.085	0.088	0.136	0.200	0.284	0.294
Human F (8.11)	0.079	0.083	0.129	0.189	0.270	0.284
Bovine (11.4)	0.094	0.097	0.142	0.206	0.293	0.307
Porcine (12.4)	0.088	0.091	0.137	0.194	0.276	0.290
Equine (13.4)	0.091	0.093	0.137	0.196	0.275	0.289
Ovine A (14.5)	**0.131**	**0.131**	**0.172**	**0.226**	**0.305**	**0.316**
Ovine F (14.11)	0.096	0.096	0.140	0.204	0.290	0.305

Standard procedure; Optica CF4.

$\varepsilon_{O_2Hb}(\lambda)$ in $L \cdot mmol^{-1} \cdot cm^{-1}$ from tables indicated in parentheses. A is adult; F is foetal.

Table 14.12 there is no difference in $\varepsilon_{O_2Hb}(\lambda)$, for example, between foetal ovine and (adult) bovine haemoglobin.

Apart from the difference in the red/infrared region, there appeared to be little difference between the HHb, O_2Hb and COHb spectra of adult and foetal sheep haemoglobin (Fig. 14.3; Table 14.9). This is in contrast with the distinct differences between human HbA and HbF, as shown in Chapter 8 (Fig. 8.4; Table 8.12). It should be kept in mind, however, that only a few specimens of foetal sheep haemoglobin have been examined. Not enough data have been obtained to confidently calculate 'standard spectra' of the common haemoglobin derivatives. The data presented for foetal sheep haemoglobin therefore should be interpreted with caution.

CHAPTER 15

COMMENTS ON THE DETERMINATION OF ABSORPTION SPECTRA OF HAEMOGLOBIN

The necessity to distinguish between the haemoglobin absorption spectra of different animal species, and within the human species between those of adult and foetal haemoglobin, has been recognised only recently. In the first of Drabkin's pioneering spectrophotometric studies [95] the results obtained for human, dog and rabbit haemoglobin differed so little that it was considered an open question whether one should use the spectrophotometric constants interchangeably from species to species. Fourteen years later, Drabkin reported on an extensive study [94] in which the crystallographic differences between human, horse and dog haemoglobin were demonstrated, but the question of any difference in spectrophotometric properties was not raised again. It seems that it was tacitly assumed that such differences would not exist because the colour of haemoglobin originates from the iron-porphyrin moiety, which does not differ between animal species. When the question was occasionally posed, it was answered in the negative: in 1972 Siggaard-Andersen *et al.* [357] found no differences between the absorption spectra of adult and foetal haemoglobin, nor between haemoglobins from rabbit, mouse, cow, pig, and monkey. This course of events is not surprising: the differences are extremely small and the problem was scientifically not exactly exciting.

It was only through the advent of spectrophotometric multicomponent analysis, which is very sensitive to small differences in the absorptivities used in the calculations, that differences between the absorption spectra of human foetal and adult haemoglobin were recognised [177, 446, 478]. Further exploration of these differences then seemed worthwhile and it appeared unavoidable to take a closer look at the obvious possibility of similar differences between animal species. Since variations in technique had been shown to cause differences in the measured absorptivity in magnitude comparable to the species differences, a more refined technique was required, especially for preparation of the samples.

In Table 8.15 the influence of various methods of preparation and measurement on the absorption spectra of the haemoglobin derivatives has been demonstrated with the β-peak of adult human O_2Hb as an example. O_2Hb is easy to prepare and quite stable; therefore, ε_{O_2Hb} is suitable for demonstrating the effect of clearing the samples before the measurement. The table shows that residual turbidity has a distinct influence on the measured values of the absorptivity. For this reason, the first step in the earlier methods for the determination of the absorptivities of haemoglobin derivatives was always to prepare a clear erythrolysate [7, 211] or even a stroma-free haemoglobin solution [472, 481]. The stock solutions thus prepared were the starting material for the preparation of solutions of the various haemoglobin derivatives.

However, in the course of time it appeared that this apparently ideal procedure had a serious disadvantage. When O_2Hb solutions were prepared of human foetal haemoglobin, a high fraction of Hi was formed. As such, this is not a serious hindrance to the determination of the absorption spectrum of O_2Hb, since a correction for any contaminating Hi can be made (Chapter 7; equation 7.2), but Hi formation is often a first step in the deterioration of haemoglobin and should be prevented as far as possible. Moreover, there were other observations that extended tonometry of stroma-free haemoglobin solutions at a high p_{O_2} might lead to changes in a small part of the haemoglobin. There was, for instance, the observation that unexplained negative COHb fractions were found when the O_2Hb solutions were subjected to multicomponent analysis (Chapter 17).

Although it had not been demonstrated conclusively that the phenomena mentioned above were due to the way in which the haemoglobin solutions were equilibrated with oxygen, we changed to a tonometry procedure in which erythrocyte suspensions instead of haemoglobin solutions were equilibrated with oxygen and with other gas mixtures. Erythrolysis was then effected by introducing a non-ionic detergent solution into the tonometer near the end of the equilibration period. This procedure forms part of the **standard procedure** as described in Chapter 6.

It should be realised that solutions of haemoglobin derivatives obtained with the standard procedure are not as clear as those obtained from a stroma-free haemoglobin solution, as used in the **SFH procedure**. This is shown, for instance, by the data given in Tables 8.8 and 8.15. It can also be concluded from Table 8.8 that by switching from the SFH procedure to the standard procedure clarity is, in a sense, traded-off for purity. The spread in the absorptivity values of O_2Hb prepared by means of tonometry is definitely smaller when the standard procedure is used, whereas the absorptivity values of Hi are more precisely determined with the SFH procedure. Complete conversion of all haemoglobin into Hi is easily accomplished by the addition of $Fe(CN)_6^{3-}$; the spread in the series of measurements thus results mainly from differences in residual turbidity, which are smallest when the SFH procedure is used.

These method-dependent differences in the absorption spectra of the haemoglobin derivatives raise the question as to how far the true values of the absorptivity can be and should be approximated. The *true absorptivity* at a certain wavelength would be the absorptivity of a pure haemoglobin derivative measured in a solution with zero turbidity by a perfect spectrophotometer. Furthermore, any differences in refractive index between solution and cuvette should be taken into account. Although all measurements are made against a blank measurement, using the same cuvette filled with water, refractive index mismatch between sample and water may introduce some deviation because of differences in the reflection losses at the sample-to-glass interfaces in the cuvette for water and sample [139], with the number of sample-to-glass interfaces depending on whether the cuvette is used with or without a glass insert.

If important information concerning structure or function of haemoglobin could be obtained by means of the true absorption spectra, it would be worthwhile to make the effort to attain true spectra by further refining the methods of preparation and measurement. However, as the knowledge of the absorption spectra of the haemoglobin derivatives mainly serves practical applications, it is more important to keep the application in mind when determining the spectra.

An example of such an approach is the procedure followed by Zwart *et al.* [444, 447] in the development of a method for multicomponent analysis of five haemoglobin derivatives using a general purpose diode array spectrophotometer. Realising that the measurements had to be made on freshly drawn blood samples of patients, the samples for the determination of the absorption spectra of HHb, O_2Hb and COHb were likewise prepared by introducing fresh whole blood into the tonometers. That absorptivity values (Appendix ref. [447]) were obtained by this procedure, which are in fair agreement with those obtained with the standard method (Table 8.1), is due to the effectiveness of the cotton wool filtration (Chapter 6; Fig. 6.3) used in both series of measurements. (For Hi there is a difference between the two series of absorptivity values, because the ε_{Hi} values of Table 8.1 are based on a new series of measurements with closer control of pH).

A practical approach has also been followed in dealing with the problems of the tendency of HHb solutions to reoxygenate, of the pH-dependency of the Hi spectrum, and of the temperature-dependency of the absorptivity values. The solutions to the latter two problems, as given in Chapter 7, although not perfect, function reasonably well in our experience. The dilemma of using or not using sodium dithionite to prevent reoxygenation, however, is indeed a difficult choice. Tables 8.8 and 8.13 show that without dithionite some degree of reoxygenation can hardly be avoided; Tables 8.13, 12.5 and 14.5, on the other hand, demonstrate that even a trace of $Na_2S_2O_4$ used as described in Chapter 6, usually causes some distortion of the HHb spectrum.

In spite of the shortcomings of the method, the absorption spectra determined with the standard procedure are well suited for multicomponent analysis and other biomedical applications. With the exception of dog haemoglobin, which was investigated before the standard procedure had been developed, and rat haemoglobin, to which the standard procedure could not be applied, all data presented are based on the measurement of samples prepared with the standard procedure. For human HbA, absorption spectra obtained using the SFH procedure have also been presented for comparison of the two procedures (Fig. 8.3; Table 8.8) and for comparison of dog and human haemoglobin (Fig. 9.1; Table 9.2).

In all measurements, care has been taken to keep the measured absorbance within limits allowing reliable measurements to be made. As explained in Chapter 7, this requires that the lightpath length is varied over a wide range, which, in turn, requires the use of different types of cuvettes, with or without glass inserts.

As described in Chapter 3, the principal difference between the two types of spectrophotometer used for the measurements concerns the spectral resolution. This difference is clearly discernible in the results presented in Chapters 8–14 and has been taken into account in the interpretation of these data. The Optica CF4 spectrophotometer was used to exactly locate the wavelengths of maximal and minimal absorption for adult human, dog and rat blood, as well as to obtain, for all kinds of haemoglobin, a series of absorbance measurements beyond 800 nm, the end of the spectrum scanned by the HP8450A. However, because a single (manual) measurement with the Optica CF4 takes 1–2 min, wavelength selection was necessary to complete the measurement series within a reasonable time. The wavelengths 775, 805, 845, and 904 nm were included because of their use in near infrared spectroscopy [68].

Table 15.1 compares the absorptivities at five wavelengths between 630 and 800 nm of four common derivatives of human, bovine, porcine, equine and ovine haemoglobin measured with the two types of spectrophotometer. The samples were all prepared with the standard method and dithionite was used to prevent reoxygenation of HHb. The measurements were made using the same cuvette for each haemoglobin derivative. Hence, there was no difference in lightpath length, nor in the influence of differences in refractive index between water and sample. Thus the only difference between the two measurements was the spectrophotometer used. The table reveals a systematic difference: the absorptivity measured with the HP8450A is generally higher than the absorptivity measured with the Optica CF4.

The difference between the absorptivities obtained with the two spectrophotometers can probably be ascribed to a difference in response of the instruments to residual scattering in the erythrolysates, *e.g.* by remnants of erythrocyte membranes, which have a refractive index that differs from that of the solution. The amount of scattered light that is captured by the detector depends on the numerical aperture of the detector, which is larger for the Optica CF4 than for the HP8450A. The observation that absorptivities < 0.010 have been measured with the Optica CF4 (Tables 8.4 and 8.12) gives confidence that practically all scattered light is captured by the detector of this instrument [140].

Table 15.1.
Absorptivities ($L \cdot mmol^{-1} \cdot cm^{-1}$) at wavelengths between 630 and 800 nm of the common haemoglobin derivatives of various animal species measured with an HP8450A diode array (HP) and with an Optica CF4 grating spectrophotometer (O)

Species	λ (nm)	ε_{HHb}		ε_{O_2Hb}		ε_{COHb}		ε_{Hi}	
		HP	O	HP	O	HP	O	HP	O
Man (HbA)	630	1.051	1.058	0.139	0.115	0.196	0.187	3.905	3.914
	660	0.815	0.814	0.100	0.080	0.069	0.061	0.945	0.819
	700	0.447	0.437	0.105	0.088	0.028	0.021	0.222	0.204
	750	0.388	0.387	0.148	0.136	0.019	0.013	0.261	0.251
	800	0.215	0.206	0.208	0.200	0.015	0.010	0.406	0.399
Cow	630	1.117	1.105	0.185	0.167	0.248	0.218	3.972	3.900
	660	0.842	0.845	0.108	0.099	0.103	0.087	0.957	0.871
	700	0.472	0.467	0.102	0.097	0.050	0.037	0.224	0.198
	750	0.391	0.406	0.144	0.142	0.039	0.028	0.260	0.247
	800	0.226	0.221	0.206	0.206	0.033	0.024	0.409	0.403
Pig	630	1.150	1.148	0.212	0.186	0.281	0.258	4.020	3.921
	660	0.841	0.855	0.111	0.088	0.112	0.095	1.014	0.883
	700	0.475	0.477	0.109	0.091	0.061	0.045	0.233	0.197
	750	0.397	0.427	0.149	0.137	0.047	0.035	0.267	0.245
	800	0.230	0.231	0.204	0.194	0.039	0.029	0.413	0.402
Horse	630	1.090	1.066	0.173	0.147	0.255	0.221	3.961	3.954
	660	0.837	0.830	0.119	0.099	0.122	0.102	0.933	0.855
	700	0.466	0.456	0.110	0.093	0.066	0.050	0.180	0.162
	750	0.386	0.403	0.151	0.137	0.054	0.041	0.266	0.218
	800	0.227	0.218	0.208	0.196	0.048	0.037	0.381	0.378
Sheep (adult)	630	1.171	1.146	0.233	0.201	0.267	0.226	4.012	4.015
	660	0.882	0.873	0.164	0.136	0.121	0.092	1.021	0.889
	700	0.508	0.494	0.156	0.131	0.072	0.047	0.244	0.210
	750	0.418	0.423	0.188	0.172	0.059	0.038	0.277	0.257
	800	0.245	0.230	0.239	0.226	0.050	0.033	0.423	0.411

Absorptivity values from Tables 8.2 and 8.4 (man), 11.2 and 11.4 (cow), 12.2 and 12.4 (pig), 13.2 and 13.4 (horse), 14.2 and 14.4 (sheep).

Table 15.1 shows one peculiar irregularity. Although the absorptivities measured with the HP8450A are generally higher than those measured with the Optica CF4, it is the other way round at 750 nm for HHb, for all animal species. At this point in the HHb spectrum the absorptivity measured with the Optica CF4 exceeds that measured with the HP8450A. This difference cannot be easily explained. It is definitely not caused by the use of dithionite, since it is also present in measurements made without dithionite. The absence of measurements with the Optica CF4 near 750 nm makes the interpretation still more difficult.

Because of the differences in the measured absorptivities between different types of spectrophotometer, it is mandatory to use in each application absorptivity values obtained with the same type of instrument. For this reason Optica CF4 measurements have been made over the range of 600–1000 nm, although reliable measurements up to 800 nm had already been made with an HP8450A. The results have, for example, been used in the error analysis of the influence of the presence of HbF, COHb and Hi on the oxygen saturation as measured with a pulse oximeter using spectral bands at 660 and 940 nm [294, 480] (Chapter 18).

The interpretation of the differences between the standard Hi spectra of different kinds of haemoglobin is complicated by the fact that Hi, in the absence of other anions that bind more readily, binds OH^- ions in dependence of the pH of the solution. A solution of Hi thus actually is a solution of Hi and HiOH (Fig. 2.2), and the absorption spectrum is dependent on the [Hi]/[HiOH] ratio. The [Hi]/[HiOH] ratio is, in turn, dependent on the OH^- affinity of Hi as well as the pH of the solution. The OH^- affinity of Hi may vary between different kinds of haemoglobin. In the interpretation of the spectral absorption curves of Hi one should therefore distinguish between differences in pH, differences in OH^- affinity, and differences in molecular structure.

If identical absorption spectra are obtained at similar pH, there is no spectroscopically discernible difference between the Hi derivatives of the two kinds of haemoglobin. If a difference in pH is necessary to have the absorption spectra coincide, the difference between the Hi derivatives of the two kinds of haemoglobin is a difference in OH^- affinity. If there is no pH at which the absorption spectra can be made to coincide, a difference in molecular structure is the cause of the spectral differences between the Hi derivatives of the two kinds of haemoglobin, possibly in addition to a difference in OH^- affinity.

To distinguish between these possibilities, absorption spectra at various pH values were recorded in addition to the standard Hi spectra of the various kinds of haemoglobin (Figs 8.2, 8.5, 9.2, 10.2, 11.2, 12.2, 13.2, 14.2, 14.4, 14.5). To facilitate the interpretation of the spectral absorption curves by visual inspection, the curves were normalised by reducing $\varepsilon_{Hi}(630)$ to the same numerical value, most often the corresponding value of the standard Hi spectrum of human HbA: 3.895 $L \cdot mmol^{-1} \cdot cm^{-1}$ (Table 8.1). All absorptivities were then multiplied by $3.895/\varepsilon_{Hi}(630)$. If, as in the case of human HbF, $\varepsilon_{Hi}(630) = 4.021$ (Table 8.9), the multiplication factor was $3.895/4.021 = 0.969$.

Table 15.2 consolidates some absorptivities at wavelengths near the principal absorption maxima and minima of HHb, O_2Hb and COHb of all types of haemoglobin of which absorption spectra have been measured. This gives a general impression of the differences between the animal species and between adult and foetal haemoglobin. The species are arranged, from top to

Table 15.2.
Absorptivities ($L \cdot mmol^{-1} \cdot cm^{-1}$) near some absorption minima and maxima of HHb, O_2Hb and COHb of various animal species measured with an HP8450A diode spectrophotometer

Species	$\varepsilon_{HHb}(556)$	$\varepsilon_{O_2Hb}(578)$	$\varepsilon_{O_2Hb}(560)$	$\varepsilon_{O_2Hb}(542)$	$\varepsilon_{COHb}(570)$	$\varepsilon_{COHb}(538)$
Dog*	12.90	14.93	8.67	14.45	14.16	14.27
Man (A)*	12.99	15.13	8.72	14.41	14.30	14.22
Horse	13.24	15.14	8.69	14.41	14.29	14.18
Pig	13.29	15.32	8.62	14.49	14.26	14.36
Sheep (F)	13.30	15.51	8.68	14.64	14.30	14.32
Cow	13.31	15.33	8.68	14.53	14.31	14.27
Rat**	13.35	15.87	8.87	14.81	14.37	14.51
Man (A)	13.36	15.36	8.77	14.52	14.46	14.30
Sheep	13.39	15.59	8.66	14.57	14.42	14.44
Man (F)	13.53	15.45	8.60	14.62	14.42	14.47

Samples prepared with the standard procedure, except * = SFH procedure and ** = Rat procedure (Chapter 10). A = adult Hb; F = foetal Hb). Data (top to bottom) from Tables 9.1, 8.7, 13.1, 12.1, 14.7, 11.1, 10.1, 8.1, 14.1 and 8.9.

bottom, according to the values of $\varepsilon_{HHb}(556)$ that are placed in order of increasing magnitude. It can be seen that the difference in spectral absorbance that was first observed between types of haemoglobin, the difference between human HbA and HbF [177, 446, 478], is among the most distinct differences hitherto observed.

It should be noted that the differences between the absorptivities of dog and human haemoglobin, both measured in samples prepared with the SFH method, are smaller than the differences found between the absorptivities of human HbA prepared either with the standard method or with the SFH method. This again demonstrates that in choosing the appropriate absorptivity values for a particular application, the procedure and the circumstances of the measurement should be taken into account as well as the type of spectrophotometer.

CHAPTER 16

HAEMOGLOBINOMETRY

The determination of the absorption spectra of haemoglobin derivatives as presented in Chapters 8–14 was only possible through the availability of an accurate method for the determination of total haemoglobin, ct_{Hb}. Moreover, the suitability of the absorptivities of the various haemoglobin derivatives for applications yielding good interlaboratory comparability required that the determination of ct_{Hb} was based on an internationally accepted standard. As described in Chapter 4, the ICSH-recommended HiCN method fulfilled this requirement. This method is based on the absorptivity of HiCN at 540 nm, $\varepsilon_{HiCN}(540)$, which has, for human haemoglobin, been determined by various methods (Table 4.1), and found to be $11.0\ L \cdot mmol^{-1} \cdot cm^{-1}$. In Chapter 5 it has been shown that for the investigated animal haemoglobins $\varepsilon_{HiCN}(540)$ was not significantly different from the human value. Hence $\varepsilon_{HiCN}(540) = 11.0\ L \cdot mmol^{-1} \cdot cm^{-1}$ also became the anchor value for the absorptivities of the derivatives of haemoglobin of various animal species.

The intensive investigation that has led to standardised haemoglobimetry was motivated by the importance of the rapid and accurate determination of ct_{Hb} in clinical medicine. This determination remains one of the most frequently performed in medical practice, and thus still is one of the most important applications of the spectrophotometry of haemoglobin. This chapter describes the standard HiCN method in greater detail and gives a survey of past and present procedures for the routine determination of ct_{Hb} and their relationship with the reference method.

A SHORT HISTORY

Among the first methods described for the clinical estimation of blood haemoglobin is a device consisting of a small stand holding two small tubes of equal bore [134]. One tube contained a reference composed of gelatine coloured with carmine and picrocarmine; the other was graduated and, in it, dilutions of the blood sample were made until the colour visually matched that of the reference. Since then, a great many methods have been developed to determine ct_{Hb}. These methods include measurements based on physical properties of blood, *e.g.* specific gravity [17] and refractive index [376], measurements based on chemical properties, *e.g.* iron content, the ability to reversibly unite with oxygen [392] or carbon monoxide [393], and measurements based on the spectral characteristics of haemoglobin derivatives.

Among the first methods based on the spectral characteristics of haemoglobin derivatives were determinations as carboxyhaemoglobin. In these methods the colour of diluted blood saturated with CO was compared with a diluted COHb reference solution [150]. The simplest technique ever devised is undoubtedly the one of Tallqvist [379] where a drop of blood is absorbed by a piece of filter paper and its colour then visually compared with a lithographed, graded colour scale. Between 1920 and 1940 numerous visual and photoelectric colorimetric methods were introduced, based on measurement of O_2Hb [337], HHb [51], Hi [152], acid haematin [333], alkaline haematin [431], and a haemochromogen formed from haem and pyridine [321].

Of these methods, Sahli's acid haematin method [333] has been used most extensively. Using a special pipette ('Sahli pipette') 20 μL of blood is mixed with 0.1 mol/L hydrochloric acid in a graduated glass tube and, after 1 min, diluted with water until the colour visually matches the colour of a second tube containing a brown coloured substance. The colour however, develops gradually, then fades; substantial errors are rule rather than exception. When in the 1950s the HiCN method began to win ground, Sahli's method lost its popularity and the acid haematin and other methods based on the measurement of haematin complexes dropped out of use. In 1984 Zander *et al.* [439–441] proposed a revival of this approach by describing a spectrophotometric determination of ct_{Hb} as a haematin-detergent complex, called alkaline haematin D-575. The method however, has disadvantages and was no match for the methods available at the time [14].

The HiCN method was introduced by Stadie [371] in 1920, who used a potassium hexacyanoferrate(III) (potassium ferricyanide; $K_3Fe(CN)_6$) solution to convert all haemoglobin to Hi, followed by a solution of potassium cyanide (KCN) to form HiCN from Hi. Later a solution of 200 mg $K_3Fe(CN)_6$, 50 mg KCN, 1.0 g $NaHCO_3$ made up with water to 1000 mL ('Drabkin's solution') came into use [208]. The measurement was first performed by visual, later by photoelectric spectrophotometry. In the 1960s the standardisation of haemoglobinometry was effected on the basis of the HiCN method as described in Chapter 4. In 1975 the HiCN method was almost generally used in clinical chemical laboratories, and had become the first internationally accepted standardised clinical chemical method [205, 210].

With the advent of automation, modification of the standard method was required for incorporation of the measurement of ct_{Hb} in automated haematology analysers. The increased use of multipurpose haematology analysers in turn induced a need for simple and accurate instruments for single determinations of ct_{Hb}. Hence, a new diversity of methods evolved. This, however, has not produced a reversion to the chaotic situation existing before 1950: the standardised HiCN method has been converted from a routine method for general use to the reference method for haemoglobinometry [194].

The principal methods now in use for the routine measurement of ct_{Hb} and their relation to the reference method are described further on in this chapter.

THE REFERENCE POINT OF STANDARDISED HAEMOGLOBINOMETRY

The first step required for standardisation was to establish a reference point for haemoglobinometry [486]. This primarily involved two decisions to be made. The first decision concerned the selection of a physical or chemical property of haemoglobin that could be measured accurately and easily; the second decision was to adopt one or more methods to exactly correlate the result of the physical or chemical measurement with a certain amount of haemoglobin.

Of the physico-chemical properties of haemoglobin that can be measured accurately and easily, the absorption of visible light has been most widely used. With modern spectrophotometers the absorbance of haemoglobin in solution can easily be measured within 0.5%. It needs no argument, of course, that in the solution all forms of haemoglobin must have been converted into one and the same haemoglobin derivative, or at least into derivatives that are photometrically indistinguishable at the wavelength of measurement. The absorbance of HiCN solutions at 540 nm fulfilled all requirements (Chapter 4) and therefore was selected as the quantity to be measured.

For the second decision a choice had to be made between either a method based on some functional property of haemoglobin, *e.g.* oxygen or carbon monoxide binding capacity, or an analytical method based on the chemical composition of the haemoglobin molecule. From a physiological point of view, standardisation of haemoglobinometry on the basis of the oxygen binding capacity of haemoglobin would have been desirable. However, the inevitable presence of some dyshaemoglobin, as well as the technical difficulties and uncertainties of blood gas analysis, favoured the use of an analytical method based on the composition of the haemoglobin molecule. The complete elucidation of the composition of human haemoglobin A [39, 159] added a strong argument in favour of chemical analysis.

Before the chemical composition of HbA had become completely known, it had already been demonstrated that each haemoglobin monomer contained only a single iron atom. Measuring the iron concentration of a haemoglobin solution thus was the obvious choice for the determination of the haemoglobin concentration of the solution by chemical analysis. As described in Chapter 4, $\varepsilon_{HiCN}(540)$ has largely been determined on the basis of iron measurements. Although the results have been confirmed by measurements of nitrogen and carbon (Table 4.1), iron remained the atom of choice (Chapter 5).

THE HAEMIGLOBINCYANIDE METHOD

Specimens for the determination of ct_{Hb} are preferably obtained by venipuncture and the blood is anticoagulated with appropriate salts of EDTA, or heparin. When blood cell counting and sizing are also contemplated, ICSH recommends dipotassium EDTA, 1.5–2.2 mg (3.7–5.4 μmol) per mL of blood as the anticoagulant of choice [188]. When liquid anticoagulants, *e.g.* solutions of tripotassium EDTA or heparin, are used, there is a dilution of the specimen of 1–2%. Before sampling, the specimen must be mixed well by completely inverting the specimen container at least 8 times.

Best results are obtained when fresh blood is analysed; ct_{Hb} should be determined on the day the specimen is drawn. It has been shown, however, that overnight storage of blood in a properly stoppered container at room temperature does not affect ct_{Hb} obtained using the HiCN method.

When the specimen has been obtained, a known volume of blood is mixed with a known volume of reagent solution in order (1) to lyse the cells, (2) to dilute the haemoglobin solution, and (3) to convert all haemoglobin derivatives to HiCN. Ideally the result is a perfectly clear solution of HiCN of which the absorbance at 540 nm is determined using a spectrophotometer properly calibrated as to wavelength and absorbance, and glass cuvettes providing an exactly known lightpath length. The extent to which these requirements should be met depends on the purpose of the ct_{Hb} measurement: routine haemoglobinometry or reference procedure. In the following description this difference is taken into account.

The **conversion** of the common haemoglobin derivatives present in blood to HiCN takes place in two steps: (1) oxidation of haemoglobin iron by potassium hexacyanoferrate(III), converting all haemoglobin to Hi, and (2) the addition of cyanide, CN^-, to form HiCN.

$$\mathrm{Hb(Fe^{2+}) + Fe(CN)_6^{3-} \rightarrow Hi(Fe^{3+}) + Fe(CN)_6^{4-}}$$
$$\mathrm{Hi + CN^- \rightarrow HiCN}$$

The oxidation is rather slow, with the reaction velocity dependent on the pH of the solution. At pH = 8.5 and a room temperature of 20–25 °C, the reaction takes some 30 min; at pH = 7.2 this is < 5 min. Oxidation of COHb takes considerably longer, some 90 min at pH = 8.5, some 30 min at pH = 7.2 [7, 381]. The binding of CN^- to Hi is instantaneous.

COHb is completely converted to HiCN, but when the longer conversion time is not taken into account an erroneous result may be obtained. Because the absorptivity of COHb at 540 nm is 14.3 $\mathrm{L \cdot mmol^{-1} \cdot cm^{-1}}$ (Table 8.1), non-conversion to HiCN may lead to an overestimation of ct_{Hb}. Errors may be easily prevented by increasing the time between dilution of the suspect sample and measurement.

The only Hb derivative occurring in patients that is not converted to HiCN is SHb; SHiCN is formed instead (p. 13). Since the absorptivity of SHiCN at 540 nm is 8.0 $L \cdot mmol^{-1} \cdot cm^{-1}$ [103, 210], a SHb fraction of 1% causes an underestimation of ct_{Hb} of 0.27%. Since SHb fractions exceeding 5% are rarely encountered in clinical practice [442, 489], the possible underestimation is limited to 1.5%.

The blood sample must be **diluted** in order to obtain a HiCN solution of which the absorbance is within the range of 0.090–0.700, a range that can be measured accurately with conventional spectrophotometers with a lightpath of 1.0 cm (Chapter 7). For routine determination of ct_{Hb} a dilution of 1 : 251 is recommended: 20 μL blood and 5 mL reagent, using calibrated glassware. The highest accuracy is attained with calibrated 0.400 or 0.500 mL to-contain pipettes and 100 mL Class A volumetric flasks; this virtually eliminates all variance caused by dilution errors. In practice, accurate dilutions with a reproducibility $< 0.5\%$ may be obtained by diluting 0.100 mL well-mixed blood (calibrated to-contain pipette) with 25.0 mL of reagent (calibrated Class A volumetric flask) [49] or by diluting 40 μL well-mixed blood (calibrated positive displacement pipette) with 10.0 mL reagent (calibrated Class A volumetric flask) [176]. After aspirating the blood, wipe the outside of the pipette. Ensure that the meniscus at the tip of the pipette is flat. Deliver the pipette contents into the reagent and rinse the pipette by aspirating and delivering the blood-reagent mixture at least five times.

Once the sample is mixed with reagent, the resulting HiCN solution is stable and can be measured when convenient. Storage for periods longer than 6 h should be at 4–10°C, in tightly stoppered containers, in the dark.

Reagent solutions used in the HiCN method generally contain $K_3Fe(CN)_6$, usually 200 mg/L, and KCN, usually 50 mg/L. Because HCN is a weak acid ($pK_a = 9.1$), $K_3Fe(CN)_6$/KCN solutions have a high pH, around 9.6, and oxidation of O_2Hb and HHb proceeds slowly. To lower the pH and, consequently, decrease the time required for oxidation of haemoglobin, 'Drabkin's solution' includes $NaHCO_3$ (p. 198) and has a pH of 8.6. This solution was quite popular but still requires conversion times in excess of 15 min. To further lower the pH and decrease the conversion time, $NaHCO_3$ was replaced by KH_2PO_4.

Lowering the pH, however, favours precipitation of plasma proteins, primarily gamma globulins with an isoelectric point near the pH of the reagent. This required the addition of a non-ionic detergent to effectively prevent turbidity in the majority of cases. Thus the reagent solution first described by van Kampen and Zijlstra [208] and recommended by ICSH [190–192, 194] contains $K_3Fe(CN)_6$, 200 mg (0.607 mmol), KCN, 50 mg (0.769 mmol), KH_2PO_4, 140 mg (1.028 mmol), non-ionic detergent, 0.5–1.0 mL, and deionised or distilled water to 1000 mL. The reagent is a clear, pale yellow solution with a pH of 7.0–7.4, an osmolality of 6–7 mmol/kg (as determined by freezing point depression), and does not absorb light above 480 nm; the conversion time of most haemoglobin derivatives to HiCN is 3–5 min. Stored at room temperature in a tightly capped, brown borosilicate glass bottle, the shelf life is 6–8 weeks. However, the reagent is unstable when exposed to light and $K_3Fe(CN)_6$ is destroyed by freezing.

On freezing and subsequent thawing, $Fe(CN)_6^{3-}$ is converted to $Fe(CN)_6^{4-}$ and therefore oxidation of haemoglobin to Hi does not take place [409, 453]. The haemoglobin in the solution remains mainly as O_2Hb and, because the absorptivity of O_2Hb at 540 nm is 14.3 $L \cdot mmol^{-1} \cdot cm^{-1}$ (Table 8.1), ct_{Hb} values up to 30% too high may be measured. Decomposition on freezing can be prevented by the addition of 20 mL/L ethanol, methanol, or ethylene glycol, or 5 mL/L glycerol to the reagent solution [453].

Loss of CN^- from the reagent when stored in plastic containers has been described [7]. Haemoglobin is then oxidised to Hi, but the subsequent formation of HiCN is incomplete or absent. Because the absorptivity of human Hi at 540 nm is 6.3 $L \cdot mmol^{-1} \cdot cm^{-1}$ (Table 8.1), ct_{Hb} values up to 40% too low may result. The cyanide concentration of reagent solutions can be measured by means of an ion-selective electrode [469] or, when in doubt, ct_{Hb} can be remeasured after the addition of some extra KCN to the reagent solution.

Optically clear, pure HiCN solutions are characterised by $A(750) \leqslant 0.003$ and $1.59 \leqslant A(540)/A(504) \leqslant 1.63$ (p. 205). The recommended reagent [190, 191, 208] contained 0.5 mL Sterox SE as non-ionic detergent; this detergent, however, is no longer available. Nonidet P-40, 1.0 mL/L, or Triton X-100, 1.0 mL/L, are now recommended [194]. However, neither of these is as effective in preventing turbidity as Sterox SE was.

Therefore, even in measuring normal blood specimens, some residual turbidity may be encountered because of protein aggregates and/or erythrocyte stroma remnants. Many HiCN solutions prepared from normal human blood are actually slightly turbid, as shown by $A(750)$ values up to 0.005. Because an error of 0.001 in the measurement of $A_{HiCN}(540)$ equates to an error of 0.37 g/L in ct_{Hb}, the reference method requires membrane filtration of the HiCN solutions before measuring.

It is recommended to filter the HiCN solutions with a syringe and a disposable filter cartridge containing a low-binding, low-release, 25-mm diameter membrane filter with a mean pore diameter of 0.20–0.25 μm. The filtered solution is dispensed directly into the measuring cuvette and should stand for at least 1 min to allow air bubbles to escape. Good results have been obtained with 0.22 μm Millex-GV and with 0.20 μm Puradisc 25 PP membrane filters [176].

In some pathological conditions, *e.g.* severe lipaemia or proteinaemia, white cell counts $> 20 \times 10^9$/L, or platelet counts $> 700 \times 10^9$/L, turbidity of the resulting HiCN solutions may cause a significant overestimation of ct_{Hb}. This had already been observed when Sterox SE was still consistently used. To combat turbidity in such cases several modified reagents with increased ionic strength have been proposed [253, 254, 397]. These reagents have not been widely used and it has proven more practical to clear turbid HiCN solutions by centrifugation, membrane filtration, or by the addition of a drop of 25%(v/v) ammonia.

The **absorbance** of the diluted, filtered blood-reagent mixture is measured at 540 nm against water or reagent as blank, using matched 1.000-cm glass cuvettes and a spectrophotometer calibrated as to wavelength and absorbance. The natural bandwidth of HiCN around 540 nm of 70 nm dictates that the spectral bandwidth at the wavelength of measurement should be $\leqslant$ 6 nm (Chapter 3). Although the reagent does not absorb light above 480 nm, some instruments have been identified that show apparent light absorption in the 500–600 nm range [176]. Water can therefore be used as a blank only when the laboratory has verified that, for a particular instrument–cuvette combination, the reagent does not display light absorption above 500 nm.

HiCN solutions strictly follow Lambert–Beer's law [481] and ct_{Hb}, expressed in g/L, is calculated from $A_{HiCN}(540)$ by means of the following equation:

$$ct_{Hb} = [A_{HiCN}(540) \times 16114.5 \times f]/(11.0 \times l \times 1000), \qquad (16.1)$$

where $A_{HiCN}(540)$ is the absorbance of the HiCN solution at 540 nm; 16114.5 (g/mol) is the molecular mass of the haemoglobin monomer (64458/4) [38]; f is the dilution factor used in the preparation of the HiCN solution; 11.0 ($L \cdot mmol^{-1} \cdot cm^{-1}$) is the absorptivity of HiCN at 540 nm (p. 28); l (cm) is the lightpath length, usually 1.000 cm; and 1000 is a factor to convert mg/L to g/L.

For a dilution of 1 : 251 and 1.000-cm cuvettes:

$$ct_{Hb}\ (g/L) = 367.7 \times A_{HiCN}(540). \qquad (16.2)$$

If the results are to be expressed as substance concentration based on the haemoglobin monomer:

$$ct_{Hb}\ (mmol/L) = 22.82 \times A_{HiCN}(540). \qquad (16.3)$$

HAEMIGLOBINCYANIDE STANDARDS

A **reference material** is defined [193] as a material or substance, one or more properties of which are sufficiently well established to be used for calibration of an instrument, assessment of a measurement method, or for assigning values to other materials. If one or more values are assigned to the material by means of a collaborative study in several designated centres in accordance with a specified protocol and controlled by a certifying body, the reference material is usually called a **certified reference material**, CRM. In daily usage, reference materials are often called standards, while CRMs may be called certified standards.

ICSH International HiCN Standards have been prepared, on behalf of ICSH, at the Dutch Institute of Public Health (RIV) from 1965 to 1986 and have been designated *International Standard for the Measurement of Haemoglobin in Blood* by the World Health Organisation (WHO) [427]. These standards are used to assign values to secondary HiCN standards (reference materials, calibration solutions) and as a reference against which to judge the purity of such HiCN solutions. The standards were prepared at the RIV using a procedure described by Zijlstra and van Kampen [485] and by Holtz [167].

Fresh donor blood, since the 1980s tested for the absence of hepatitis and HIV antibodies, is collected in 1/5 the volume of 3.2% (w/v) sodium citrate, centrifuged and the plasma removed. The cells are washed twice with NaCl, 9 g/L, and twice with NaCl, 12 g/L. An equal volume of twice distilled or deionised water is added to the cells, and 0.4 volume reagent grade toluene. The mixture is stirred thoroughly and allowed to stand, at 4–6°C, for 12 h. The mixture is then centrifuged, resulting in the formation of three clearly distinct layers. The upper layer consists of toluene, the middle layer is a turbid suspension of erythrocyte stromata, the third a clear solution of haemoglobin. The first two layers are removed and the haemoglobin solution is filtered through ash-free filter paper. The haemoglobin concentration of the solution is determined and a calculated volume of reagent solution containing 200 mg $K_3Fe(CN)_6$, 50 mg KCN, and 1.0 g $NaHCO_3$ per litre is added to a given amount of haemoglobin solution to obtain a final HiCN solution at a concentration of approximately 600 mg/L. This solution is sterilised by means of membrane filtration, mean pore size $< 45\ \mu$m, and dispensed into brown, borosilicate glass ampoules.

Figure 16.1 shows the absorption spectrum of a HiCN Standard; it is characterised by a maximum at 540 nm and a minimum at 504 nm. The ratio of the absorbances at these two wavelengths, $A(540)/A(504)$, is used as a criterion of the purity and the clarity of the HiCN Standard, and of any HiCN solution as well. Since at 750 nm the absorptivity of HiCN is zero, $A(750) \leqslant 0.003$ is used as a criterion for the absence of turbidity.

The HiCN concentration of the standards is calculated from absorbance measurements at 540 nm with a spectrophotometer of which the wavelength and absorbance scales have been checked, and a slitwidth chosen so that the spectral bandwidth is $\leqslant 2$ nm.

$$c_{\text{HiCN}} = [A_{\text{HiCN}}(540) \times 16114.5]/(11.0 \times l), \qquad (16.4)$$

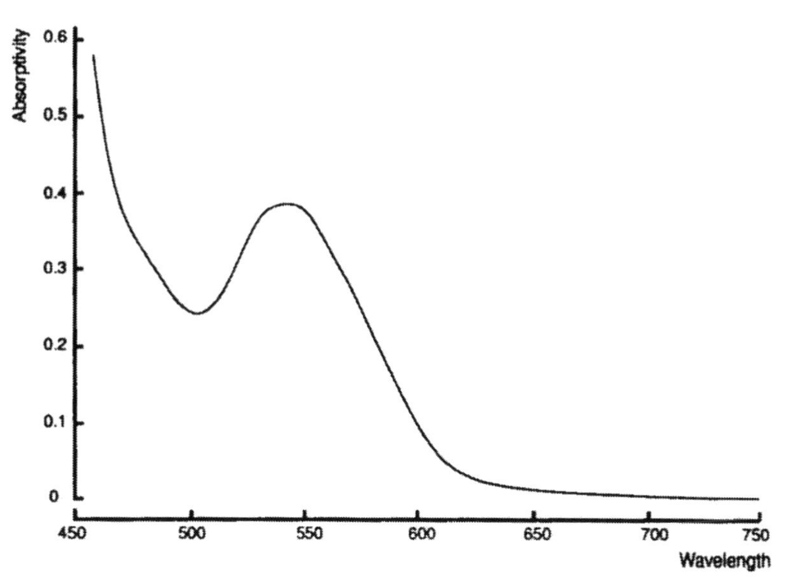

Figure 16.1. Absorption spectrum of HiCN standard (RIVM No. 20600).
$A(540) = 0.378$, corresponding to $ct_{HiCN} = 553.8$ mg/L; $A(504) = 0.235$; $A(540)/A(504) = 1.61$; $A(750) = 0.001$. (Compare Table 16.2, batch 20600 at 2–2.5 years.)

where c_{HiCN} is expressed in mg/L; $A_{HiCN}(540)$ is the absorbance of the HiCN solution at 540 nm; 16114.5 (g/mol) is the molecular mass of the haemoglobin monomer (64458/4) [38]; 11.0 ($L \cdot mmol^{-1} \cdot cm^{-1}$) is the absorptivity of HiCN at 540 nm (Table 4.1); and l (cm) is the lightpath length, usually 1.000 cm.

The concentration of each batch of ICSH International HiCN Standard is determined in at least five laboratories nominated by the ICSH Board. The determination is performed at 20–25°C by means of a spectrophotometer, properly calibrated as to wavelength and absorbance scale. Plan parallel glass cuvettes are used with an inner wall-to-wall distance of 1.000 ± 0.0005 cm. The HiCN concentration is calculated with equation 16.4. Table 16.1 shows an example of the results of measuring a batch of ICSH International HiCN Standard in nine ICSH-appointed laboratories.

The stability of ICSH international HiCN standards was checked by the ICSH-appointed reference laboratories at 3-month intervals before issue of the standards, then at 6-month intervals until the assigned expiration date. The long-term stability has been monitored in selected ICSH reference laboratories [8, 9, 10, 15, 210]. Table 16.2 summarises the results of monitoring the stability of the HiCN standards prepared by the RIV and stored in one of the reference laboratories at 4–6°C. Although the standards at first were prepared yearly and issued for a period of two years, stability

Table 16.1.
Results from ICSH-appointed laboratories obtained for ICSH International HiCN Standard No. 20600

Laboratory	$A(750)$	$A(540)$	$A(504)$	R
Atlanta	0.000	0.382	0.2365	1.62
Bilthoven	0.0005	0.3815	0.237	1.61
Cleveland	0.0005	0.382	0.2365	1.62
Groningen (VK)	0.0005	0.383	0.238	1.61
Groningen (Z)	0.001	0.3805	0.2365	1.61
Kumamoto	0.0005	0.384	0.2395	1.60
London (Ph)	0.0005	0.382	0.2375	1.61
London (W)	0.0005	0.3825	0.2385	1.60
Rome	0.0005	0.382	0.237	1.61

$R = A(540)/A(504)$. With Bilthoven (RIVM) excluded, the mean value of $A(540)$ was 0.382 ± 0.001 (*SD*), corresponding to $c_{HiCN} = 559.6$ mg/L (equation 16.4).

data indicated that the standards could be prepared at 3-year intervals and could be issued for a 6-year period.

To monitor the stability of the standards, spectrophotometric measurements were made with a Beckman DU, spectral bandwidth 2.0, 0.52 and 0.42 nm at 750, 540 and 504 nm respectively, and/or an Optica CF4, spectral bandwidth 0.5, 0.13 and 0.13 nm at 750, 540 and 504 nm respectively, and/or a HP8450A, spectral band width 2 nm at all wavelengths of measurement, and/or a Varian Cary 219, spectral band width 1.0 nm at 750 nm and 0.5 nm at 540 and 504 nm. All measurements were made using certified or calibrated 1.000-cm glass cuvettes against water as blank, and at room temperature (20–25°C). The wavelength scale was checked regularly by means of the mercury or deuterium emission lines and the absorbance scale by means of a certified Corning HT glass filter or Scott neutral density filters (p. 23).

Each value in Table 16.2 is the mean of two or more measurements by one or two operators measuring a single vial on one to three instruments. Occasionally a second vial was measured for verification. The measured absorbances at 750 nm have not been included because, as long as $A(540)/A(504)$ remained $\geqslant 1.59$, $A(750)$ was consistently $\leqslant 0.003$. As $A(540)/A(504)$ was found to decrease, $A(750)$ increased. Ongoing stability studies have shown that properly prepared, sterile HiCN solutions stored at 4–6°C are stable, *i.e.* the decrease in $A(540)$ is $\leqslant 2\%$, for a period of 15 years.

A recent development in the manufacture of HiCN standards has been the use of *bovine* erythrocytes as source material (Chapter 9). Accelerated thermal degradation studies [215, 216] have been performed on a batch of bovine HiCN standard prepared by Diagnostic Reagents and one prepared by Euro-Trol, with ICSH HiCN standard batch 70 600 as control. Vials were stored at 20, 30 or 37, and 45 or 50°C and measured weekly for a period

Table 16.2.
Long-term stability of ICSH International HiCN Standards stored at 4 °C

Batch No.	40400		60400		70400		80400	
Year	1964		1966		1967		1968	
Age (yr)	*A*(540)	*R*	*A*(540)	*R*	*A*(540)	*R*	*A*(540)	*R*
0	0.391	1.61	0.386	1.61	0.405	1.62	0.383	1.60
0.25			0.387	1.61			0.381	1.60
0.5			0.387	1.61	0.404	1.62	0.386	1.61.5
0.75			0.387	1.61			0.383	1.61
1	0.389	1.61	0.388	1.61	0.405	1.595		
2–2.5	0.391	1.61	0.387	1.61	0.404	1.62	0.386	1.61
3–3.5	0.391	1.62	0.388	1.61	0.404	1.61	0.386	1.61
5–5.5	0.393	1.61	0.386	1.61	0.404	1.62	0.384	1.61
9–10	0.390	1.60	0.386	1.61	0.404	1.61	0.381	1.60
13–15			0.381	1.60	0.405	1.585	0.3865	1.605
16–17	0.3875	1.59	0.380	1.60	0.401	1.595	0.382	1.59
18–19	0.3915	1.58	0.3805	1.595	0.3995	1.60	0.3805	1.605
20–21	0.382	1.565					0.3795	1.595
22–25	0.380*	1.57	0.380	1.58	0.3985	1.59	0.378	1.58
26–27			0.372	1.62	0.4015	1.60	0.378	1.595

Batch No.	90400		00400		10400		20400	
year	1969		1970		1971		1972	
Age (yr)	*A*(540)	*R*	*A*(540)	*R*	*A*(540)	*R*	*A*(540)	*R*
0	0.389	1.61	0.395	1.61	0.413	1.61	0.408	1.61
0.25	0.389	1.61	0.395	1.61	0.414	1.60	0.408	1.61
0.5	0.388	1.62	0.396	1.61	0.415	1.61	0.408	1.61
0.75								
1	0.388	1.61	0.400	1.62			0.408	1.62
2–2.5	0.390	1.61	0.393	1.60	0.413	1.61	0.404	1.61
3–3.5	0.387	1.61	0.393	1.61	0.412	1.61	0.407	1.60
5–5.5	0.387	1.60	0.393	1.60	0.411	1.61	0.402*	1.60
9–10			0.389	1.59	0.405	1.61		
13–15	0.384	1.59	0.391	1.58	0.405	1.605		
16–17	0.389	1.59	0.386	1.585				
18–19					0.403	1.595		
20–21	0.383*	1.60	0.386	1.57				
22–25			0.384*	1.56	0.3975	1.535		

Batch No.	30400		40500		50500		60500	
Year	1973		1974		1975		1976	
Age (yr)	*A*(540)	*R*	*A*(540)	*R*	*A*(540)	*R*	*A*(540)	*R*
0	0.403	1.61	0.3825	1.60	0.379	1.61	0.3925	1.60
0.25	0.402	1.61	0.380	1.61	0.381	1.61	0.394	1.61
0.5	0.403	1.62	0.381	1.60	0.3815	1.61	0.392	1.61
1	0.402	1.61	0.378	1.61	0.3795	1.60	0.3935	1.60
2–2.5	0.401	1.61			0.380	1.60	0.391	1.61

Table 16.2.
(Continued)

Batch No.	30400		40500		50500		60500	
Year	1973		1974		1975		1976	
Age (yr)	*A*(540)	*R*	*A*(540)	*R*	*A*(540)	*R*	*A*(540)	*R*
3–4	0.4035	1.605					0.391	1.62
5–6					0.3795	1.61	0.3875	1.605
8–11			0.376	1.61	0.3775	1.61	0.393	1.61
11.5–13			0.3775	1.61	0.379	1.605		
13–15	0.400	1.605			0.376	1.595	0.393	1.58
16–19	0.3965	1.58	0.3765	1.595	0.3805	1.595	0.3895	1.59
20–21	0.3935	1.535						

Batch No.	70500		80500		90500		20600	
Year	1977		1978		1979		1982	
Age (yr)	*A*(540)	*R*	*A*(540)	*R*	*A*(540)	*R*	*A*(540)	*R*
0	0.3945	1.60	0.396	1.60	0.3925	1.61	0.3825	1.60
0.25	0.396	1.61	0.393	1.61	0.392	1.61	0.3805	1.61
0.5	0.395	1.61	0.394	1.60	0.3935	1.61		
0.75					0.3925	1.61		
1	0.395	1.61	0.3935	1.61	0.3925	1.60	0.3815	1.59
2–2.5	0.392	1.61	0.393	1.61	0.3915	1.60	0.378	1.60
3–4			0.3925	1.605	0.3905	1.60	0.3795	1.59
5–6	0.3915	1.61	0.3935	1.60	0.3865	1.59	0.377	1.60
8–11	0.3925	1.61	0.3955	1.605	0.386	1.595	0.376	1.58
11–13	0.392	1.595	0.3905	1.60			0.3745	1.595
13–15			0.3955	1.59	0.384	1.59		
16–19	0.3965	1.605			0.377	1.56	0.372	1.58

Batch No.	40600		70600	
Year	1984		1987	
Age (yr)	*A*(540)	*R*	*A*(540)	*R*
0	0.3995	1.60	0.404	1.61
0.25			0.404	1.61
0.5	0.402	1.605	0.400	1.595
0.75	0.402	1.605		
1	0.401	1.61	0.403	1.61
2–2.5	0.3965	1.605	0.402	1.61
3–4	0.398	1.60	0.402	1.60
5–6	0.399	1.595	0.4015	1.60
8–11	0.3975	1.585	0.400	1.61
11–13				
13–15	0.391	1.58		

* Last ampoule of batch.

of 50, respectively 12 weeks. The absorbance at 540 nm of the samples at the given temperature, relative to that of the samples stored at 4°C, was determined. This yields the reaction constant at various temperatures for the decomposition, which is assumed to be a first order reaction. The Arrhenius equation is then used to extrapolate the relation between reaction constant and temperature to a storage temperature of 4°C. The results indicate that HiCN standards prepared from bovine blood are at least as stable over time as the HiCN standards traditionally prepared from human blood.

ROUTINE HAEMOGLOBINOMETRY

Although the HiCN method has been universally accepted as the most accurate method for the determination of ct_{Hb}, and although all newer methods of determination are either based on and/or calibrated to yield results commensurate with this method, some of the older methods remain firmly entrenched for routine laboratory determination and screening for anaemia. Physician's offices and small clinics, for example, continue to use the AO Hb-Meter (American Optical) to screen patients for anaemia. A drop of skin puncture blood is allowed to flow into a small glass chamber and is haemolysed by gentle agitation with a 'hemolysis applicator'. The chamber is covered with a cover plate and inserted into the instrument. The haemoglobin concentration is determined by visual comparison of the haemoglobin (mainly O_2Hb) solution with a green split field; a slide is moved until the two halves of the field are equally light and ct_{Hb} is read from a scale.

New routine methods which have largely superseded the manual HiCN method fall into three main categories: (1) automated systems which simultaneously determine other haematological quantities; (2) simple haemoglobinometers for single measurements for office and point-of-care testing; (3) multiwavelength haemoglobin photometers, which also measure oxygen saturation and dyshaemoglobins (Chapter 17).

A majority of haemoglobin measurements in the clinical laboratory are presently carried out by **haematology analysers**, which are designed for the rapid processing of large series of blood samples. These measuring systems either use a modified HiCN method, or convert all haemoglobin in the sample into other haemoglobin species, or into haem derivatives through alkaline denaturation of the globin moiety [250]. The ensuing products are then measured by photometry. An additional motive for changing to other haemoglobin species is avoiding the use of toxic reagents and generating cyanide-containing waste.

Comparability of haematology analyser ct_{Hb} results to HiCN method results remains assured through the analyser calibration process, because the calibrators used generally have assigned ct_{Hb} values obtained with the HiCN reference method [192]. Analyser precision is reflected by results from proficiency assessment programs where, for haemoglobin determination in the reference range a coefficient of variation $\leqslant 1.5\%$ is consistently found.

Over the years, several photoelectric haemoglobinometers have been recommended for the determination of ct_{Hb} on the basis of O_2Hb or HiCN [209, 240]. None, however, has found wide acceptance until, in the mid-1980s a new instrument developed by Leo Diagnostics, the **Hemocue B-hemoglobin photometer**, became available. The Hemocue makes use of disposable cuvettes of small volume (10 μL) and a short lightpath (0.13 mm). Dry reagents, sodium desoxycholate for haemolysis, sodium nitrite to oxidise all haemoglobin derivatives to Hi [12], and sodium azide to convert Hi to HiN_3 (Figs 2.1 and 2.2) [11, 396] are deposited on the inner walls of the cuvette. The blood sample — capillary blood from a skin puncture or blood from venipuncture — is drawn into the cuvette by capillary action and spontaneously mixes with the reagents. Once filled, the cuvette should be kept in a horizontal position and inserted, within 10 min, into the Hemocue photometer, which measures at 565 and 880 nm. The instrument displays ct_{Hb} as digital readout within 45 s.

The factory calibration of the instrument is checked daily with a dummy cuvette containing an optical filter ('red control cuvette') certified by the manufacturer to be equivalent to a specified haemoglobin concentration. During an evaluation of 100 instruments and 5 batches of cuvettes, a systematic error of about -3.5% was found in comparison with the HiCN method [229]. By introducing an improved calibration procedure, this error was reduced to about -1%. This improved factory calibration procedure, which is a good example of coupling a routine method to the ICSH international reference method, consists of the following steps [229]. (1) Accurate determination by means of the HiCN reference method of ct_{Hb} of a fresh, properly anticoagulated human blood specimen from a healthy non-smoking donor. (2) Calibration of a single haemoglobinometer by measuring ct_{Hb} of the same specimen with this instrument ('key instrument'); in case of the Hemocue this means that the ct_{Hb} value represented by the dummy cuvette is established. (3) Calibration of serial instruments with the aid of the key instrument. (4) Checking each batch of cuvettes as to lightpath length and reagent filling by measuring ct_{Hb} on the key instrument.

The Hemocue method has been shown to be sufficiently accurate for all clinical and experimental applications [41, 339, 445] and to be only marginally sensitive to the presence of COHb and low oxygen saturation [451].

A variant of the Hemocue method, the **Hemosite**, has been developed by GDS Diagnostics. It allows the use of a first drop of blood from a skin puncture and is based on dry chemistry blood separation technology. The sample is applied to the well of a test card consisting of multilayered filtration membranes and reagent membranes. Colour development of the erythrocyte–sodium nitrite–sodium azide mixture takes place in the final layer and is measured with a small reflectance photometer. The precision, expressed as coefficient of variation, has been reported by the manufacturer to be 3.5%. Comparison of 372 specimens with results obtained with a Coulter haematology analyser reportedly gave a mean difference of −0.2%. A published evaluation of the method has not yet come to our attention.

The use of **cyanide** in haemoglobinometry has recently become a matter of concern, not so much as to the reference method but increasingly as to the large volumes of waste produced by the routine measurement of ct_{Hb}. Generally, there are three concerns regarding use and disposal of CN^- containing laboratory reagents: environmental impact, laboratory employee safety, and financial burden of adhering to imposed regulations promulgated by central or local authorities. Governmental regulations, *e.g.* Belgium, Japan, USA, are usually directed toward the environmental impact of toxic chemicals that may leach into ground water when disposed of in landfills. Laboratory worker safety is an issue of protection of employees from accidental exposure through ingestion of cyanide or to hydrogen cyanide gas liberated by degradation of bulk cyanide reagents or through inadvertent admixture of reagent with acids. However, it requires at least 4 L of the conventional CN^- containing HiCN reagent solution to ingest a lethal dose of cyanide.

Cyanide waste released from typical automated haematology analysers ranges in concentration from 6.8 to 65.8 mg/L [46]. Disposal of such cyanide waste is theoretically simple: alkaline solutions of sodium or calcium hypochloride (NaOCl, CaOCl) will oxidise cyanide to cyanate; alternatively, the laboratory may use a commercial waste handling and disposal company. The latter, especially, may prove prohibitively expensive. Thus, manufacturers of haematology analysers have investigated the use of non-cyanide-containing reagents for the determination of ct_{Hb}.

Oshiro *et al.* [300] were among the first to describe the use of an anionic detergent, **sodium laurylsulphate** (sodium dodecylsulphate) for the determination of ct_{Hb}, naming the end product measured 'SLS-hemoglobin'. The reaction mechanism of this method has been studied by Matsubara and Minura [252] using a reagent consisting of 0.7 g sodium laurylsulphate (SLS) and 1.0 mL Triton X-100 in 1000 mL 0.033 mol/L phosphate buffer ($pH = 7.35 \pm 0.15$). They hypothesise that the quaternary structure of the haemoglobin tetramer is disrupted by the hydrophobic group of SLS. This allows molecular oxygen to oxidise the haemoglobin iron. The hydrophilic group of SLS then combines with the oxidised haemoglobin iron to form SLS-haemoglobin. The non-ionic detergent Triton X-100 is included in the reagent to protect SLS from precipitating at low temperature and to promote rapid and complete lysis of the red blood cells.

It would be more correct to characterise the end product as ***haemiglobin-laurylsulphate*** (HiLS) for the following reasons. The absorption spectrum of HiLS shows a characteristic peak at 534–536 nm with a shoulder near

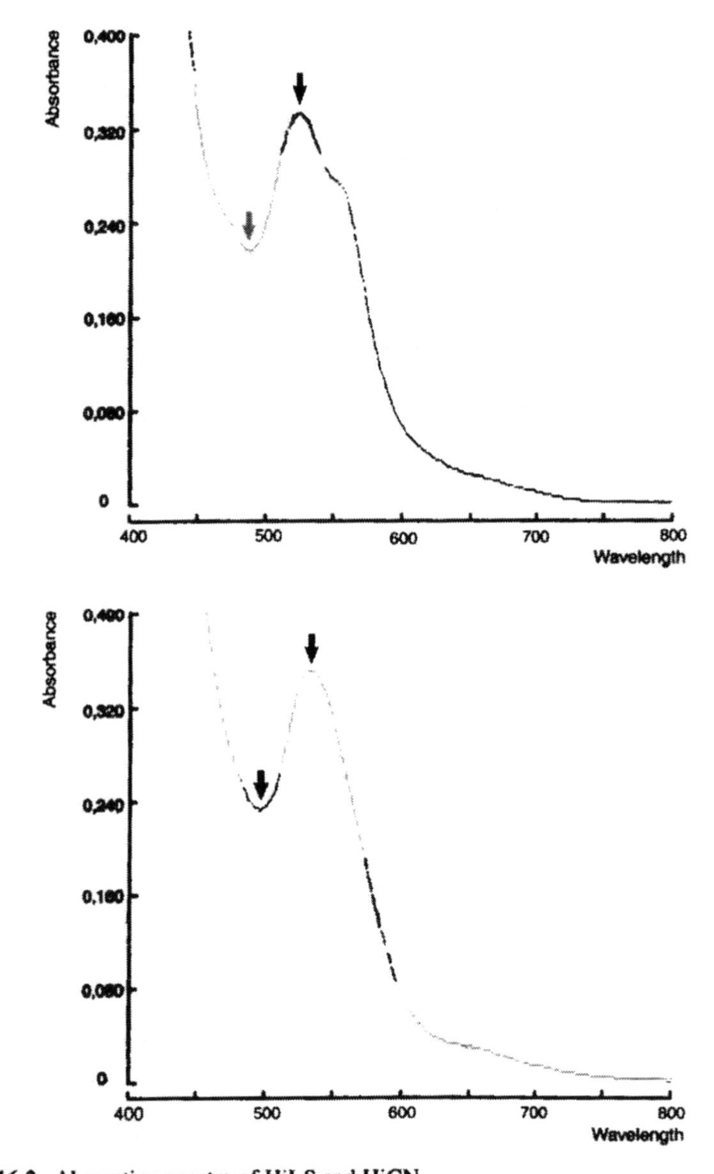

Figure 16.2. Absorption spectra of HiLS and HiCN.

Upper panel: Spectrum of a 250-fold dilution of whole blood with SLS reagent. Maximum at 534 nm, minimum at 502 nm; lower panel: spectrum of the HiLS solution after adding excess KCN. Maximum at 542 nm, minimum at 506 nm.

570 nm, and a minimum at 502 nm (Fig. 16.2, upper panel). On addition of CN^- to an HiLS solution (250-fold dilution of whole blood sample), the absorption maximum shifts to 540–542 nm and the spectrum becomes almost identical with that of HiCN [470]. The addition of excess SLS reagent to an HiCN solution (250-fold dilution of whole blood sample) resulted in a shift

of the absorption maximum from 540 nm to 536 nm with a spectrum in between that of pure HiCN and pure HiLS [470]. Hence, it appears that LS^- can partly displace CN^- from HiCN, and excess CN^- can displace LS^- from HiLS. The displacement series shown (p. 12) thus can be extended to

$$OH^- < F^- < NO_2^- < N_3^- < LS^- < CN^-$$

and the term haemiglobinlaurylsulphate (HiLS) appears to be correct.

A comparison of the HiLS method with the reference HiCN method has shown that the HiLS method is equivalent with the HiCN method as to precision and accuracy [241]. The method has been adapted for use on Sysmex automated haematology analysers with a proprietary reagent 'sulfolyser'. A good correlation between the automated HiLS method and the HiCN reference method has been obtained by Fujiwara *et al.* [124] ($N = 60$; $r = 0.999$; $y = 1.001x - 0.033$), by Oshiro *et al.* [299] ($N = 120$; $r = 0.998$; $y = 1.025x - 0.332$), and by Saito *et al.* [334] ($N = 100$; $r = 0.998$; $y = 1.004x + 0.007$). An automated HiLS method has also been compared to a similar automated HiCN method with mouse, dog and rat blood samples [113]. Good correlation was found with dog and rat samples ($r = 0.99$), poor correlation with mouse samples ($r = 0.84$). Using dog, monkey, mouse and rat blood specimens, the automated HiLS method has been compared to a HiCN-based method with an Ortho ELT-8 haematology analyser [281]. For all the 60 animal specimens studied the correlation was 0.999 with $y = 1.069x - 0.375$.

Although determination of correlation coefficients in and of themselves is not adequate to evaluate one method against the other [31], it appears that the HiLS method, as applied in automated haematology analysers, provides results of ct_{Hb} in accordance with those obtained with the HiCN method.

CHAPTER 17

MULTICOMPONENT ANALYSIS (MCA) OF HAEMOGLOBIN DERIVATIVES

The difference between the absorption spectra of the various haemoglobin derivatives has, from the very beginning of haemoglobin spectrophotometry [178, 180, 467], been used for the identification of haemoglobin derivatives in solution and for the measurement of their concentration. The most important clinical application has always been the measurement of the oxygen saturation of the blood, *i.e.* the concentration of O_2Hb as a fraction of the sum of the concentrations of O_2Hb and HHb. This application is discussed extensively in Chapter 18, but there the emphasis is on measurement *in vivo*. In the present chapter the measurement of the oxygen saturation is treated on a par with the spectrophotometric methods for the determination of the concentrations or the fractions of the common haemoglobin derivatives in a mixture.

The principles underlying these methods have been given in Chapter 3. The main requirement to be fulfilled in all applications obviously is that Lambert–Beer's law is valid. This not only means that equation 3.3 can be applied, but also that all absorbing solutes act independently of each other as expressed in equation 3.6. That Lambert–Beer's law holds over an extensive range of concentrations for O_2Hb and HHb had already been shown by Drabkin and Austin [94, 97]. When, half a century later, this finding was challenged [52], we confirmed Drabkin's results for O_2Hb and HiCN [481].

THE LENGTHY ROAD TO MCA

As mentioned, the analysis of multicomponent systems presupposes the availability of accurate values for the absorptivity of all components at all wavelengths used in the measurement. The determination of reliable absorptivity values for the common haemoglobin derivatives [7, 211], however, was not immediately followed by straightforward multicomponent analysis of mixtures of haemoglobin derivatives. This is not surprising in light of the influence of the method of sample preparation and instrumentation on the absorptivity values measured (Chapter 15). Especially, the light scattering effect of plasma proteins in the haemolysed blood samples usually employed in the clinical application of the methods introduced errors and negated the use of absorptivity values determined in partly purified haemoglobin solutions. This only changed after effective means for removing most of the scattering material had been introduced (Chapters 6 and 15) and thus methods that could be calibrated under the conditions prevailing in the measurement remained in use for a considerable period.

As explained in Chapter 3, the fractions of two haemoglobin derivatives present in a *two-component system* can be determined by measuring at two wavelengths without knowledge of the absorptivities, provided that samples containing the pure haemoglobin derivatives can be prepared and measured under similar conditions (equations 3.7 and 3.8). The requirement that only two components are present can only be fulfilled to a certain extent for a blood specimen, but a fair approximation is possible.

With regard to measuring **oxygen saturation**, in most of the blood specimens taken from patients, the fractions of the common dyshaemoglobins are but slight [447]; moreover, wavelengths can be selected where the absorbance of COHb, Hi and SHb has little influence on the absorbance ratio, $A(\lambda_1)/A(\lambda_2)$. Fully saturated and desaturated samples can be obtained by tonometry as described in Chapter 6, but more simply by gently shaking a sample in a flask with oxygen or room air, and by addition of sodium dithionite, respectively. When using room air, it may be that the oxygen saturation stays a few tenths of a percent below 100; the possible influence of the added $Na_2S_2O_4$ on the absorbance at the wavelengths of measurement is not taken into account. A series of measurements of $A(\lambda_1)/A(\lambda_2)$ with samples containing 0 and 100% O_2Hb enables the constants a and b of equation 3.7 to be calculated.

Numerous two-wavelength methods have been described for measuring oxygen saturation, using various combinations of wavelengths and multiple types of cuvettes with different lightpath lengths [7, 116, 128, 132, 166, 202, 210, 211, 219, 272, 284, 285, 318, 458]. In most of the methods an isosbestic wavelength was used, but a few used two wavelengths at which there is a significant difference in absorbance between O_2Hb and HHb [219, 272]. These methods were quite accurate and rather easy to carry out with a conventional general purpose spectrophotometer such as, for example, the Beckman DU and the Optica CF4.

Generally, wavelengths > 600 nm were preferred because the lower absorptivity in this region allowed the use of cuvettes with 0.1–0.2 cm lightpath length. These are easier to handle than the thin layer cuvettes necessary for measurements at $\lambda < 600$ nm. Using the 500–600 nm region continued to be advocated because of the lower non-haemoglobin absorption in this region and the elusive character of the isosbestic wavelength near 800 nm [211]. However, the shift of the 800 nm isosbestic wavelength was shown to be caused by the use of excess sodium dithionite for deoxygenation (p. 43), and Mook *et al.* [271] demonstrated that the sensitivity of the measurements, expressed as dA/dS_{O_2}, *i.e.* the change in absorbance per unit change in oxygen saturation, is much higher when measuring between 600 and 730 nm than in any other region of the visible spectrum. An additional argument to prefer the red/near infrared region is that in this region a pair of wavelengths can be chosen in such a manner that the influence of the possible presence of COHb and Hi in the blood sample on the measured value of S_{O_2} is slight.

In the determination of the **fraction of COHb** the presence of other dyshaemoglobins has usually been neglected. A two-component system can be attained either by complete deoxygenation, the COHb/HHb system, or by full oxygenation, the COHb/O_2Hb system [207, 210, 211]. Complete deoxygenation can be achieved by the addition of $Na_2S_2O_4$, the possible influence of which on the absorbance measured may be disregarded because it is added to all samples including those used for calibration. When the COHb/HHb system is used, samples containing 100% HHb and 100% COHb are necessary for calibration. Freshly drawn blood from a non-smoking donor may, after addition of $Na_2S_2O_4$, be considered to contain only HHb. Blood with 100% COHb can be obtained by equilibrating a sample of the same blood with a gas mixture containing a high fraction of CO. A disadvantage of methods in which $Na_2S_2O_4$-containing blood is brought into contact with CO gas, is that some COSHb may be formed [207, 211].

In methods for measuring the COHb fraction of blood in the COHb/O_2Hb system all haemoglobin in the samples must be saturated with oxygen. This must be achieved by tonometry with room air, with the risk that some CO dissociates from the COHb present in the blood. However, it has been demonstrated that, even when oxygen-enriched air is used, full saturation of all HHb in the solution can be attained with little or no conversion of COHb into O_2Hb [210, 211]. Samples containing 100% O_2Hb for calibration may be obtained by tonometry with oxygen of freshly drawn blood from a non-smoking donor, 100% COHb samples by tonometry with a CO-containing gas mixture. Measuring $A(\lambda_1)/A(\lambda_2)$ for samples with 0 and 100% COHb then yields the values of a and b of equation 3.7.

Many two-wavelength methods have been described for measuring carboxyhaemoglobin, using various combinations of wavelengths and different procedures to arrive at a two-component system [7, 29, 102, 206, 207, 210, 211, 326, 458, 479]. Some authors prefer to use the Soret band for the determination of COHb. Because of the high absorptivity in this region ($\varepsilon_{COHb}(420) = 192.0\ L \cdot mmol^{-1} \cdot cm^{-1}$) [7], the blood sample has to be diluted by about a factor 1000. Rodkey *et al.* [326] and Beutler and West [29] use the COHb/HHb system and measure at the absorption maxima of COHb and HHb, 420 and 432 nm, respectively. As no isosbestic point of the two components is used, an equation of the type of equation 3.8 is used in the calculation.

The **fraction of Hi** can be determined reliably with the KCN addition method, originally devised by Evelyn and Malloy [114] and slightly modified by Dijkhuizen *et al.* [103], but other spectrophotometric methods have also been described [7, 151, 211, 283, 326, 458]. The KCN addition procedure has the advantage of being an absolute method in the sense that it is not dependent on constants that must be determined using samples of known concentration. For that reason it can be used as a reference method [443, 444].

In the **KCN addition method**, the absorbance difference at 630 nm is measured between the Hi present in the sample and the HiCN formed from it; this difference is divided by the absorbance difference at 630 nm occurring when all haemoglobin present is converted from Hi to HiCN. Tables 8.1 and 8.2, and Fig. 8.1 show that the absorptivity of Hi at 630 nm is considerably higher than that of HiCN; hence, the absorbance decreases when Hi is converted to HiCN. The blood specimen is mixed with four different reagent solutions; the dilution factor is 51. Solution 1 contains a phosphate buffer (Na_2HPO_4 27.50 mmol/L; KH_2PO_4 13.16 mmol/L; pH = 7.4) and 0.5 mL of Sterox SE per L (or an other non-ionic detergent; p. 202). Solutions 2 and 3 contain, in addition, 3.84 mmol/L KCN and 3.04 mmol/L $K_3Fe(CN)_6$, respectively. Solution 4 contains, in addition, both 3.84 mmol/L KCN and 3.04 mmol/L $K_3Fe(CN)_6$. The Hi fraction is calculated with the equation:

$$F_{Hi} = (A_1 - A_2)/(A_3 - A_4), \qquad (17.1)$$

where A_1, A_2, A_3, and A_4 represent the absorbances of the four blood-reagent mixtures at 630 nm. The time interval between dilution of the blood specimen and measurement of the samples should be 10 min. The method can be carried out with any general purpose spectrophotometer.

Two-component analysis thus proved to be reasonably accurate and not too difficult to perform. *Systems with more components*, however, remained difficult to analyse. Yet, methods for the simultaneous determination of more than two haemoglobin derivatives in human blood appeared to be a useful objective and almost from the beginning of photoelectric spectrophotometry attempts have been made to devise suitable methods for the analysis of systems containing more than two haemoglobin derivatives [114, 175].

For a long time the approach to analysing multicomponent systems was to divide the procedure into separate steps, each comprising the measurement of not more than two components. Sometimes it is possible to determine in one of the steps the concentration of one of the components by measuring the change in absorbance at a certain wavelength resulting from the addition of an agent that only affects that component, as in the KCN addition method for measuring Hi. By using an isosbestic wavelength two components can first be treated as a single one. In the next step, in which other wavelengths are used, the two components are then differentiated from each other. The addition of chemicals between the measurements makes the procedure more complicated. A more straightforward single step procedure to analyse a three-component system through measurement at four wavelengths [494], however, proved to be beset by cumbersome calculations.

Most methods concerning more than two components are limited to the analysis of *three-component systems*, either HHb, O_2Hb and COHb [30, 156] or O_2Hb, COHb and Hi [363, 494]. A few procedures have been developed for analysing systems of *four components* [329], and even *five components* [291, 357]. Rossi-Bernardi *et al.* [329] used a set-up including a p_{O_2}-electrode and a photometer for absorption measurement in wavebands around 497, 565 and 620 nm (interference filters). A 10-μL blood sample was injected into a closed cuvette containing a buffer solution (100-fold dilution). The fall in p_{O_2} occurring in the solution, measured by the electrode, was assumed to be proportional to the amount of HHb present. The HHb becomes oxygenated and the number of components in the solution thus is reduced to three. The resulting three-component system (O_2Hb, COHb, Hi) was then analysed by photometry at three wavelengths, assuming that 497 nm is isosbestic for O_2Hb and COHb and that $\varepsilon_{COHb}(620) \approx \varepsilon_{O_2Hb}(620)$.

The method developed by Siggaard-Andersen *et al.* [291, 357] for analysing the system HHb, O_2Hb, COHb, Hi, and SHb is a purely spectrophotometric procedure that can be executed by means of a general purpose spectrophotometer, only needing a special thermostatted microcuvette with a lightpath length of 0.01 cm. Haemolysis is effected by freezing and thawing. By addition of dithionite O_2Hb and Hi are reduced to HHb, leaving the system HHb, COHb and SHb, which is analysed by measurements at 620, 600 and 470 nm. By addition of CO and CN^- a three-component system of COHb, HiCN and COSHb is formed that is analysed by measuring at 620, 570 and 470 nm. The rather tedious calculations are facilitated by the use of nomograms. A problem is the estimation of the fraction of O_2Hb in the unmodified specimen, but this may be done by using a simple two-wavelength method and correcting the measured absorbances on the basis of the measured fractions of the dyshaemoglobins in the specimen.

SELECTED TWO-WAVELENGTH METHODS FOR O_2Hb AND COHb

Although most determinations of the oxygen saturation and the COHb fraction of the blood are now performed with (semi-)automated analysers, mainly multiwavelength haemoglobin photometers, it is worthwhile to describe in more detail a robust, thoroughly tested two-wavelength method for the determination of each of these quantities that can be carried out on any conventional, general purpose spectrophotometer.

The best choice for measuring **oxygen saturation**, S_{O_2}, is the wavelength pair 680/800 nm [210]. The absorbance can then be measured with a lightpath length of 0.20 cm. Heparinised venous or arterial blood is collected in a glass syringe; for a determination of S_{O_2} 2 mL of blood is transferred anaerobically to a 2-mL syringe, which contains a mixing ring and the dead space of which is filled with a 10% solution of a non-ionic detergent. After thoroughly mixing and discarding the first three drops, the haemolysate is injected into a 0.20-cm cuvette with the help of a blunt needle. The absorbance is measured at 680 and 800 nm, using as a blank a similar cuvette filled with water. The oxygen saturation is calculated with an equation of the type of equation 3.7.

The constants a and b have been determined by measuring the ratio $A(680)/A(800)$ for a series of samples of completely oxygenated and completely deoxygenated blood. This yielded mean values of 0.4276 and 3.0461 for $S_{O_2} = 1$ and 0, respectively, from which the constants of equation 17.2 were calculated:

$$S_{O_2} = -0.3819 \times [A(680)/A(800)] + 1.1633. \qquad (17.2)$$

If one calculates the values of a and b from the absorptivities given in Table 8.4, the following equation is obtained:

$$S_{O_2} = -0.3943 \times [A(680)/A(800)] + 1.1676. \qquad (17.3)$$

Substitution of the $A(680)/A(800)$ values corresponding with $S_{O_2} = 1$ and 0, then gives $S_{O_2} = 0.999$ and -0.034 (99.9 and -3.4%). The cause of the slight difference obviously is that the absorptivities of Table 8.4, based on haemoglobin solutions prepared with the standard procedure that yields quite clear solutions, are not identical with the corresponding quantities pertaining to haemolysed fresh blood.

An advantage of the method is that the measured oxygen saturation is relatively insensitive to the presence of COHb and Hi, especially in the range of arterial S_{O_2}-values occurring in the majority of patients (Table 17.1).

The absorption maximum of *indocyanine green*, a dye used in clinical tests, for example for measuring cardiac output by dye dilution, coincides with the isosbestic point of O_2Hb and HHb at 800 nm [237]. This coincidence was actually the reason for the introduction of indocyanine green by Fox *et al.* [119], because it allows the recording of dye dilution curves without

Table 17.1.
Errors in the determination of S_{O_2} measured at 680 and 800 nm caused by the presence of COHb and Hi

S_{O_2} (%)	**Measured values of S_{O_2} (%)**			
	$F_{COHb} = 5\%$	$F_{COHb} = 10\%$	$F_{Hi} = 5\%$	$F_{Hi} = 10\%$
100	99.8	99.5	99.3	98.6
80	80.0	79.8	81.3	82.3
60	60.5	60.3	63.6	66.2
40	41.2	41.0	46.0	50.3

interference by variations in oxygen saturation (p. 255). The high absorbance of indocyanine green at 800 nm, however, renders this wavelength unfit for measuring S_{O_2} after indocyanine green has been injected. A longer wavelength, where light absorption by indocyanine green is absent or slight should be used instead. By using the wavelength pair 660/880 nm, S_{O_2} can be measured virtually independent of the presence of indocyanine green in the blood. As no isosbestic point of the two components is used, an equation of the type of equation 3.8 must be used in the calculation of S_{O_2}.

For measuring the **fraction of COHb**, F_{COHb}, in human blood, the system COHb/O_2Hb and the wavelength pair 562/540 nm is used, where 540 is an isosbestic point of the two components [210, 479]. A fresh, heparinised blood sample is first oxygenated by rotation in a small open, cylindrical tonometer flushed with oxygen before use. This is sufficient to saturate all HHb in the sample with oxygen, while the influence on COHb is insignificant for $F_{COHb} < 40\%$ [102].

1 mL of tonometered blood is transferred to a 1-mL syringe, which contains a mixing ring and the dead space of which is filled with a 10% solution of a non-ionic detergent. After thoroughly mixing and discarding the first three drops, the haemolysate is injected, with the help of a blunt needle, into a plan parallel glass cuvette with a lightpath length of 0.100 cm. A 0.095-cm glass plate is inserted into the cuvette, leaving a lightpath of 0.005 cm (Fig. 6.4). The absorbance is measured at 562 and 540 nm, using a similar cuvette filled with water as blank. F_{COHb} is calculated with an equation of the type of equation 3.7.

The constants a and b have been determined by measuring the ratio $A(562)/A(540)$ for 46 human blood specimens containing various fractions of COHb. The blood was obtained from 22 healthy donors and contained heparin as anticoagulant. Part of the blood was oxygenated, part of it was equilibrated with a CO-containing gas mixture, and samples with various F_{COHb} where prepared by mixing various volumes of O_2Hb and COHb containing blood. F_{COHb} was determined by the titrimetric method described in Chapter 19, and $A(562)/A(540)$ was plotted against F_{COHb} [102, 210, 479]. The linear regression line yielded absorbance ratios of 0.598 and 0.909 for $F_{COHb} = 0$ and 100%, respectively, from which the constants of equation 17.4 were calculated:

$$F_{COHb} = 3.215 \times [A(562)/A(540)] - 1.923. \qquad (17.4)$$

If one calculates the values of a and b from the absorptivities given in Table 8.1, the following equation is obtained:

$$F_{COHb} = 3.5137 \times [A(562)/A(540)] - 2.1518. \qquad (17.5)$$

Substitution of the $A(562)/A(540)$ values corresponding to $F_{COHb} = 0$ and 1, then gives $F_{COHb} = -0.051$ and 1.042 (−5.1 and 104.2%). The cause of the difference is that the absorptivities of Table 8.1, based on haemoglobin solutions prepared with the standard procedure, which yields quite clear solutions, are different from the corresponding quantities pertaining to haemolysed fresh blood.

MULTICOMPONENT ANALYSIS

Direct spectrophotometric **multicomponent analysis** is the determination of the concentration of each of several substances in a solution by measuring the absorbance at several wavelengths and calculating the concentrations on the basis of Lambert–Beer's law (equation 3.6), knowledge of the lightpath length, and of the absorptivity of all components at all wavelengths used. Application of this technique to the determination of haemoglobin derivatives in blood has long been impeded by two difficulties. First, the measurements have to be made on freshly drawn blood without air contact, lest a change in one of the most important quantities, the oxygen saturation, occurs. However, freshly drawn blood, after haemolysis, contains a considerable amount of scattering material that must be removed without causing any change in the fractions of the haemoglobin derivatives in the specimen. Second, until the 1970s, equipment for rapidly solving equations with more than three unknowns was not generally available. The latter problem has been solved completely by progress in computing techniques. The former, however, is still with us, although it has been solved to a considerable extent.

Spectrophotometric MCA of haemoglobin strictly complying with the definition given above would require the use of the 'true' absorptivities of the haemoglobin derivatives and perfect measurement of the absorbance of the blood samples as defined in Chapter 15. Neither of these requirements is fulfilled in the methods presently available. The five-wavelength method for measuring HHb, O_2Hb, COHb, Hi, and SHb developed by Zwart *et al.* [443] is a fair approximation as far as absorbance measurement is concerned. It is performed on an Optica CF4 grating spectrophotometer with a very small spectral band width, and error-free calibration of the wavelength and absorbance scales (p. 23). As to clearing the samples of scattering material, the introduction of cotton wool filtration (Fig. 6.3) was a significant improvement; for the first time it actually made MCA of haemoglobin possible. But perfect clearance was not achieved and, therefore, the determination of the absorptivities was adapted to the way in which the absorbance of the blood samples was measured (p. 191).

After the introduction of **diode array spectrophotometers** the determination of complete absorption spectra and the performance of MCA of haemoglobin became much easier, but at the price of an increase in spectral band width and some loss in the accuracy of wavelength calibration. Using absorptivity values from the standard spectra as given in Chapters 8–14, MCA of haemoglobin derivatives can be performed easily, with good precision and accuracy. However, using lyophilised bovine haemoglobin as test material, it has been shown that identical absorbances on different types of spectrophotometer can only be obtained after normalisation of the measured absorbances on the basis of measuring holmium oxide solutions [7, 210] along with the haemoglobin samples [249].

Thus it seems that the present methods for MCA of haemoglobin are not yet completely method- and instrument-independent. This means that a particular method works best when absorptivity values are used that have been determined under similar conditions as those that prevail in the measurement of the blood samples. This probably holds to a still higher degree for the special multiwavelength haemoglobin photometers used for routine MCA in clinical practice. Figure 17.1 shows that a computer simulation of some of these instruments in quite a few cases produces results different from those of the actual instruments, although these instruments have been shown to generally give quite accurate results. This means that the methods do not yet conform in every respect to the ideal MCA concept defined above, and that the non-ideal aspects of a measuring system are compensated for by system-specific measures based on empirical study of that system. This also means that, although MCA of haemoglobin has become accepted practice in clinical chemistry, there is still room for improvement.

In the method of Zwart *et al.* [443] for the simultaneous determination of HHb, O_2Hb, COHb, Hi, and SHb on a *conventional spectrophotometer* five wavelengths were used, chosen in such a manner that for each haemoglobin derivative a local maximum was included: 760 nm for HHb, 577 nm for O_2Hb, 569 nm for COHb, 500 nm for Hi, and 620 nm for SHb. The absorptivities of ε_{SHb} at 500, 569, 577, 620 nm were taken from ref. [103], the other absorptivities from ref. [7]. At 500, 577 and 569 nm the absorbance was measured with $l = 0.007$ cm (0.100-cm cuvettes with 0.093-cm glass inserts; Fig. 6.4); at 620 and 760 nm the measurements were made at $l = 0.200$ cm. Measurements were made on freshly drawn, heparinised blood. From the specimen, 2 mL was transferred without air contact to a 2-mL glass syringe containing a mixing ball and with the dead space filled with a solution of a non-ionic detergent. After mixing, a filter unit containing cotton wool was placed on the syringe and, after discarding the first 10–20 drops to flush needle and filter, the two cuvettes were filled; a glass plate was inserted into the 0.100-cm cuvette.

The absorbance at the five wavelengths was measured with an Optica CF4 with similar cuvettes filled with water in the reference channel. The wavelength calibration had been checked with mercury emission lines, the absorbance scale with a Corning HT yellow filter (p. 23). The concentration of the haemoglobin derivatives was calculated from the absorbance values by matrix calculation with the help of a HP9845A desk top calculator (Appendix ref. [443]). The total haemoglobin concentration was determined by adding the concentrations of the individual haemoglobin derivatives. Oxygen saturation and dyshaemoglobin fractions were calculated with equations as given in Chapter 2.

This five-wavelength method showed the feasibility of straightforward multicomponent analysis by means of a conventional spectrophotometer, but the procedure was still rather tedious. This changed when diode array spectrophotometers became available.

The HP8450A diode array spectrophotometer has the additional advantage of a built-in microprocessor with a 16K random access memory, capable of storing about 50 complete spectra from 200 to 800 nm; for permanent storage the spectra can be recorded on magnetic tape. The processor can perform mathematical operations on spectra, store calibration data for the calculation of concentration from absorbance, and perform multicomponent analysis of up to 12 components in a mixture. To this end, the spectra of the individual pure components are introduced into the work space of the processor and fitted on the measured spectrum of the unknown sample by a matrix calculation procedure for an overdetermined system (p. 20). The concentration of each component that corresponds to the best fit of the spectra is displayed on a built-in cathode ray tube. This type of instrument has been the general purpose spectrophotometer of choice for MCA of haemoglobin derivatives since 1984.

The procedure of Zwart *et al.* [444] for MCA of haemoglobin derivatives by means of a *diode array spectrophotometer* is analogous to the procedures described in Chapters 6 and 7 for the determination of the spectra of pure haemoglobin derivatives. Of a freshly drawn heparinised blood specimen, 2 mL is transferred to a 2-mL glass syringe containing a 5-mm-diameter glass mixing ball and a 100 mL/L solution of Sterox SE or an other non-ionic detergent in the syringe's dead space. Blood and reagent solution are thoroughly mixed and ejected through a cotton wool filter (Figs 6.3 and 6.4). After discarding the first 10–20 drops, which serve to flush filter space and needle, a 0.100-cm cuvette is filled with the haemolysate. A plan parallel glass plate is then inserted into the cuvette leaving a lightpath of 0.010 cm (Fig. 6.4). The cuvette is placed in the spectrophotometer and the absorbance spectrum is determined over the interval of 480–650 nm. The absorbance is divided by the lightpath length and the MCA program is used to calculate the concentrations of the individual haemoglobin derivatives.

The absorptivities of the haemoglobin derivatives originally used in the calculation were determined as mentioned in Chapter 15; they are given in the Appendix of ref. [447] and are in reasonable agreement with those of the standard spectra (Table 8.1). Oxygen saturation and dyshaemoglobin fractions are calculated with the equations given in Chapter 2; ct_{Hb} is equal to the sum of the concentrations of the haemoglobin derivatives present in the sample.

The method has been evaluated by measuring series of blood samples containing known fractions of the various haemoglobin derivatives [444]. Samples with different ratios of HHb and O_2Hb were prepared by tonometry. To measure S_{O_2} in these samples, a Radiometer OSM2 hemoximeter [355] was used. The instrument was recalibrated for the blood of each donor and the method had previously been checked by comparison with the 680/800 nm method described on page 219. In a series of 115 comparative determinations the mean difference between MCA and OSM2 ($X_{MCA} - X_{OSM2}$) was $-0.32\%S_{O_2} \pm 1.21\%S_{O_2}$ (SD).

Blood samples containing different concentrations of COHb were prepared by tonometry for a few minutes in a revolving glass tonometer (Fig. 6.1) with a humidified gas mixture of 5% CO and 95% N_2. This procedure yielded blood samples containing HHb, O_2Hb and COHb. To a few samples some Hi was also added. The COHb concentration was determined with the chemical (titrimetric) method described in Chapter 19. In a series of 42 comparative determinations the mean difference between MCA and chemical analysis ($X_{MCA} - X_{Chem}$) was 0.53% ± 1.31% COHb (SD).

To prepare blood containing various Hi concentrations, one drop of undiluted Sterox SE was added to the blood specimen for haemolysis. The specimen was distributed over several test tubes containing different volumes of a 90 mmol/L $K_3Fe(CN)_6$ solution and kept at room temperature for 90 min. The blood samples then contained various concentrations of HHb, O_2Hb and Hi. F_{Hi} was measured with the KCN addition method and calculated with equation 17.1. In a series of 56 comparative determinations, the mean difference between MCA and KCN addition method ($X_{MCA} - X_{KCN}$) was 0.62% ± 1.00% Hi (SD).

Blood samples containing SHb were prepared by a procedure analogous to the one described in Chapter 6. This yielded samples containing SHb and O_2Hb. F_{SHb} was determined by measuring ct_{Hb} and $A(620)$ of the sample, correcting for $A_{O_2Hb}(620)$, and calculating the SHb concentration assuming $\varepsilon_{SHb}(620) = 20.8\ \mathrm{L \cdot mmol^{-1} \cdot cm^{-1}}$. Some of the samples were then subjected to tonometry for a few minutes with a humidified gas mixture of 94.4% N_2 and 5.6% CO_2 to obtain samples containing SHb, HHb and O_2Hb. In a series of 42 comparative determinations the mean difference between MCA and the 620 nm method ($X_{MCA} - X_{620}$) was $-0.09\% \pm 0.30\%$ SHb (SD).

Most of the samples used in the comparative measurements of S_{O_2} contained only two haemoglobin derivatives: HHb and O_2Hb. To a few samples some COHb or Hi, or both had been added. In these cases the OSM2 was recalibrated with the blood containing COHb and/or Hi. The agreement between the two methods was not impaired by the presence of COHb and Hi.

The COHb fractions in the samples used in the comparative measurements were mostly determined in the presence of two other haemoglobin derivatives, HHb and O_2Hb, but occasionally a considerable amount of Hi had been added. The titrimetric method is insensitive to any other haemoglobin derivative. The comparative measurements showed that MCA yielded correct values for F_{COHb} irrespective of the presence of other haemoglobin derivatives.

The Hi fraction in the samples used in the comparative measurements was kept $< 30\%$, because the KCN addition method tends to overestimate F_{Hi} for values $> 40\%$. For a few samples with $F_{Hi} > 40\%$ the value determined by MCA was checked by converting all non-Hi in the samples to COHb and measuring F_{COHb} by means of the chemical method. F_{Hi} calculated as $1 - F_{COHb}$ was in good agreement with F_{Hi} measured by MCA.

For **routine application** of the method in the clinical laboratory, the procedure was slightly modified [447]. An HP8451A diode array spectrophotometer, equipped with an HP98155 keyboard and an HP89057-60100 keyboard module was used, and the workspace was extended with an HP82903A 16K memory module. The results were printed on the internal thermal printer. Program and spectra of the pure haemoglobin derivatives were stored on a floppy disc by using an HP9121 flexible disc drive unit and an HP82937A-IB interface. The spectra (Appendix ref. [447]) were introduced using the keyboard. An HP89052A peristaltic pump was connected to the spectrophotometer through an HP89053A sipper/sampler interface module. The absorbance was measured in a Hellma 170.032.01 flow-through cuvette with a lightpath length of 0.008 cm. The supply tube of the pump was immersed in a bottle containing isotonic saline solution, while the outlet tube from the cuvette, ended in a waste container.

Isotonic saline solution was used to clean as well as to blank the system between two consecutive measurements. For a duplicate determination, 5 mL of venous or arterial blood was collected in glass syringes containing heparin to prevent coagulation. For each measurement, 2 mL of blood was transferred without air contact to a 2-mL glass syringe containing a 5-mm-diameter glass mixing ball and a 100 mL/L solution of the non-ionic detergent Nonidet P40 in the syringe's dead space. Blood and reagent solution were thoroughly mixed and the syringe was connected to the peristaltic pump.

This adaptation of the original procedure for routine use proved to fulfil the requirement that the method be simple enough to be executed by inexperienced personnel and rapid enough to process a considerable number of specimens. The modifications described also had some analytical advantages. An 0.008-cm flow-through cuvette that remains in its place in the spectrophotometer is superior to a 0.10-cm cuvette with glass inserts, which have to be manipulated for every measurement. Also, the automatic check of the blank spectrum after the cuvette has been rinsed, immediately before the next measurement, may have contributed to the precision of the measurements.

It proved worthwhile to make a systematic use of the performance checks provided by the microprocessor, especially the relative fit error, which gives an indication of how well the algorithm was able to fit the measured spectrum on the spectra of the pure haemoglobin derivatives. Without this check, errors due to a high concentration of bilirubin or lipids, or in case of paraproteinaemia, would not be detected. Bilirubin shows significant absorption below 510 nm. Because Hi has an absorption maximum near 500 nm, it is not surprising that bilirubin simulates Hi. However, in these cases the relative fit error is abnormally high. When this occurs, the measurement may be repeated leaving out the spectral range of 480–510 nm. Another overall performance check is to periodically compare ct_{Hb} measured by MCA with ct_{Hb} measured by the HiCN method.

The method has been used in specimens of 4066 patients in a Dutch general hospital, 3863 patients (both sexes; 12–85 years of age) awaiting major surgery for various ailments and 203 patients (both sexes; 25–70 years of age) visiting the outpatient department for pulmonary diseases. All results are given in ref. [447]. A few remarkable findings are the following. Not a single case of sulfhaemoglobinaemia was detected; F_{SHb} was always $< 0.4\%$. This does not confirm the view of Park and Nagel [303] that sulfhaemoglobinaemia is probably underdiagnosed. However, the examined patient population may not have been representative, because sulfhaemoglobinaemia is associated with chronic multiple-drug use and abuse. On the other hand, it may be that sulfhaemoglobinaemia is indeed declining as a result of the discontinuation of the use of acetanilide and phenacetin.

About 63% of all patients had $F_{COHb} \leqslant 1.5\%$; when F_{COHb} was increased, it was usually between 1.5 and 10%; in a few cases values as high as 15% were observed. The percentage of patients with $F_{COHb} > 1.5\%$ (37%) agrees very well with the percentage of smokers in the Dutch population in 1985 (36%). There was a weak, but significant correlation between F_{COHb} and ct_{Hb}; in people with $F_{COHb} \leqslant 1.5\%$, $ct_{Hb} > 150$ g/L occurred in 17.4% of all cases, whereas this occurred in 41.7% of all cases in people with $F_{COHb} > 5.0\%$.

There were two cases of haemiglobinaemia due to food poisoning with $F_{Hi} = 32$ and 24%. In the surgical patients, F_{Hi} was $0.4 \pm 0.2\%$ (*SD*), but in the outpatients with pulmonary disease F_{Hi} was $1.5 \pm 0.8\%$ (*SD*), with occasional values around 5%. In this group of patients, long-term treatment with theophylline and bromhexine was customary at the time. These drugs, however, had up to then not been incriminated as a possible cause of Hi formation.

The method for MCA of haemoglobin of Zwart *et al.* [444] has been used continually since 1984 in our laboratory for various research purposes. The methods used for the determination of the absorptivities of pure haemoglobin derivatives of different kinds of haemoglobin gradually improved to the level of the standard procedures, described in Chapters 6 and 7, and used for most of the absorption spectra presented in Chapters 8–14. These spectra are currently used with the method to determine either five (HHb, O_2Hb, COHb,

Table 17.2.
Results of MCA of hypothetical blood samples of different species, when calculated on the basis of the absorption spectra of adult human haemoglobin

Haemoglobin	ct_{Hb}	S_{O_2}	F_{HHb}	F_{O_2Hb}	F_{COHb}	F_{Hi}	F_{SHb}
Man HbA							
5-MCA	1.000	90.0	8.0	72.0	10.0	10.0	0.0
4-MCA	1.000	90.0	8.0	72.0	10.0	10.0	
Man HbF	0.998	94.3	4.3	71.2	13.6	10.9	
Cow	1.005	90.4	7.6	71.5	9.3	11.7	
Pig	1.003	90.5	7.3	69.9	9.8	12.9	
Horse	1.002	88.6	9.8	75.7	3.7	10.8	
Sheep HbA	1.013	90.0	8.0	71.8	7.8	12.5	
Sheep HbF	1.006	94.1	4.3	69.1	14.6	12.0	
Man HbA (SFH)	0.993	91.2	6.7	69.6	12.4	11.3	
Dog (SFH)	0.997	92.0	5.9	67.8	13.6	12.7	
Rat (Rat proc)	1.022	93.4	5.2	73.1	10.4	11.3	

ct_{Hb} in mmol/L; S and F in %; 5-MCA is five-component analysis; 4-MCA is four-component analysis; SFH is SFH procedure (Chapter 8); Rat proc is rat procedure (Chapter 10). When no procedure is mentioned, the standard procedure was used for the determination of the spectra of the pure haemoglobin derivatives.

Hi, SHb) or four components (SHb left out). In the following this method is referred to as **standard multiwavelength method** (MWM). Addition of other components can be made as is shown in the case of HiCN (p. 236).

The effect the differences between the absorption spectra of various kinds of haemoglobin may have on the results of a multicomponent analysis can be illustrated by a mental experiment. First, calculate for each kind of haemoglobin, on the basis of the standard spectra, the absorbances for the spectral range of 480–650 nm of a hypothetical haemolysed blood sample containing various fractions of haemoglobin derivatives, assuming a certain ct_{Hb} and lightpath length. Second, introduce this absorption spectrum into the work space of the spectrophotometer microprocessor, as if it were a measured spectrum. Third, calculate with the usual algorithm, the fractions of the haemoglobin derivatives, S_{O_2}, and ct_{Hb}, on the basis of the standard spectra of adult human haemoglobin (Tables 8.1 and 8.2). Table 17.2 shows the results obtained for a hypothetical blood sample with $ct_{Hb} = 1.000$ mmol/L and 8% HHb, 72% O_2Hb, 10% COHb, and 10% Hi. S_{O_2} is $72/(72 + 8) = 90\%$.

Table 17.2 mainly shows differences between kinds of haemoglobin, but some differences due to differences in preparation of the haemoglobin solutions are also seen. There is, as expected in a theoretical exercise, no difference between 5-MCA and 4-MCA. Remarkable is the sensitivity of the COHb fraction to spectral differences, while, as expected, ct_{Hb} is hardly affected. It should also be noted that the MCA results of the two foetal haemoglobins differ considerably from those of the corresponding adult haemoglobin derivatives, but that there is less difference between the MCA results of the two foetal haemoglobins.

The differences due to a difference in procedure used for the determination of the spectra of the haemoglobin derivatives can be seen by comparing the MCA results of human HbA prepared with the standard procedure with those prepared with the SFH procedure.

The MCA results of canine and human haemoglobin both prepared with the SFH procedure differ little. This confirms that only slight differences exist between the human and canine haemoglobin spectra, as is shown in Fig. 9.1 and Table 9.2. Since these differences are smaller than those between MCA results of human HbA prepared with the standard spectra and human HbA prepared with the SFH procedure, and since no standard spectra are available for dog haemoglobin, it seems indicated to use the standard spectra of human HbA for MCA of dog blood.

Ideally, MCA of haemoglobin should be carried out at *constant temperature*. When equipment for thermostatted diode array spectrophotometry became available, changing over to this equipment was deemed too expensive compared to the possible slight increase in accuracy. Keeping room temperature within narrow limits and avoiding series of measurements during extreme weather conditions indeed proved sufficient to keep the influence of changes in temperature at a negligible level. However, it would have been advantageous if the absorption spectra could also have been determined at temperatures other than $\sim 22°C$.

Another only loosely controlled, and only loosely controllable, factor is the *pH of the haemoglobin solutions*. The effect of pH on the spectra of O_2Hb and COHb is but slight [155, 415, 416], but that on Hi is considerable. Therefore, much attention has been paid to the changes in the Hi absorption spectrum. However, the only feasible measure to diminish the effect of differences in pH was to select, as standard Hi spectrum, a spectral absorption curve within the range of pH values that can be expected to result from the haemolysis of a freshly drawn, normal blood sample of the species. In view of the range in which blood pH may vary in patients, this will not completely eliminate the effect of pH on MCA of haemoglobin. Table 11.5 shows that for bovine haemoglobin, for which a pH range of 7.24–7.32 had been chosen as acceptance criterion for the Hi spectra underlying the standard Hi spectrum, MCA indeed indicates 100% for a pure Hi solution at pH = 7.28.

For a pure Hi solution MCA should give 0 for the other haemoglobin derivatives. Table 17.3 gives the results of MCA at various pH values for a human blood sample containing 100% Hi. Note the considerable influence that the pH-dependent changes in the Hi spectrum have on the fractions measured for the other haemoglobin derivatives. For human HbA the acceptance criterion for Hi spectra underlying the standard Hi spectrum was pH = 7.14–7.30. At pH = 7.23, F_{Hi} indeed hardly differs from 100%

Table 17.3.
Results of MCA of a human blood sample containing 100% Hi at various pH values

pH	F_{HHb} (%)	F_{O_2Hb} (%)	F_{COHb} (%)	F_{Hi} (%)	F_{SHb} (%)
6.74	−8.0	−2.8	4.8	107.0	−1.0
7.05	−4.2	−1.4	2.4	103.7	−0.4
7.23	0.4	0.3	−0.2	99.4	0.1
7.35	4.9	1.6	−2.6	95.8	0.3
7.48	9.6	3.3	−4.7	91.7	0.1
7.54	11.6	3.7	−5.9	89.9	0.7

and the fractions of the other haemoglobin derivatives are virtually zero. It should be noted that Table 17.3 exaggerates the error pH variations may cause in MCA of haemoglobin. In actual practice Hi fractions will be lower by a factor of about 10 and the possible error reduced accordingly.

MULTIWAVELENGTH HAEMOGLOBIN PHOTOMETERS

Multiwavelength haemoglobin photometers are automated photometers for absorbance measurement at a limited number of wavelengths for multicomponent analysis of specified haemoglobin derivatives. The work cycle of the instrument usually includes haemolysis of the blood sample, blanking of the photometer, and filling and cleaning of the (flow-through) cuvette. The results are displayed on a screen and/or printed out. Only μL-samples are needed in most of the instruments.

The first instrument to possibly belong in this category of photometers is the **IL182 CO-Oximeter**, although it is no more than an automated version of a variant of the spectrophotometric three-wavelength methods for HHb, O_2Hb and COHb developed in the nineteen sixties (p. 219). It determined ct_{Hb} and the fractions of O_2Hb and COHb by measurement at wavebands around 548, 568 and 578 nm. Whole blood was introduced into the instrument and the results were displayed digitally. It was shown to be a useful and fairly accurate instrument [247].

Around 1980, the IL182 was replaced by the **IL282**, the first real multiwavelength haemoglobin photometer [47, 92]. The instrument measures HHb, O_2Hb, COHb and Hi in an exactly determined system at 535.0, 585.2, 594.5, and 626.6 nm. The wavelength setting for the four absorbance measurements is obtained by means of a servo-controlled rotating wheel with four interference filters, each selecting a particular line from the emission spectrum of a thallium-neon hollow cathode lamp. A distinct advantage of this construction is the accurate, stable and reproducible wavelength setting, which obviates the necessity of regular wavelength calibration. The measurements are made in a cuvette with a lightpath length of 125 μm; 175 μL whole blood is introduced into the instrument. A built-in microprocessor calculates the concentrations of the four haemoglobin derivatives from the measured absorbances, the lightpath length and the absorptivities, sums the four concentrations to find ct_{Hb}, and divides the four concentrations by ct_{Hb} to obtain the fractions of the four haemoglobin derivatives.

Zwart *et al.* [446] compared IL282 results with their manual five-wavelength method (p. 223) and found good agreement for F_{O_2Hb}, F_{COHb} and ct_{Hb}. There was a tendency for the IL282 to overestimate F_{Hi}. The presence of SHb gave a clear, concentration-dependent overestimation of F_{Hi} and underestimation of F_{COHb}, thus allowing detection of any significant fraction of SHb in the blood. The instrument proved to be robust, fast and easy to handle.

The IL282 has exerted significant influence on the development of haemoglobin spectrophotometry in several respects. Firstly, it showed the suitability of the automated special purpose photometer for the rapid determination of haemoglobin derivatives and related quantities in clinical conditions. Consequently, an avalanche of technical development in this field followed.

Secondly, use of the IL282 led to the recognition of the differences between the absorption spectra of foetal and adult human haemoglobin, and thus to the search for similar differences between animals, because it yielded abnormally high COHb fractions in foetal cord blood [177, 446, 478]. Interestingly, it had already been observed by the designers of the IL282 that for measurements with non-human blood the absorptivity matrix had to be adapted [47, 92]. It seems that the general significance of this observation was not immediately recognised.

Thirdly, the preference of the designers of the IL282 of F_{O_2Hb} over S_{O_2} and, hence, their decision to display F_{O_2Hb}, calling it O_2Hb saturation [47], immediately led to confusion of the two quantities [92] and soon to a serious misunderstanding in pulse oximetry (p. 271).

Fourthly, because the IL282 was, like its predecessor, called CO-Oximeter although it served for measuring four haemoglobin derivatives, the mental association of CO with carbon monoxide faded. CO-Oximeter became 'co-oximeter' and subsequently all kinds of multiwavelength photometers were called by this name. This, in turn, led to use of the term 'co-oximetry' as synonymous with MCA of haemoglobin.

The **Hemoximeter OSM3** (Radiometer) is a six-wavelength haemoglobin photometer for measuring ct_{Hb}, S_{O_2}, and the fractions of HHb, O_2Hb, COHb and Hi, while it indicates the presence of SHb [117]. The instrument is equipped with a mode of operation for dealing with the presence of HbF [118]. The optical system of the OSM3 consists of a 20-W halogen lamp, a heat-absorbing filter, several lenses, a glass cuvette, a diffraction grating, and six photodiodes located behind slits that select six wavelengths from the continuous spectrum. The light source is equipped with stabilising circuitry. The blood is haemolysed in the thermostatted cuvette (37 °C) by vibration at a frequency of about 40 kHz. The required sample volume is 35 μL, of which only 1 μL is in the cuvette during the measurement. The result is displayed after some 20 s; a whole measuring cycle, including automatic rinsing, takes about 1 min.

The six wavelengths used in the OSM3 are 535, 560, 577, 622, 636 and 670 nm. Five wavelengths are necessary to account for HHb, O_2Hb, COHb, Hi and SHb, while the sixth wavelength serves to minimise interference by turbidity. The instrument thus performs MCA in an exactly determined system, and the calculations are made accordingly. This yields the concentrations of the haemoglobin derivatives present in the sample; ct_{Hb} is calculated by adding the concentrations of the haemoglobin derivatives, and S_{O_2} and the fractions of HHb, O_2Hb, COHb and Hi are calculated according to equations 2.9–2.13. To measure samples containing a high fraction of foetal haemoglobin, the calculations can be made on the basis of absorptivity values of human HbF [118]. No value for F_{SHb} is displayed on the screen, but an indication of the presence of SHb is given. Values < 0 and $> 100\%$ are not displayed on the screen, but a printer can be added to the instrument to record the absorbances and the concentrations of the haemoglobin derivatives present, including SHb.

We compared the performance of the OSM3 with MCA carried out with the semi-automated multiwavelength method using an HP8451A described in the preceding section (p. 226). Most of the comparative measurements were made with patients' blood drawn for pre-operative screening. Some blood specimens from healthy donors were used to make several series of samples containing considerable amounts of dyshaemoglobins in various combinations. Most of the results of the OSM3 measurements were read from the screen, but a printer had been included, and for part of the samples we recalculated S_{O_2} and dyshaemoglobin fractions from the printed concentrations [482].

For 100 freshly drawn blood specimens with S_{O_2} between 23 and 100%, the mean difference between OSM3 and multiwavelength method (MWM), $X_{OSM3} - X_{MWM}$, was $0.90\%S_{O_2} \pm 1.14\%S_{O_2}$ (SD). Some of the samples contained up to 12% COHb. In 23 samples containing up to 60% COHb the mean difference was $-0.47\%S_{O_2} \pm 1.66\%S_{O_2}$ (SD). In 32 samples containing up to 70% Hi the mean difference was $-0.85\%S_{O_2} \pm 2.83\%S_{O_2}$ (SD).

Comparison of F_{COHb} in 214 samples as measured by the two methods yielded a mean difference, $X_{OSM3} - X_{MWM}$, of 0.03% ± 0.47% COHb (SD). In 37 samples containing 1–60% COHb the mean difference, $X_{OSM3} - X_{MWM}$, was 0.29% ± 0.96% COHb (SD). In 24 samples containing up to 70% Hi in addition to COHb the mean difference was 1.42% ± 1.26% COHb (SD).

In the same 214 samples in which F_{COHb} was measured, F_{Hi} was between 0 and 1.4%; the mean values were 0.46% (MWM) and 0.81% (OSM3). In

Table 17.4.
Comparison of results of MCA of 16 contrived blood samples with an OSM3 haemoglobin photometer and with a multiwavelength procedure carried out with an HP8451A diode array spectrophotometer

S_{O_2} (%)		F_{COHb} (%)		F_{Hi} (%)		F_{SHb} (%)	
MWM	OSM3	MWM	OSM3	MWM	OSM3	MWM	OSM3
28.9	31.0*	5.6	16.5	1.5	−9.4	15.9	17.0
23.6	43.1	0.3	9.8	5.0	−3.3	15.9	15.4
5.8	11.4*	8.2	14.4	−1.2	−6.3	15.7	15.0
55.3	65.6	−0.1	7.5	6.6	−2.4	9.7	9.3
25.7	32.0	5.2	32.0	25.8	4.5	5.2	7.8
70.3	75.0	−0.3	3.7	4.0	−0.5	4.3	3.6
66.7	77.8	−1.4	5.3	27.3	23.3	2.5	2.8
49.5	58.8	13.7	37.3	40.7	32.7	1.8	3.2
91.9	94.6	29.7	31.6	15.2	14.2	0.7	0.6
98.9	99.4	16.9	17.3	12.6	11.7	0.3	0.0
98.5	99.3	23.4	24.6	14.5	15.0	0.2	0.1
68.8	68.4	6.2	6.8	7.3	5.8	0.3	0.0
76.1	76.9	5.4	6.8	13.5	11.2	0.3	0.0
69.9	67.2	5.3	6.6	16.7	14.2	0.3	0.1
91.4	94.7	22.2	20.8	13.2	12.2	0.3	−0.1
94.0	98.5	17.8	19.2	37.6	43.8	0.2	0.1

* Displayed on OSM3 screen with question mark.

34 samples containing 1–70% Hi, the mean difference, $X_{OSM3} - X_{MWM}$, was −0.29% ± 2.29%Hi (SD). In 29 samples containing up to 60% COHb in addition to Hi, the mean difference, $X_{OSM3} - X_{MWM}$, was 0.16% ± 3.24%Hi (SD).

To put the instrument to a more severe test, 16 blood samples were made containing exceptional combinations of all five haemoglobin derivatives, including considerable amounts of SHb. The results of comparative measurements with the OSM3 and the multiwavelength method are listed in Table 17.4.

Taking into account that comparative measurements contain systematic and random errors of both procedures, there is fair agreement between the two methods for all samples within the performance specifications of the OSM3. When the sample contains an appreciable fraction of SHb, the OSM3 overestimates S_{O_2} and F_{COHb}, and underestimates F_{Hi}. It appears that the overestimation of S_{O_2} and F_{COHb} is increased through the presence of a considerable amount of Hi. The measurement of F_{SHb} is surprisingly accurate.

Other types of multiwavelength haemoglobin photometer, such as the IL482 and IL682, the CCD270 (Chiron) and the AVL912 and AVL Omni, are not essentially different from the earlier instruments, although there is a tendency to more overdetermined systems, the CCD270 using 8, the AVL912 17 and the AVL Omni 64, and the photometry module of the Radiometer ABL725 128 wavelengths. The PolyOx module of the AVL Omni actually contains an array of 512 photodetectors scanning the wavelength range of 520–650 nm in steps of 0.25 nm. For MCA of five haemoglobin derivatives, 64 wavelengths are selected. In the photometry module of the Radiometer ABL725, an array of 128 photodiodes in the wavelength range of 478–672 nm is used.

Through the remarkable progress in the development of dedicated instruments for MCA of haemoglobin, the determination of ct_{Hb}, S_{O_2}, and dyshaemoglobin fractions has become easy and reliable. However, the more these methods are used in, and beyond, the clinical laboratory, the greater is the need for reliable **quality control materials**. For ct_{Hb}, all requirements can be fulfilled easily (Chapter 16). For S_{O_2}, on the other hand, no reliable quality control material has been produced. The only adequate quality control material would be blood with different, exactly known levels of S_{O_2}. This can hardly be produced because there is no simple, generally accepted reference method for S_{O_2}, and, if produced, it cannot be kept constant for more than a few days. For the dyshaemoglobins, the situation is intermediate: quite stable solutions of COHb and Hi can be made.

Most quality control materials for MCA of haemoglobin consist of dye solutions simulating certain mixtures of haemoglobin derivatives. These solutions are suitable for testing photometer performance, but the spectral properties are different from those of the haemoglobin solutions to be measured. A stable solution of haemoglobin derivatives, the fractions determined by several reliable multiwavelength methods, would be preferable.

It has been demonstrated that **lyophilised bovine haemoglobin** (LBH) containing various fractions of O_2Hb, COHb and Hi is stable for more than a year, and that it yields, after reconstitution, a haemoglobin solution that is spectrophotometrically equivalent to a fresh bovine haemoglobin solution containing the same fractions of O_2Hb, COHb and Hi [248]. For a study of the suitability of this material for quality control in MCA of haemoglobin, four LBH preparations of different composition were prepared; F_{O_2Hb} varied between 88.8 and 42.6%, F_{COHb} between 6.9 and 19.8% and F_{Hi} between 4.2 and 37.6%. The fractions of the haemoglobin derivatives were determined by the standard multiwavelength method (p. 227) on the basis of the absorption spectra of pure LBH derivatives. These spectra did not significantly differ from the absorption spectra of the corresponding bovine haemoglobin derivatives (Tables 11.1 and 11.2). The composition of the four preparations is given in the first two columns of Fig. 17.1; these data are called the 'true values' of the four preparations.

A computer program was developed to calculate the composition of mixtures of hemoglobin derivatives by simulating MCA [248]. The program was used either with a set of absorptivities from 480 to 650 nm with intervals of 2 nm (general model), or with specific sets of absorptivities for wavelengths as applied in various types of multiwavelength haemoglobin photometers (specific models). When the whole spectral range of 480–650 nm was used the computer program gave results corresponding with those of the standard multiwavelength method.

The computer program was first used to estimate the effect of wavelength selection on the results of MCA. The open bars in Fig. 17.1 show the differences between the true values and those calculated with specific models, *i.e.* a four-wavelength model comprising the four wavelengths used by the IL282 and 482, a six-wavelength model comprising the six wavelengths used by the OSM3 and ABL520 (Radiometer), an eight-wavelength model comprising

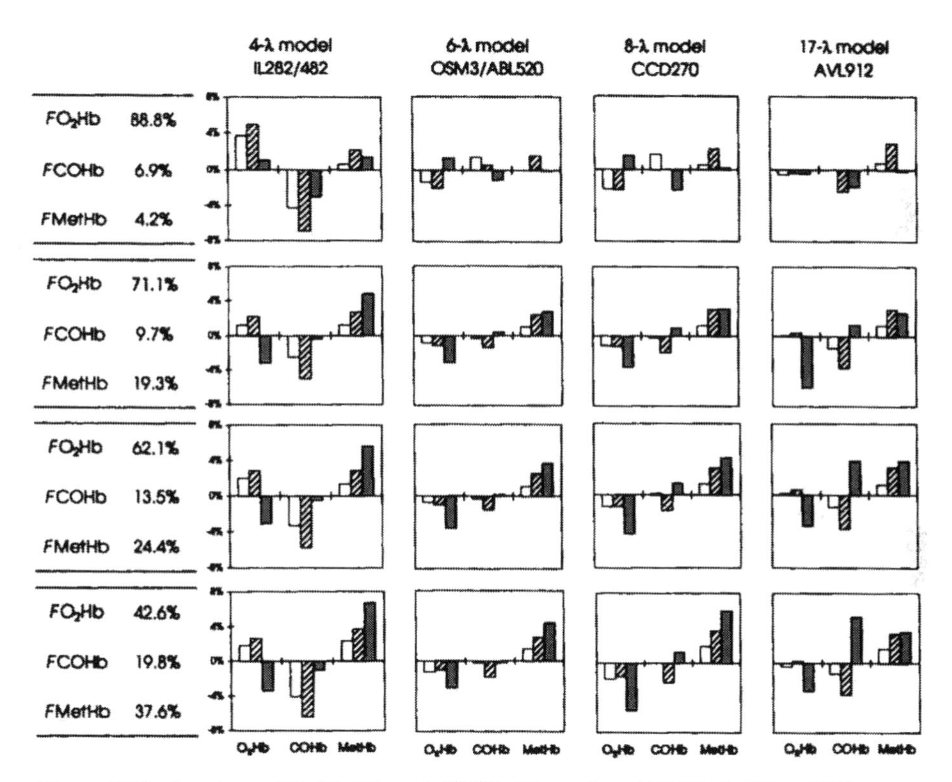

Figure 17.1. Fractions of O_2Hb (F_{O_2Hb}), COHb (F_{COHb}), and Hi (F_{Hi}) of four LBH preparations as calculated with the general model and bovine absorptivities ('true values'; second column), as calculated with the various specific models and bovine and human absorptivities, respectively, and as measured with various types of multiwavelength haemoglobin photometer.

The open bars represent the values calculated with the specific models and bovine absorptivities minus the 'true values'. The shaded bars represent the values calculated with the specific models and human absorptivities minus the 'true values'. The solid bars represent the values measured with the various types of haemoglobin photometer minus the true values (from ref. [248]).

the eight wavelengths of the CCD270 (Chiron) and a 17-wavelength model comprising the 17 wavelengths of the AVL912. It appears that the difference in comparison with the true values becomes smaller when more wavelengths are used.

In a second experiment the same calculations were made, but now using human absorptivities (Tables 8.1 and 8.2) instead of bovine absorptivities. The differences in comparison with the true values are shown by the shaded bars in Fig. 17.1. Each of the specific models used in combination with the absorptivities of human haemoglobin is the closest simulation of a particular multiwavelength haemoglobin photometer one can make without further knowledge of construction details and software.

The black bars in Fig. 17.1 resulted from measurements of the four LBH preparations with the various types of instrument in a field trial of 14 IL282/482, 13 OSM3/ABL520, 17 CCD270 and 4 AVL912 haemoglobin photometers. For each group of instruments, the mean values for the different haemoglobin derivatives in the four LBH preparations were calculated and compared with the true values. This demonstrated the potential of LBH as a reference material for MCA of haemoglobin. The differences between the results calculated for the various measuring systems on the basis of simple models and the results provided by the instruments show that the actual instruments do not conform to the straightforward MCA concept formulated earlier in this chapter (p. 222). This is not surprising in light of the complex character of the material to be analysed: freshly drawn blood includes a multitude of not completely colourless substances and scattering particles, for which empirical corrections can be incorporated in programming of the instruments.

MCA INCLUDING Hi AND HiCN

Little attention has been paid to the determination of haemoglobin derivatives other than HHb, O_2Hb and the common dyshaemoglobins, obviously because medical practice does not ask for more. However, there is no theoretical or technical impediment standing in the way of adding any number of haemoglobin derivatives or other coloured substances to the MCA of haemoglobin, provided that there is a clear difference between the spectra of the components to be measured. If there is a need to add a different component, it can easily be done. This is shown by the example of HiCN, the measurement of which appeared to be desirable when the treatment of cyanide poisoning with the help of haemiglobin-forming agents was proposed.

Acute cyanide (CN^-) poisoning can be treated by injection of Hi-forming agents, *e.g.* $NaNO_2$ or 4-DMAP [280]. Through its high affinity for cyanide [163], Hi is able to keep CN^- away from the cytochrome a_3 component of cytochrome oxidase (EC 1.9.1.3), which otherwise would become blocked.

Thus, time is gained for detoxification through the conversion of CN^- to CNS^-, mediated by rhodanase (thiosulfate cyanide sulfertransferase; EC 2.8.1.1). The enzyme can handle a considerable amount of CN^-, but the reaction is limited by scarcity of substrate, unless thiosulfate is supplied as a source of sulfur. The CNS^- formed is relatively non-toxic and easily excreted by the kidneys.

For effective treatment a considerable amount of Hi should be formed. To bind all CN^- from an ingested amount of 1 g KCN, F_{Hi} should be 30% in a 70-kg man with ct_{Hb} = 150 g/L [473]. The protection of cytochrome oxydase is achieved at the cost of a considerable decrease in oxygen carrying capacity of the blood: oxygen transport capability is traded for maintaining oxidative energy release. The precarious balance between these two factors threatening the patient's life requires strict control of the fractions of Hi and HiCN in the blood. Hence, F_{Hi} and F_{HiCN} should be measured at short intervals, in addition to ct_{Hb}. In fire victims the oxygen carrying capacity of the blood may simultaneously be diminished through the presence of COHb [20, 227]. Hence, also F_{COHb} should be measured and taken into account in the estimation of the maximum allowable rate of Hi formation.

We modified the standard multiwavelength method by adding HiCN and, for the sake of convenience, leaving out SHb [473, 474]. The method was tested by adding known amounts of KCN to blood samples in which 11.7–77.0% Hi had been induced. Because of the high value of the equilibrium association constant ($K = a_{HiCN}/(a_{Hi} \cdot a_{CN^-}) \approx 1.5 \times 10^6$) [163], $< 0.1\%$ of any CN^- added to Hi-containing blood remains unbound as long as the added amount of CN^- does not exceed the amount of Hi present. Consequently, the HiCN concentration can be calculated from the amount of KCN added. Table 17.5 shows part of the results. The mean difference between the sum of the Hi and HiCN fractions and the initial Hi fraction was $-0.1 \pm 0.3\%F$ (SD) $(n = 25)$. The measured HiCN fractions differed $-0.5 \pm 0.8\%F$ (SD) from the calculated HiCN fractions. To make the method applicable to any type of spectrophotometer an analogous five-wavelength procedure (500, 540, 554, 560, 578 nm) was developed, which matched the original procedure as to accuracy and precision [474].

Even if the five-wavelength method is used, special equipment (thin layer cuvettes) is necessary and the procedure takes more time than is available in the emergency of acute cyanide poisoning. Only an automated measuring system requiring the injection of only a small volume of blood can meet the requirements of the occasion. The feasibility of such a system has been investigated by analysing blood samples containing various fractions of Hi, HiCN and COHb with a modified version of the Radiometer OSM3 multiwavelength haemoglobin photometer [475].

The **modified version of the OSM3**, referred to as OSM3M, differed from the standard instrument only in that the measurement at 622 nm, which corresponds with the absorption maximum of SHb, was replaced by

Table 17.5.
Hi and HiCN fractions determined with MCA (HHb, O_2Hb, COHb, Hi, HiCN; 480–650 nm; HP8450A) and HiCN fraction calculated from added amounts of KCN

$F_{Hi} + F_{HiCN}$ (%)	F_{Hi} (%)	F_{HiCN} (%)	F_{HiCN} (calc%)*
11.7	10.0	1.7	1.7
12.5	9.4	3.1	3.5
11.8	6.8	5.0	5.1
11.7	5.7	6.1	6.9
11.8	4.0	7.8	8.6
12.0	3.5	8.5	10.3
12.2	1.3	10.9	12.0
22.5	20.6	1.9	2.1
22.3	19.0	3.3	4.2
22.8	17.0	5.8	6.3
23.0	14.9	8.1	8.4
23.3	13.6	9.7	10.5
23.6	11.9	11.7	12.6
23.7	10.0	13.7	14.7
75.1	73.1	2.0	2.1
75.0	70.9	4.1	4.3
74.9	69.2	5.7	6.4
74.8	66.5	8.3	8.5
74.5	62.7	11.8	12.8
74.7	58.6	16.1	17.1
74.6	50.6	24.0	25.6

* F_{HiCN} (calc) is calculated HiCN fraction.

Table 17.6.
Absorptivity matrix for automated five wavelength MCA (HHb, O_2Hb, COHb, Hi, HiCN) (OSM3M)

Component	$\varepsilon(518)$	$\varepsilon(535)$	$\varepsilon(560)$	$\varepsilon(577)$	$\varepsilon(636)$	$\varepsilon(670)$
HHb	6.25	9.15	13.09	9.85	0.98	0.73
O_2Hb	5.55	12.81	8.77	15.31	0.11	0.10
COHb	7.63	13.83	12.25	11.00	0.15	0.05
Hi	8.00	6.89	4.43	4.20	3.60	0.45
HiCN	8.37	10.89	9.54	6.82	0.49	0.19

ε in $L \cdot mmol^{-1} \cdot cm^{-1}$.

a measurement at 518 nm. This wavelength is nearly an isosbestic triple point of COHb, Hi and HiCN (Fig. 8.1). The absorptivity matrix used in the calculations is shown in Table 17.6. The absorptivity data are from ref. [480] and correspond, except for Hi, with those of Tables 8.1 and 8.2. The operating procedure was as with the standard OSM3: a 35-μL blood

Table 17.7.
COHb, Hi and HiCN fractions determined by MCA (HHb, O_2Hb, COHb, Hi, HiCN) by means of a multiwavelength method (MWM) and with a modified OSM3 haemoglobin photometer (OSM3M)

KCN added (mL)	0.1	0.2	0.4	0.6	0.8
Multiwavelength method					
F_{COHb} (%)	40.5	40.6	40.9	41.0	41.8
F_{Hi} (%)	45.9	40.4	30.4	21.6	13.2
F_{HiCN} (%)	6.6	12.5	23.0	31.4	40.0
$F_{Hi} + F_{HiCN}$ (%)	52.5	52.9	53.4	53.0	53.2
Modified OSM3					
F_{COHb} (%)	44.0	45.0	44.7	44.5	43.3
F_{Hi} (%)	44.5	39.2	27.9	17.6	8.2
F_{HiCN} (%)	8.1	13.5	26.1	37.1	46.6
$F_{Hi} + F_{HiCN}$ (%)	52.6	52.7	54.0	54.7	54.8
Calculated from KCN added					
F_{HiCN} (%)	6.0	12.1	24.2	36.3	48.4

MWM (480–650 nm) carried out with HP8450A; KCN added as 1 g/L solution.

sample is introduced into the instrument, of which only 1 μL is used in the measurement after it has been haemolysed by vibration at a frequency of 40 kHz [482].

The results obtained with the OSM3M were compared with those of the multiwavelength method (480–650 nm) carried out with an HP8450A. Heparinised blood from healthy donors was used to prepare samples containing different haemoglobin derivatives, largely following the procedures of Chapter 6. Samples containing 50–70% Hi were prepared by injecting ~ 2 mL of a 7 g/L $NaNO_2$ solution into the revolving tonometer. To obtain samples with a known fraction of HiCN, increasing volumes (0.1–1.0 mL) of a 1 g/L KCN solution were added to 4-mL aliquots of blood. From 11 specimens of fresh human blood from different donors a total of 125 samples of different composition were prepared.

Table 17.7 shows the results obtained for a single specimen of which a series of samples had been prepared containing considerable fractions of Hi and COHb to which subsequently increasing amounts of KCN had been added. Upon the addition of KCN, HiCN increases and Hi decreases, while the sum of the fractions of Hi and HiCN remains constant. The formation of HiCN does not interfere with the determination of COHb, which appears to be measured with reasonable precision. The complete results [475], which agree with those of Table 17.7, demonstrate the feasibility of a practical instrument for the determination of F_{Hi} and F_{HiCN} in rapid succession with sufficient accuracy to monitor the treatment of cyanide poisoning with Hi-forming agents.

The maximum allowable fraction of Hi that may be induced primarily depends on the remaining oxygen capacity of the blood. Because there may be a concomitant CO intoxication [20, 227, 280], it is important that the method also gives a reliable value of F_{COHb}. In this investigation the OSM3M measured the COHb fractions some 5% F too high over the whole range, but with good precision. It should be noted that in this case the wavelengths, and the number of wavelengths, could not be chosen freely, but that the possible selection was limited by the specifications of the original instrument. There is no doubt that a combination of wavelengths can be found that allows accurate determination of COHb in the presence of Hi and HiCN. However, it may be that more than five wavelengths are necessary for optimal resolution of this combination of haemoglobin derivatives, because of the rather flat spectral absorption curve of HiCN (Fig. 8.1).

CURRENT DEVELOPMENTS

With the present tendency to use more overdetermined systems in instruments for MCA it will be possible to use different combinations of wavelengths for analysing different combinations of components chosen at will. The powerful new photometric devices, such as the AVL PolyOx module and the Radiometer photometer module of the ABL725 analyser which scan the relevant part of the spectrum with high spectral resolution, especially, will allow the addition of other components and the detection of interfering coloured substances with little more than extended software, provided the absorption spectra of these substances are accurately known. Even the absence of reliable absorptivity values seems to become less of a problem, since the photometry systems in the analysers have now evolved to a stage that they seem to be excellently suited to determine the absorption spectrum of any candidate substance for inclusion in the MCA. A step in the direction of extended MCA appears to be the inclusion of bilirubin, fat particles ('Intralipid') and interfering dyes in the MCA as carried out with the photometer module of the ABL725 analyser [360].

An interesting development is the construction of multiwavelength haemoglobin photometers measuring the light extinction in a layer of unhaemolysed whole blood [130]. Dispensing with haemolysis not only simplifies the procedure, it is also an analytical improvement because any addition may change the composition of the sample. The main problem, also the main possible advantage, of these measuring systems is that, in calculating concentration from light extinction, scattering as well as absorption should be taken into account. Properly solving this problem eliminates an important potential source of error in absorbance measurement in haemolysed blood, namely, the interference by residual scattering after haemolysis.

An adequate test of instruments for multicomponent analysis of haemoglobin in unhaemolysed whole blood is the absence of interference by fluorocarbon emulsions added to the blood as oxygen carrier. Shepherd and Steinke have compared the interference by perflubron emulsion in haemolysing and non-haemolysing instruments [351]. Generally, as expected, there is much less interference in the non-haemolysing systems. However, the paper is not quite convincing because there is little difference in results between the IL682, a haemolysing multiwavelength haemoglobin photometer, and the IL Synthesis 35, a non-haemolysing one. The paper does not contain enough technical details to speculate on the cause of this finding.

CHAPTER 18

OXIMETRY AND RELATED TECHNIQUES

The oxygen saturation of blood, especially the arterial blood, is one of the most important quantities in clinical (emergency) medicine. The observation of an abnormal oxygen saturation, *i.e.* a *too low* oxygen saturation (the normal value is near 100%), has always been intimately connected with the spectral properties of haemoglobin. The symptom of **cyanosis**, a blue discoloration of the skin, was known as a danger signal long before the concept of oxygen saturation, S_{O_2}, was thought of. Although (chronic) cyanosis may occur through the presence of Hi or SHb in the blood (p. 288), the symptom is usually associated with a fall in S_{O_2} in the skin capillaries. This may be caused by a decreased arterial S_{O_2} (**central cyanosis**) or by a slowing of the cutaneous circulation so that the oxygen supply does not come up to the cutaneous oxygen consumption (**peripheral cyanosis**).

However, cyanosis is an unreliable symptom. In cases of anaemia or cutaneous vasoconstriction, a deep fall in arterial S_{O_2} may occur unnoticed. Moreover, the individual ability to observe cyanosis is quite variable. In the classical experiments of Comroe and Botelho [75] this was demonstrated by having physicians and medical students observe the colour of experimental subjects breathing different gas mixtures with high and low oxygen fractions, while S_{O_2} was monitored with the aid of an ear oximeter. Their main conclusions were:

1. The majority of 127 observers were unable to detect the presence of definite cyanosis until the arterial oxygen saturation fell to approximately 80%; 25% of observers did not note definite cyanosis even at arterial saturation levels of 71 to 75%.

2. There were marked variations in the ability of an observer to note cyanosis in different subjects or even in the same subject at different times. There were wide variations in color estimations when 5 to 10 observers watched cyanosis develop in the same subject at the same time [75].

Therefore, a rich variety of instruments has been constructed for the rapid determination of S_{O_2}: bench instruments for use in the clinical laboratory or near the patient, instruments containing a cuvette connected to an intravascular catheter, fibre optic instruments for measuring inside the patient and instruments for measuring non-invasively on the patient. The common feature of these widely different devices, however, is that they are photometers measuring light extinction at wavelengths at which there is a considerable spectral difference between HHb and O_2Hb.

The term **oximeter**, coined by Millikan [265] for his instrument for measuring S_{O_2} of arterial blood in man by transillumination of the ear, was first used only for similar instruments for the non-invasive measurement of S_{O_2}. In the course of time, almost any device for the determination of S_{O_2} has been called 'oximeter', including the peculiar term 'co-oximeter' for multiwavelength photometers for measuring haemoglobin derivatives *in vitro* (p. 231). In the following, the term 'oximeter' is restricted to instruments for measuring S_{O_2} in or on the body (*in vivo*) or in a cuvette connected to an intravascular catheter (*ex vivo*). These instruments are the main subject of this chapter.

DEVELOPMENT OF OXIMETRY

With Vierordt's observation of the spectral changes in a transilluminated finger when the blood flow through the finger was stopped [403], the principle on which *in vivo* oximetry rests had been discovered. Application of this principle, however, had to wait for the availability of practical devices for converting light to electricity, such as the selenium barrier layer photocell. This photocell, introduced into physiological instrumentation around 1930, has, in a low resistance circuit, a current output that is proportional to the intensity of the illumination and high enough to be measured by the dependable and sensitive galvanometers then available. During the 1930s, German investigators (Nicolai [286], Kramer [224, 225], Matthes [255], Matthes and Groß [256, 257]) performed the fundamental experiments that paved the way for the construction of the first practical oximeters.

Squire [370, 454] described, in 1940, a fairly advanced instrument to be applied to the web of the hand that was primarily intended for the study of the peripheral circulation, yet exhibited various features of later oximeters. There are no reports on the application of the instrument, probably because it was a cumbersome device, but it may have had considerable influence on the development of more practical methods for measuring S_{O_2} *in vivo*.

In 1942 Millikan published his oximeter [265], which is generally acknowledged as the beginning of oximetry in physiology and in clinical medicine, although Goldie [129, 454] had described a similar device earlier that year. It may be that the name of the instrument has contributed to the success of Millikan's oximeter, its use has certainly been promoted by the fact that an improved version became commercially available (Coleman).

The Millikan oximeter consisted of an earpiece, a switchbox and a sensitive galvanometer. The earpiece contained a small 6–8 V miniature incandescent lamp that transilluminated the pinna of the ear and the light was detected by

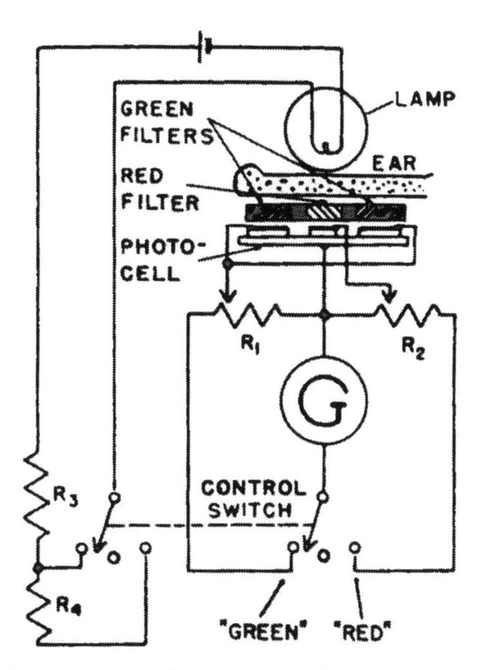

Figure 18.1. Circuit diagram of the original version of the Millikan oximeter.

The red and green transmissions are measured independently. The galvanometer was equipped with a multiple scale calibrated in % S_{O_2} which was read while the red transmission was measured. The proper scale was indicated by a preceding reading of the green transmission. The later versions of the Millikan oximeter and most other early oximeters are a variation on this theme (from ref. [265]).

a barrier layer photocell divided into three functional parts: a central strip covered with a red filter and two lateral strips covered with green filters. The two photocell currents were measured by the galvanometer. The current from the red part of the photocell decreased when the oxygen saturation fell, because in this range of the spectrum the absorptivity of HHb is higher than the absorptivity of O_2Hb (Fig. 8.4); the current from the green part of the photocell was assumed to be related to the total amount of haemoglobin in the lightpath. Figure 18.1 shows a circuit diagram.

Calibration of the oximeter with the aid of arterial samples analysed by Van Slyke's method [394] yielded for the S_{O_2} range of 50–100% a linear relationship between S_{O_2} and the logarithm of transmitted red light. There was no provision to distinguish between light extinction by the blood and by the tissue, *i.e.* no device for adjusting the oximeter on the bloodless ear. The measuring system thus lacked a blank; therefore, the instrument had first to be adjusted with an optical filter in the lightpath, and after the earpiece had been placed on the shell of the ear the instrument had to be set to a known value of S_{O_2}. Montgomery *et al.* [267] found in 514 measurements with 10 earpieces in 25 white subjects an average difference of 4.7% S_{O_2}, range

-11 to 23% S_{O_2}, in comparison with arterial blood samples analysed by Van Slyke's method.

At the time the Millikan oximeter was a real asset, although in its construction only poor solutions had been found for **the three basic problems of oximetry**: (1) the absence of a suitable theory of light absorption and scattering in living tissues; (2) the necessity to differentiate between blood and non-blood in the lightpath, complicated by the fact that the amount of blood can vary considerably; (3) the presence in the lightpath of arterial, capillary and venous blood, while the *arterial* S_{O_2} is the relevant quantity to be measured [454]. It is the effort to solve these problems that to a large extent determined the further course of oximeter development.

In the Millikan oximeter the third problem was solved by having the lamp produce enough heat to keep the ear 'fully flushed', *i.e.* in a state of maximum vasodilation. The oxygen supply to the tissue in the lightpath then exceeds by far the oxygen consumption so that there is hardly any difference in S_{O_2} between the arterial and the venous blood. Apart from application of heat, 'arterialisation' of the blood in the vessels of the skin can also be achieved by the local application of vasodilating substances [108, 289, 361, 424].

The second problem was successfully dealt with by Wood and co-workers [419–421, 423] when they tried to eliminate the necessity to set the oximeter on a known value at the beginning of each series of measurements. In aviation physiology, for which the Millikan oximeter was primarily developed, this can be easily accomplished because in normal humans $S_{O_2} = 100\%$ after a few minutes of oxygen breathing, but in patients this cannot be relied upon. To provide a means of measuring the transmission of the bloodless ear, the lamp housing of the Wood oximeter was equipped with a pressure capsule with a glass surface towards the light source and with the surface towards the ear covered with a thin translucent rubber membrane. The pressure capsule was connected to a hand bulb and a manometer by means of vinyl tubing. When such an earpiece is placed on the shell of the ear and the capsule is inflated to arterial occlusive pressure, the blood is squeezed from the tissue in the lightpath. This allows measurement of the incident light intensity, analogous to I_0 in equation 2.1. When the pressure is released, the blood enters the tissue and the intensity of the transmitted light, I, is measured. Hence, the absorbance, $\log(I_0/I) = A$, of the blood can be calculated.

In the Wood oximeter an infrared filter was used instead of the green filter in the Millikan oximeter. Functionally this may have made little difference. Figure 8.4 shows that in the green part of the spectrum the absorptivity of HHb and O_2Hb is so high that hardly any light of these wavelengths passes through the ear. Millikan's green filter probably transmitted some infrared light and since most of the barrier layer photocells then in use were sensitive in the near IR region, this may have produced an appreciable signal.

The oximeter thus measured the absorbance of the blood in the shell of the ear in two wavebands, one in the red ($\sim$620 to $\sim$680 nm) and the other in the infrared around the isosbestic point of HHb and O_2Hb at 800 nm (Fig. 8.4). If Lambert–Beer's law could be applied in the system, an equation for calculating S_{O_2} from the absorbance ratio, $A(R)/A(IR)$, analogous to equations 3.7 and 17.2 should be valid. The equation predicts a straight line when $A(R)/A(IR)$ is plotted against S_{O_2}. However, when 211 oximeter readings were plotted against the results of Van Slyke analysis of simultaneously drawn arterial blood samples, a curved line was obtained with a declining increase in $A(R)/A(IR)$ per % S_{O_2} when S_{O_2} decreased. Between 50 and 90% S_{O_2}, however, the calibration line could be reasonably approximated by a straight line. At S_{O_2} between 15 and 100%, the standard deviation of the difference between S_{O_2} determined with the oximeter and determined by Van Slyke's method was 2.9% S_{O_2} [423].

These results demonstrated that S_{O_2} can be determined *in vivo* by a method analogous to the spectrophotometric procedure described on p. 219, but that the relation between $A(R)/A(IR)$ and S_{O_2} does not exactly comply with equation 17.2. This is obviously due to the fact that the light during its passage through the ear is attenuated by scattering as well as by absorption.

Various versions of the Wood oximeter have become commercially available (Waters). The original version was a double-scale procedure, preferably using separate galvanometers for measuring the incident and transmitted light in the two wavebands. This procedure has the highest accuracy, but it requires some calculations to convert the double-scale readings to S_{O_2}. In newer versions of the instrument, a single scale procedure was used allowing continuous reading or recording of S_{O_2}. An oximeter employing simultaneous double- and single-scale procedures has been devised to combine the advantages of the single-scale procedure with the accuracy of the double scale method [425].

In later oximeter development, electronic amplification replaced the use of sensitive galvanometers and resulted in instruments allowing direct reading of S_{O_2} from a linear scale [287, 288, 304]. Basically there was little change in design. All instruments were two-colour photometers employing broad bands of red and infrared light and contained devices for compression of the tissue to measure the transmission of the bloodless ear, or, as in the Nilsson oximeter [288] (Atlas) to obtain desaturated values for calibration, assuming that when the tissue is isolated from the circulation S_{O_2} rapidly falls to 40%. The problem of continuously discriminating between light absorption by blood and by non-blood was reasonably solved in most procedures. The instruments also contained many technical improvements that made them more dependable and easier to use, but the procedures remained complicated [289, 425].

The easy and accurate, absolute-reading oximeter, clinical practice was waiting for, did not appear. Even the HP47201, an eight-wavelength spectrophotometer using optic fibres for transmitting light to and from the earpiece, did not come up to the expectations. It was a precalibrated instrument that gave fairly accurate results [338], but practical disadvantages such as the large earpiece and the high price impeded general use. Oximetry remained a research tool: for clinical practice it was too complicated and too difficult to apply.

In this respect **reflection oximetry** was no exception. Reflection oximetry was introduced by Brinkman and Zijlstra around 1950 [42, 43, 45]. It started with the observation that the oxygen saturation of (non-haemolysed) blood could be determined by measuring light reflection in *a single band of red light*. The amount of red light scattered backwards by a layer of blood proved to be independent of ct_{Hb} and layer thickness within wide limits, and there was an almost linear relationship between the logarithm of the amount of reflected light ($\log R$) and S_{O_2} [268, 454, 459, 462, 487, 490]. On this basis a simple method was developed to determine S_{O_2} in blood samples [43] (p. 251).

In developing a reflection method to measure S_{O_2} *in vivo*, several problems had to be solved. First the measurement had to have a zero point. In transmission photometry *in vitro*, the zero point of the absorbance measurement is found by measuring the transmission of a cuvette containing the solvent (water); in the Wood oximeter it was found by measuring the transmission of the bloodless ear. In the Haemoreflector (p. 251) it was obtained by measuring the reflection of a cuvette filled with Indian ink, which has zero reflection. For a reflection oximeter to be used on the human skin the best solution proved to be a measurement with a waveband in which light absorption by the blood is so high that the amount of back-scattered light is practically zero. Thus, in the oximeter, at least two wavebands had to be used, one in the red, preferably 600–700 nm, and one in a spectral region where the absorptivity of HHb and O_2Hb is high, 510–590 nm (Fig. 8.4).

As to the 'three basic problems of oximetry', the third one could be solved by arterialisation of the spot on the skin where the reflection photometer would be applied and the best method to arterialise a spot on the skin proved to be the application of histamine through iontophoresis [108, 361, 454, 490].

The first of the basic problems could only be approached empirically because, for light remitted by a suspension, there was no simple relationship such as Lambert–Beer's law which, although not exactly valid under the circumstances, at least could lead the way. For the second problem, how to separate blood reflection from non-blood reflection, a technical solution analogous to Wood's compression manoeuvre had to be found.

The result was a small skin reflection photometer, soon called the **Cyclops oximeter** [380, 425, 454, 457, 459, 462, 464, 490], because it was usually placed on the forehead, although it could be used almost anywhere on the skin (Fig. 18.5). Figure 18.2 is a diagram of the Cyclops. It contains two selenium barrier layer photocells in back-to-back position, insulated by a thin sheet of mica. Light from one red and two green light sources can reach the skin through holes in the photocells and is reflected to the sensitive side

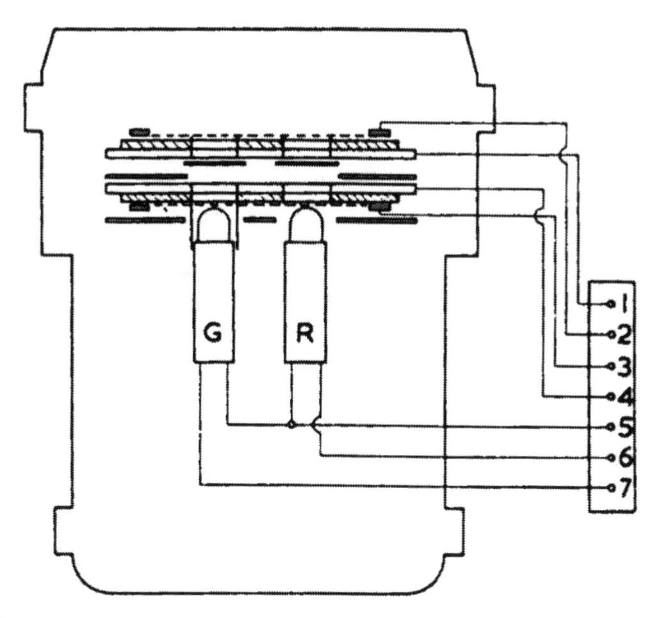

Figure 18.2. Diagram of the Cyclops oximeter. Terminals 1–4 are connected to the galvanometer, terminals 5–7 to a stabilised power supply. The connections are made through a switchbox allowing various adjustments to be made (from ref. [490]).

of the anterior photocell. The light sources are endoscope lamps covered with gelatine filters; the red one is an Ilford 281 filter cutting off sharply at 600 nm, the green one is a specially prepared filter with a transmission maximum that coincides with the β peak of O_2Hb (Fig. 8.1; Table 8.1). The posterior photocell is directly illuminated by the red lamp; it is covered by a grey filter that reduces the illumination to the average level of that of the anterior photocell.

The Cyclops was used in combination with a switchbox and a galvanometer (Kipp A70HC; sensitivy 2×10^{-9} A per mm deflection, response time 1.5 s, internal resistance 350 Ω). The photocells were connected in opposition to the galvanometer and by a potentiometer in the circuit of the posterior ('compensating') photocell, the photocurrents could be balanced perfectly. The switchbox enabled the following combinations of light sources, photocells and galvanometer sensitivity to be used:

1. Green light, anterior photocell, galvanometer at 0.1 of full sensitivity;
2. Red light, anterior photocell, galvanometer at 0.1 of full sensitivity;
3. Red light, both photocells, galvanometer at full sensitivity.

Figure 18.3 shows how this system was used to separate blood reflection from non-blood reflection. On bloodless skin the red and green reflections are made equal. On the perfused skin, the difference between the red and green reflections (Rr − Rg) then equals the blood reflection of red light, since the green reflection comes solely from non-blood. The relationship

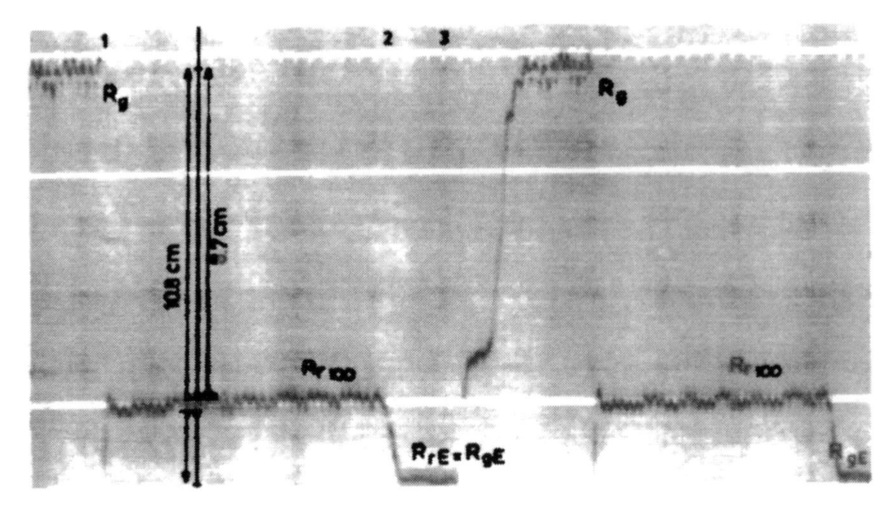

Figure 18.3. Record of light reflection measurements on an 'arterialised' spot on the human forehead.

The reflection decreases from bottom to top. With a perspex disc between the Cyclops oximeter and the skin the spot can be made bloodless by compression. As the subject breathes oxygen, S_{O_2} = 100%. The measurements are made alternately with combination 1 and 2, thus at 0.1 of full sensitivity. Rg = green reflection of the perfused skin; RrE and RgE = red and green reflection of the bloodless skin; $R\text{r}_{100}$ = red reflection of the perfused skin with S_{O_2} = 100%. The blood reflection, $R\text{r}_{100} - R\text{g}$ = 8.7 cm galvanometer deflection. At full sensitivity (combination 3) the S_{O_2} scale extends over 87 cm and the reading spot is kept on the visible part of the scale by adjusting the output of the compensating photocell (from ref. [490]).

between (Rr − Rg) and S_{O_2} had been determined in calibration experiments that showed log(Rr − Rg) to be linearly related to S_{O_2} over a wide range. During the actual S_{O_2} measurements, Rr was observed at full sensitivity and through the predetermined relationship between log(Rr − Rg) and S_{O_2}, converted to % S_{O_2} [380, 459, 462, 464, 490].

The Cyclops oximeter has for many years been commercially available (Kipp). A practical problem was that the slope of the calibration line was subject to variation due to differences between instruments and differences between patients. The former source of error could be largely eliminated by proper selection of photocells. The latter, also present in most other types of oximeter, was mainly due to the use of a broad spectral band necessary to attain a sufficient level of illumination. It could only be eliminated by individual calibration. To this end, S_{O_2} was measured in two or more arterial samples taken during the observation period. When this procedure was used, an excellent accuracy could be obtained. Using a single Cyclops oximeter and a single Haemoreflector (see below) there was in a series of 104 determinations, no mean difference and the standard deviation of the difference between the two methods was only 1.22% S_{O_2}. A series of experiments in which a Nilsson oximeter was used simultaneously with a Cyclops oximeter demonstrated the equivalence of transmission and reflection oximetry [491]. With both techniques accurate results could be obtained during observation periods of several hours under different circumstances, including surgical operations [454, 459, 490], but the procedures remained unfit for general application. This only changed through the advent of the pulse oximeter (p. 263).

In contrast with oximeters for use *in vivo*, the reflection method for measuring S_{O_2} *in vitro* proved to be quite suitable for routine application. This method was implemented in a practical instrument, the **Haemoreflector** (Kipp) [425, 455–457, 459, 464, 490, 492], that became widely used, especially for the rapid analysis of multiple blood samples taken during cardiac catheterisation and in pulmonary function testing.

The original reflection photometer contained as light source a neon tube which emits a band of red light suitable for oximetry [43]. The necessary high voltage (1000 V), however, made this light source impractical for instruments to be used in the clinical laboratory. Therefore, it was replaced in the Haemoreflector by a small 4-V incandescent lamp with a red filter. Light from this source passed through a hole in an assembly of two selenium photocells in back-to back position and reached the underside of a cylindrical cuvette placed a little distance above the photocells. Light scattered backwards by the bottom and the contents of the cuvette thus reached the photocell facing the underside of the cuvette; the other photocell was illuminated directly by the light source. The photocells were connected in opposition to a similar galvanometer as used with the Cyclops oximeter.

The Haemoreflector was provided with a revolving cuvette compartment containing three cuvettes — one with Indian ink, one with a red glass disc, and one containing the blood sample. With the Indian ink cuvette in the lightpath the galvanometer was set to zero by adjusting the output of the directly illuminated ('compensating') photocell. With the cuvette containing the red disc over the photocells, the sensitivity of the system was adjusted to a constant value. With the cuvette containing the blood sample over the photocells, the galvanometer deflection was read.

The blood samples were diluted 1 : 1 with a solution of 2% NaCl, 0.3% Na-salicylate and 0.05% NaCN. This makes the erythrocytes smaller, crenated, some even globular; the light reflection increases considerably. The cyanide served to convert any Hi to HiCN which has a lower absorptivity in this region (Table 8.2; Fig. 8.1). The blood reflection thus measured is independent of layer thickness and ct_{Hb} within wide limits; therefore measurement in a single waveband is enough. The relationship between the amount of light scattered back by the blood (R) and S_{O_2} was determined empirically: a slightly curvilinear relationship was found between log R and S_{O_2} that could be reasonably approximated by two straight lines. The Haemoreflector could be easily calibrated with the help of samples of fully saturated and fully desaturated blood.

Numerous comparative measurements using as reference methods Van Slyke's manometric method [394] and various spectrophotometric procedures showed that the Haemoreflector was sufficiently accurate for clinical use, and even for calibration of oximeters used *in vivo* and *ex vivo* [221, 454, 459, 490].

Rodrigo [327] slightly modified the method and found a linear relationship between $1/R$ and S_{O_2}. On this basis Polanyi and Hehir [308, 408] constructed a new reflection photometer for measuring S_{O_2} in blood samples using two Cd-Se photoconductor cells covered with interference filters transmitting narrow bands around 660 and 805 nm. On the basis of this prototype, a direct reading reflection photometer for measuring S_{O_2} in whole blood was constructed (American Optical). This instrument became widely used for clinical and research purposes next to the various versions of the Haemoreflector made by Kipp and, with slight modifications, by others.

CUVETTE OXIMETRY AND DENSITOMETRY

When cardiac catheterisation came into routine use as a diagnostic procedure in cardiology, especially in patients with congenital cardiac malformations, the determination of S_{O_2} at various sites in the heart and great vessels proved to be an important source of information. This created the need for serial S_{O_2} determinations, preferably during the catheterisation procedure in order that further steps in the examination could be made, taking the results into consideration. This induced Wood and co-workers to construct an oximeter that could be connected to a cardiac catheter [149, 421].

This **ex vivo oximeter** was an oblong cell containing as a cuvette a polyethylene tube with an internal volume of 0.5 mL, compressed to a lumen depth of 0.58 mm. A series of 10 tungsten filament bulbs served as a light source for transilluminating the blood sample in the polyethylene tube via a cylindrical perspex lens. The transmitted light was measured by selenium barrier layer photocells: a central photocell covered by a red filter, flanked by two photocells covered with infrared filters connected in parallel. The measurements were made either with a double scale procedure using separate galvanometers for the red and infrared absorbance or with a single scale procedure.

The double scale procedure was analogous to the one used in the Wood oximeter described on page 247, which in principle corresponded with the spectrophotometric two-wavelength method of page 219 (equation 17.2). The main differences from the latter were that broad bands of light around 680 and 800 nm were used and that the blood was not haemolysed so that Lambert–Beer's law was only valid to a limited extent. I_0 was measured with a 9 g/L NaCl solution in the cuvette, then blood from the catheter was drawn into the cuvette and I was measured. The two galvanometer readings yielded $\log(I_0/I)$ for the two wavebands and the ensuing absorbance ratio $A(\mathrm{R})/A(\mathrm{IR})$ was converted to S_{O_2} by means of a calibration graph, which, of course, did not exactly comply with equation 17.2 but showed a similar deviation from linearity as that of the Wood oximeter for use *in vivo*. Also similar to the latter method, the single scale procedure gave a direct S_{O_2} reading but was less accurate. The single scale procedure was, however, accurate enough for rapid direct intracardiac and intravascular oximetry during cardiac catheterisation. Later it was shown that the single scale procedure gave fairly accurate results because of the very fact that Lambert–Beer's law did *not* strictly apply [278].

In order to devise a simple **reflection cuvette oximeter** to measure S_{O_2} directly in samples from anywhere in the heart and great vessels, Mook and Zijlstra adapted a Cyclops oximeter to a cuvette that could be connected to a cardiac catheter [276]. It soon appeared that light reflection by blood in the cuvette was strongly dependent on the flow velocity of the blood and that stable measurements could be obtained only with a small magnetically driven mixing rod in the cuvette. The spinning rod keeps the blood in the cuvette in rapid motion and prevents rouleaux formation of the red cells. This is necessary because light reflection strongly decreases when the erythrocytes arrange themselves in rouleaux. When rouleaux formation is prevented, the amount of light scattered backwards by the blood is solely dependent on S_{O_2}, because, also in this configuration, the reflection is independent of ct_{Hb}.

Based on this prototype, the **CC-oximeter**, an *ex vivo* reflection oximeter for use during cardiac catheterisation was developed (Kipp) [268, 459, 462, 464, 490, 492, 493]. A reflectometer similar to the Cyclops oximeter (Fig. 18.2), but with only a single red light source, was placed on a cuvette assembly consisting of a cuvette filled with Indian ink and a blood cuvette with a volume of 1 mL containing a mixing rod composed of an iron core in a small perspex cylinder. The cuvette assembly was placed on top of a magnetic stirrer. On one side the blood cuvette was connected to an intravascular catheter, on the other to a syringe for drawing blood into and through the cuvette. The reflectometer, which could easily be shifted between the two cuvettes, was connected, via a switchbox, to a similar galvanometer as used with the Cyclops oximeter.

At the start of each measurement the reflection of the Indian ink filled cuvette was first used to adjust the sensitivity of the measuring system and the corresponding galvanometer deflection was subsequently reduced to zero by adjusting the output of the compensating photocell. The reflectometer was then shifted to the blood cuvette and the galvanometer deflection due to the light reflection by the blood was read. A calibration graph, $\log R$ *vs* S_{O_2}, was made during a series of cardiac catheterisations using a Haemoreflector or a spectrophotometric procedure for measuring S_{O_2} of the samples. On the ground of the calibration line a galvanometer scale reading in % S_{O_2} could be constructed.

The accuracy of the method has been determined in extensive control experiments covering a total of 1077 blood samples with, as reference methods, either a Haemoreflector or a spectrophotometric method. The overall standard deviation of the differences between the CC-oximeter and the reference method was 2.1% S_{O_2}. A comparison was also made between a Wood cuvette oximeter (double scale procedure) and a CC-oximeter by placing the two blood cuvettes in tandem. The average difference was only 0.8% S_{O_2} with a standard deviation of 1.7% S_{O_2}, thus demonstrating the equivalence of the two techniques [277].

At the time, direct oximetry during cardiac catheterisation was an important improvement. The immediate availability of S_{O_2} values together with pressure tracings and X-ray images led to a more dynamic catheterisation procedure in which each step was, at least in part, determined by the data already obtained. In most cases the diagnosis could be ascertained during the catheterisation and the number of necessary recatheterisations declined.

The immediate availability of S_{O_2} values enabled intracardiac shunts to be rapidly calculated. It can be demonstrated [490] that a left-to-right shunt (Q_y) as a fraction of the pulmonary blood flow (Q_p), $Y = Q_y/Q_p$, is given by

$$Y = [S_{O_2}(\text{PA}) - S_{O_2}(\text{SV})]/[S_{O_2}(\text{PV}) - S_{O_2}(\text{SV})], \qquad (18.1)$$

where S_{O_2}(PA), S_{O_2}(SV) and S_{O_2}(SA) are pulmonary arterial, systemic venous and pulmonary venous oxygen saturation, respectively. A right-to-left shunt (Q_x), as fraction of the systemic blood flow (Q_s), $X = Q_x/Q_s$, is given by

$$X = [S_{O_2}(\text{PV}) - S_{O_2}(\text{SA})]/[S_{O_2}(\text{PV}) - S_{O_2}(\text{SV})], \qquad (18.2)$$

where S_{O_2}(PV), S_{O_2}(SA) and S_{O_2}(SV) are pulmonary venous, systemic arterial and systemic venous oxygen saturation, respectively.

An additional advantage of direct oximetry, especially valued in paediatrics, was that after the measurement the blood was immediately reintroduced into the patient so that a virtually unlimited number of measurements could be may with very little loss of blood. Direct oximetry has been widely used [37, 268, 269, 277, 490, 492], until echocardiography made catheterisation in most cases of congenital heart disease unnecessary. Moreover, modern multiwavelength haemoglobin photometers, yielding accurate S_{O_2} values in about 1 min, while requiring < 0.1 mL of blood (Chapter 17) can easily meet all requirements when a series of S_{O_2} measurements in rapid succession are required.

Methods and instruments for oximetry *ex vivo* proved, with only small modifications, to be suitable for the application of **indicator dilution methods**. Various kinds of indicator dilution methods had already been used in physiology and medicine for many purposes, among them methods for the determination of **cardiac output**. Through the fundamental contributions of

Stewart [374] and Hamilton [214], indicator dilution methods for the determination of cardiac output had gradually improved, but it was through the application of techniques developed for *ex vivo* oximetry that dye dilution methods for the determination of cardiac output became suitable for use under clinical conditions [169, 368, 400, 401].

At first, no special equipment was used: dilution curves of blue dyes (Evans blue, Patent blue V) were recorded with the oximeter as measuring instrument connected to a peripheral artery and with injection of the dye somewhere in the venous system [490, 493]. These dyes have their absorption maximum between 600 and 700 nm, *i.e.* in the spectral region used by the oximeter to measure S_{O_2}. Hence, the oximeter is quite sensitive to the passage of the dye, but also to simultaneous changes in S_{O_2}. To eliminate the influence of changes in S_{O_2} on the dye dilution curve, Fox and Wood introduced **indocyanine green**, a dye with its maximum absorptivity near the isosbestic point of HHb and O_2Hb at 800 nm (Fig. 18.4) [119, 122, 237]. Recording indocyanine green dilution curves, however, required a densitometer using a waveband around 800 nm [120, 121, 170, 290, 344, 490, 493]. A method to measure S_{O_2} in the presence of indocyanine green has been described on p. 220.

To calculate cardiac output from the dye dilution curve the y-axis should be calibrated in units of dye concentration and the x-axis in units of time. The latter simply follows from the paper velocity of the recorder, but the former requires a calibration procedure. Moreover, the injected amount of dye has to be known exactly. After various practical devices to meet these requirements had been developed [110, 170, 171, 353, 368, 490, 493], dye dilution techniques became an established method for measuring cardiac output in man [148, 160, 172, 262, 264, 386].

Dye dilution techniques were also used for the determination of other circulatory quantities such as intracardiac shunts and valvular regurgitation [223, 251, 422, 463, 490]. When the dynamic response of the measuring system is high enough, an excellent agreement between the results of shunt determinations by dye dilution and by oximetry (equations 18.1 and 18.2) can be obtained and left-to right shunts of only 5% of the pulmonary blood flow can be detected [279]. Yet, in clinical application, ultrasound methods have now largely superseded dye dilution for the detection of shunts and valvular incompetence, and for measuring cardiac output the easier thermodilution methods [90, 160, 406, 407] are usually preferred.

SYLLECTOMETRY

In the development of the CC-Oximeter much attention was given to the elimination of the effect of **rouleaux formation** of the erythrocytes. During this investigation it appeared that the velocity of rouleaux formation can be studied by light reflection measurement and that the velocity thus measured may be a measure of the suspension stability of the blood [460, 461]. When the mixing rod in the CC-oximeter cuvette is stopped abruptly, rouleaux formation starts immediately and light reflection decreases rapidly. A set-up for studying the time course of rouleaux formation was made, comprised of a Cyclops or a CC-oximeter as a measuring instrument, a recording microvoltmeter and a device for abruptly starting and stopping the mixing rod in the cuvette [44, 490]. The name **syllectometry** (from the Greek *συλλεγομαι*: to gather) was adopted for the technique.

The change in reflection after the mixing rod has come to a stand-still is usually composed of a swift upstroke followed by an exponential decrease. The upstroke proved to be due to the rapid relaxation of flow orientation and deformation of the erythrocytes caused by the sudden decline of shearing forces. The following decrease in light reflection is caused by the formation of **rouleaux**. From the recorded 'syllectogram' the *half-time* of the reflection decrease can be calculated.

For healthy adult humans the half-time of rouleaux formation was found to be ~6 s, for human neonates (cord blood) ~15 s. Between animal species wide differences in rouleaux formation appeared to exist, varying from no rouleaux formation at all in the blood of cattle and sheep to very rapid rouleaux formation in the horse (half-time ~1.5 s) [197, 490]. The rate of rouleaux formation proved to be strongly dependent on the number of erythrocytes per unit of volume; therefore a *corrected half time* was defined as the half-time when $ct_{Hb} = 150$ g/L [197, 198]. The corrected half time was defined on the basis of ct_{Hb} because this quantity is more easily measured than the number of erythrocytes per unit of volume.

A certain concentration of macromolecular substances in the suspension medium is necessary for rouleaux formation and the nature and concentration of these substances have a considerable effect on the (corrected) half-time. In normal human blood, the concentration of fibrinogen is the principal determinant of the rate of rouleaux formation. Rouleaux formation is slowed down by increasing viscosity of the suspension medium and by decreasing temperature, but is not influenced by changes in pH [198].

The absence of rouleaux formation in bovine blood, and in the blood of some other animals as sheep and goat, is a property of the erythrocytes. Even in horse plasma and in a solution of high molecular mass dextrans, that induce very rapid rouleaux formation of most erythrocytes, bovine erythrocytes do not aggregate [198, 490].

Rouleaux formation is a prerequisite for erythrocyte sedimentation. There is no sedimentation in bovine blood, whereas the sedimentation rate of horse blood is high. In human blood a short half-time of rouleaux formation is often associated with a high sedimentation rate, but whereas a low ct_{Hb} is accompanied by slow rouleaux formation, the sedimentation rate is high. This apparent inconsistency has been explained by the model of Thorsén and

Hint concerning the sequence of events leading to the sedimentation of the red blood cells [387].

Syllectometry has been used in a study of the influence of plasma substitutes on the stability and viscosity of blood [153, 154, 483] and of haemorheological changes during normal pregnancy, pregnancy-induced hypertension and intra-uterine growth retardation [185, 186]. The method has been adopted, modified, and combined with other methods, such as the simultaneous measurement of light transmission and viscosity, by several investigators in the study of various aspects of haemorheology [127, 218, 259, 301, 340, 341, 391].

These observations also show that in all optical measurements in which native whole blood is involved, light transmission and scattering may be influenced by rouleaux formation and by flow orientation and deformation of the erythrocytes. At high shear rate the effect of orientation and deformation is considerable, while at low or zero shear rate rouleaux formation dominates.

FIBRE OPTIC OXIMETRY

An *in vivo* **fibre optic reflection oximeter** measures S_{O_2} at the tip of a cardiac catheter inserted into a blood vessel or into the heart. The catheter contains two bundles of optic fibres, one for guiding light from an external source to the blood, and one for guiding the backscattered light to a measuring system. According to the practical definition that a reflection oximeter is a photometer with light source and light detector at the same side of the blood or tissue, as opposed to a transmission oximeter with light source and light detector opposite each other, a fibre optic oximeter is indeed a reflection oximeter. However, the difference between reflection and transmission measurement in turbid media is less fundamental than formerly supposed: in all types of reflection and transmission measurement *in vivo*, the detector can only be reached by the light via complicated pathways through blood or tissue. These lightpaths strongly depend on the geometry of the experimental set-up.

Mook *et al.* [274, 275] compared the properties of a fibre optic oximeter with those of a conventional *ex vivo* reflection oximeter and found several differences. The independence of light reflection from ct_{Hb}, for instance, proved to be less distinct and strongly influenced by the number of fibres in the catheter; with two bundles of 60 fibres in the catheter light reflection even continued to increase with increasing ct_{Hb} to 250 g/L.

The most surprising result, however, was that in the fibre optic reflection spectra of (bovine) blood with $S_{O_2} = 0$ and 100%, recorded from 600 to 1000 nm, there was no isosbestic point of HHb and O_2Hb near 800 nm (Table 11.4), but an isosbestic region extending from ~840 to ~890 nm. This was definitely not caused by selective light absorption by the optic fibres, since in transmission spectra made through the same fibres the isosbestic point was found at ~790 nm. These measurements *in vitro* were corroborated by measurements in the anaesthetised dog by recording the change in the reflection of arterial blood following short periods of N_2 breathing, using different interference filters. However, in the diffuse reflectance spectra of unhaemolysed canine blood measured by Polanyi and Hehir [309] using an integrating sphere, the isosbestic point was found at 805 nm, hardly different from the value expected for canine haemoglobin. The fibre optic reflection spectra of Mook *et al.* [275] are shown in Fig. 18.4, together with the absorption spectra of HHb and O_2Hb and indocyanine green.

Another difference between the results of Mook *et al.* [274, 275], those of Polanyi *et al.* [112, 309], and those encountered in conventional *ex vivo* reflection oximetry concerns the relationship between light reflection and S_{O_2}. Polanyi found a linear relation between $R(880)/R(640)$ and S_{O_2}, but in Mook's experiments there was a slight deviation from linearity. The relation between log $R(640)/R(880)$ and S_{O_2}, however, proved to be linear, whereas on the ground of the experience with *ex vivo* reflection oximetry some deviation from linearity had been expected [490].

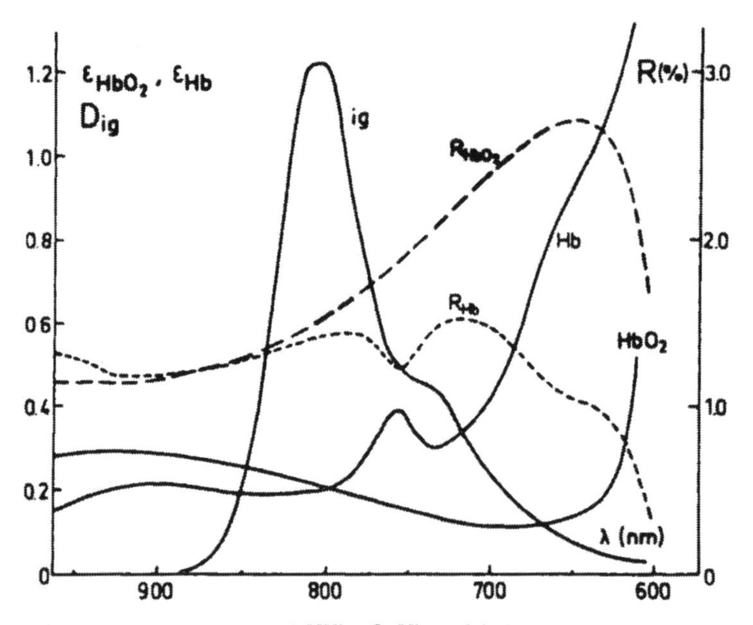

Figure 18.4. Absorption spectra of HHb, O_2Hb and indocyanine green and fibre optic reflection spectra of blood with $S_{O_2} = 0$ and 100%.

For HHb and O_2Hb the absorptivity is plotted *vs.* wavelength (ε_{HbO_2} and ε_{HHb}); for indocyanine green the absorbance of a 5 mg/L solution in plasma, measured with $l = 1.00$ cm (D_{ig}) is plotted. The fibre optic reflection spectra (R_{HbO_2} and R_{Hb}) are from Mook *et al.* [275] (from ref. [235]).

Thus it appeared that the results obtained by *in vivo* fibre optic reflection measurement of blood could not be predicted on the basis of what had earlier been observed in *ex vivo* reflection oximetry. The relation between light reflection and S_{O_2}, ct_{Hb} and blood flow proved to be dependent of the different optical geometry of, and between, fibre optic systems for oximetry *in vivo*.

The necessity in fibre optic reflection oximetry to measure at least at two wavelengths is not only due to the strong effect of variations in the haemoglobin concentration, but also to the presence of other interfering factors such as changes in blood flow. When stagnant blood starts to flow, fibre optic light reflection increases due to breaking up of rouleaux (p. 256), at high flow rate the reflection usually decreases again due to directional effects on the erythrocytes [413]. In measurements *in vivo* in arterial or venous blood the flow rate is usually well above the level where rouleaux formation of erythrocytes occurs, so that only variations in flow orientation of the erythrocytes influence the reflection measurements. These effects, however, may be considerable.

The first practical fibre optic reflection oximeter was constructed by Polanyi and Hehir [309]. The main reason to develop this instrument, primarily meant for use during cardiac catheterisation, was the low dynamic response of *ex vivo* cuvette oximeters. The continuous measurement of S_{O_2} with a dynamic response high enough to follow the variations within the cardiac cycle that may occur near a septal defect indeed seemed promising.

In their instrument, light from a tungsten filament lamp was focused on an ingoing bundle of glass fibres ($\sim$ 150 individual fibres with a diameter of $\sim$ 50 μm). The light beam was alternately intercepted by two interference filters with maximum transmission at 660 and 805 nm, with a frequency of 30–40 Hz. The light backscattered by the blood was conducted by a second fibre bundle to a single photomultiplier tube that alternately produced signals representative for the reflection of red and infrared light, from which the reflectance ratio $R(805)/R(660)$ was calculated. This ratio appeared to be linearly related to S_{O_2}. The instrument was used in patients for measuring S_{O_2} of arterial blood, as well as for the recording of dye dilution curves. To record indocyanine green dilution curves, interference filters with maximum transmission at 805 and 900 nm and the ratio $R(900)/R(805)$ were used. The wavelength 805 nm is near the absorption maximum of indocyanine green (Fig. 18.4).

Another early description of a fibre optic oximeter, operating at 640 and 805 nm, was given by Kapany and Silbertrust [204]. An improved version of the fibre optic oximeter of Polanyi and Hehir [309], with a higher dynamic response, was described by Gamble *et al.* [125]. In this instrument, white light was conducted to the blood by the efferent fibre bundle and the reflected light was divided in 540 pulse pairs per second by a rapidly rotating glass disc with alternate clear and mirrored areas. The clear areas allowed light to pass through a 660 nm filter, while the mirrored areas reflected the light to a 805 nm filter. The two beams reached the same photomultiplier tube.

Special attention was given to prevention of contact of the optic fibres with the wall of the blood vessel. This may give erroneous results through light reflected from the wall reaching the photodetector. Two devices were used; the one was a tripod-like structure, the other was similar but had, in addition, a flexible tapered tip that extended another 20–25 mm. The higher dynamic response of the measuring system was especially advantageous when it was used for recording dye dilution curves [182].

Mook *et al.* [274, 275] used for their fibre optic reflection oximeter similar optic fibres (American Optical) but with 50, 60 or 100 individual fibres in each bundle. Artefacts from wall contact of the catheter tip were avoided and flow effects were reduced by inserting the fibre optic catheter inside another catheter, so that its tip was a few mm from the tip of the outer catheter through which blood was sampled at a constant rate. The measuring system differed in several respects from the one of Polanyi and Hehir [309]. The reflected light from the afferent fibre bundle was collimated by an Abbe condenser, then split by a dichroic mirror and, after passing through different interference filters, measured by two photomultiplier tubes. To measure S_{O_2}, a filter with maximum transmission at 640 nm was used, with a filter with maximum transmission at 880 nm in the compensation channel. To record indocyanine green dilution curves, filters with maximum transmission at 800 and 920 nm were used in the detecting and the compensating channel, respectively. For oximetry the linear relationship of $\log R(640)/R(880)$ and S_{O_2} was used; for densitometry of indocyanine green, the relationship between $R(800)/R(920)$ and dye concentration, linear up to an indocyanine green concentration of 20 mg/L, was used.

A further development of the instrument was the use of silicium barrier layer photocells instead of photomultiplier tubes and the use of a low-pass electronic filter for suppression of residual noise due to flow effects [273]. Oeseburg *et al.* [298] used this fibre optic oximeter, in combination with the polarographic measurement of p_{O_2} using a catheter-mounted membrane-covered platinum electrode, for direct recording of the oxygen dissociation curve of haemoglobin in the aorta of the anaesthetised dog. The two measuring systems were calibrated *in vivo in situ* and the response times were made equal by electronic filtering. The S_{O_2} and p_{O_2} signals were fed into an *x*-*y*-recorder. S_{O_2} and p_{O_2} were made to fall and to rise again by temporarily substituting N_2 for O_2 in the gas mixture supplied to the dog. Oxygen dissociation curves were obtained down to $S_{O_2} \approx 40\%$ and a Bohr effect (Chapter 20) could be demonstrated.

A detailed description of a more advanced version of this fibre optic reflection oximeter and of the factors that, in this measuring system, determine the relation between light reflection and S_{O_2} has been given by Landsman *et al.* [234, 235]. In this instrument, the photocell signals are fed into a divider giving the ratio of measuring, $R(640)$, and compensating, $R(920)$, photocell output. The relation between $\log R(640)/R(920)$ and S_{O_2}, which is represented by a slightly curved line, is linearised by subtracting a constant voltage from the divider output before taking the logarithm. The slope of the calibration line proved to be dependent on ct_{Hb}, but for the S_{O_2} range of 70–100% an average calibration can be used. For 78 samples of pig blood with S_{O_2} between 70 and 100% the standard deviation of the difference in S_{O_2} measured with the fibre optic oximeter and with an OSM1 photometer (Radiometer) was 1.9% S_{O_2}. For 152 samples over the entire S_{O_2} range the standard deviation of the difference was 3.1% S_{O_2}. For accurate measurements below $S_{O_2} = 70\%$ the calibration must be adapted to ct_{Hb}.

Other fibre optic reflection oximeters constructed at the time [48, 217] employ pulsed light-emitting diodes and a single detector cell. The output of the detector thus contains the signals at two or more different wavelengths, which are separated electronically. Hence, the measurements at the different wavelengths are not synchronous and the compensation of non-specific effects is dependent on the pulse frequency. This is, at least theoretically, a disadvantage in the compensation of rapid flow-dependent effects. Landsman *et al.* [235] therefore preferred continuous measurement with two photocells, which gives the best possible compensation and dynamic response. Another advantage of this system is the possibility to use exchangeable light filters, so that any wavelength and spectral band width can be chosen. This is especially important for dye densitometry.

The corresponding fibre optic densitometer for indocyanine green [236] consists of the same components as the oximeter [235]. In the measuring channel a filter with maximum transmission at 800 nm is used. When fibre optic catheters with glass fibres are used, the compensation channel does not contain a filter, so that for compensation a broad spectral band is used. This proved to give the best compensation for flow effects and made the densitometer linear for indocyanine green up to a concentration of 50 mg/L. The fast dynamic response made the densitometer suited to record left and right *ventricular wash-out curves* for the determination of ejection fraction and end-diastolic volume.

The preference for using a wavelength range of 660–680 nm in fibre optic oximetry is based on the observation that in the spectrophotometric determination of S_{O_2}, the sensitivity of the measurement, expressed as dA/dS_{O_2}, *i.e.* the change in absorbance per unit change in oxygen saturation, has a maximum value near 660 nm [271]. Most commercial fibre optic oximeters operate in this wavelength range. However, it has been shown that in reflection oximetry the highest sensitivity for changes in S_{O_2} lies at a shorter wavelength; an experimental set-up using 620/810 nm proved to be considerably more sensitive than a similar device using 660/810 nm [270]. The marked increase in absorptivity of HHb and O_2Hb towards 600 nm (Table 8.2) makes wavelengths $\leqslant$ 620 nm less suitable for fibre optic oximetry, but 630 nm and 640 nm proved to be optimal for this purpose [235, 448].

The advantage of *glass fibres* is that they transmit visible and near infrared light with little loss and hardly any effect on the spectrum. The cheaper and more robust *plastic optic fibres, e.g.* polymethylmetacrylate fibres, have a considerable light filtering effect [236]. This limits the choice of wavelengths for oximetry and densitometry. For practical reasons, however, plastic optic fibres are now almost generally used for these purposes.

The Schwarzer IVH 3/4 fibre optic reflection oximeter/densitometer, a late descendant of the original instrument of Polanyi and Hehir [309], uses one (plastic) efferent fibre to guide alternating light pulses of two light emitting diodes at 660 and 805 nm (for oximetry), or at 805 and 930 nm (for measuring indocyanine green) into the blood, and two fibres to conduct the reflected light to a single photodetector. The signals are separated and processed electronically, resulting in the digital display of S_{O_2}, dye concentration and cardiac output [217, 378].

Another commercially available fibre optic oximeter (Oximetrics) uses only two optic fibres with a diameter of $\sim$0.25 mm. For measuring S_{O_2} three wavelengths are used; 1-ms light pulses from three light emitting diodes at 670, 700 and 800 nm are conducted into the blood, followed by a 1-ms dark interval. The cycle of four pulses occurs with a frequency of 250 Hz. A single afferent fibre conducts the reflected light to a photodiode. The wavelengths have been empirically selected to minimise interference through flow effects and changes in ct_{Hb}. A polynomial function was found to relate $R(670)/R(700)$ and $R(800)/R(700)$ best to S_{O_2}. The microprocessor calculating S_{O_2} also corrects for changes in ct_{Hb}, using the fact that $R(800)/R(700)$ had been found to be independent of ct_{Hb} at low values of S_{O_2}, and $R(670)/R(700)$ independent of ct_{Hb} at high values of S_{O_2} [48].

Nijland *et al.* [293] have tested this instrument in piglets for the S_{O_2} range of 15–100%. Paired data were obtained of the fibre optic oximeter with the catheter in the aorta and of arterial samples analysed by multiwavelength photometry *in vitro*. The fibre optic oximeter underestimated S_{O_2} below 78%; the underestimation worsened towards lower S_{O_2} values. The mean difference over the whole range was −3.4% S_{O_2} and the precision was 3.8% S_{O_2}. The mean difference could be eliminated by calibration *in vivo*.

An important innovation of fibre optic oximetry was that it brought monitoring of mixed venous S_{O_2} within reach of the clinician. This quantity is a measure of the difference between the oxygen supply to the tissues and the oxygen consumption by the tissues. A low mixed venous S_{O_2} may therefore be a sign of an insufficient oxygen flow to the tissues. Krauss *et al.* [226] introduced the use of a fibre optic oximeter for continuous measurement of mixed venous S_{O_2} in patients after thoracic surgery. They inserted the tip of the catheter into the pulmonary artery or left it in the superior caval vein. The measurements lasted from 20 to 60 h.

Van der Hoeven *et al.* [164] have shown that continuous measurement of venous S_{O_2} by fibre optic oximetry is also possible in newborn infants. In 24 of 36 neonates a fibre optic catheter could be inserted into the right atrium through the umbilical vein. The average observation period was 107 h.

Although measuring mixed venous or central venous S_{O_2} has been frequently applied in postoperative and intensive care of patients, its clinical significance has not yet been ascertained. Experimental evidence seems to demonstrate that only very low values of mixed venous S_{O_2} are a clear sign of oxygen-restricted metabolism. In experiments in 10- to 14-day old piglets it was observed that mixed venous S_{O_2} had to fall to 15% before clear biochemical signs of hypoxia were present [165]. However, in these experiments hypoxia was induced by decreasing inspiratory p_{O_2}; in other hypoxia models it may be different.

In biomedical research, fibre optic oximetry has found many applications. A nice example is the measurement of cardiac output in rats with the help of a 2.4F fibre optic catheter in the aortic arch and injection of indocyanine green in the right atrium. The catheter was connected to a Schwarzer IVH4 oximeter/densitometer. There was a fair agreement with electromagnetic blood flow measurement in the ascending aorta [378].

In some cases the *ex vivo* use of a fibre optic catheter may be appropriate. Woudstra *et al.* [428–430] constructed an extracorporeal system with a very small dead space for the simultaneous measurement of arterial pH, p_{CO_2}, p_{O_2} and S_{O_2} in the foetal lamb. The epoxy-resin flow-through cuvette had a central channel, 131 mm in length and 1 mm in diameter; it fitted tightly in an aluminium block which was kept at 39°C. The cuvette was connected to a foetal carotid artery and jugular vein. A roller pump maintained a steady flow of blood of 2 mL/min. The total volume of the system, including the connecting catheters was 3.25 mL. Through small holes five probes were inserted with their tips into the central channel: four electrodes (pH, pH-reference, p_{CO_2}, p_{O_2}) and a fibre optic catheter. The latter was connected to a Schwarzer IVH3 oximeter. The transport time from the arterial catheter to the sensors was 64 s. After insertion of catheters, the lamb was placed back in the uterus; the proximal parts of the catheters were then tunnelled to a bag at the flank of the ewe. For a series of measurements to be made, the foetal catheters were connected to the cuvette, with the roller pump on the catheter from the carotid artery. Reliable measurements were obtained for continuous periods of 6 h, occasionally even of 48 h.

PULSE OXIMETRY

Pulse oximetry was invented around 1972 by the Japanese bioengineer Takuo Aoyagi working in the research division of Nihon Kohden Corporation; the first publication is an abstract submitted in 1974 to the Japanese Society of Medical Electronics and Biological Engineering [349]. Although a few manufacturers of medical instruments soon started to develop non-invasive *in vivo* oximeters based on Aoyagi's idea [315, 436], it took some 10 years before it was widely perceived that through these oximeters two of the basic problems of oximetry could be solved in such a way that oximetry would, at long last, become the simple routine method clinical medicine had been waiting for since the nineteen sixties.

In spite of much research and development, oximetry had remained a complicated technique requiring special care to attain reliable results. Taking the Cyclops oximeter as an example, each series of measurements started with creating a spot with maximum vasodilation with the help of histamine iontophoresis. After the oximeter had been applied to the arterialised spot, the procedure shown in Fig. 18.3 had to be carried out to be able to distinguish, during the measurements, between reflection from blood and from non-blood. Corrections for changes in the amount of blood in the reflecting field had to be made at intervals and the oxygen saturation was calculated with a special slide rule. All this could be obviated in a single stroke by the introduction of the pulse principle. By measuring the changes in light reflection corresponding in frequency with the heart beat, a signal is obtained that originates solely from the arterial blood in the reflecting skin area.

A heart-synchronous pulsation had been observed from the very beginning in the output of conventional oximeters. Figure 18.5 shows an experiment with three Cyclops oximeters at different spots of the human body. S_{O_2} is recorded with high dynamic response and a high paper velocity. The clearly visible pulsations would correspond with a fluctuation in S_{O_2} of 1–2%, when the system had been calibrated in the conventional way. In the usual

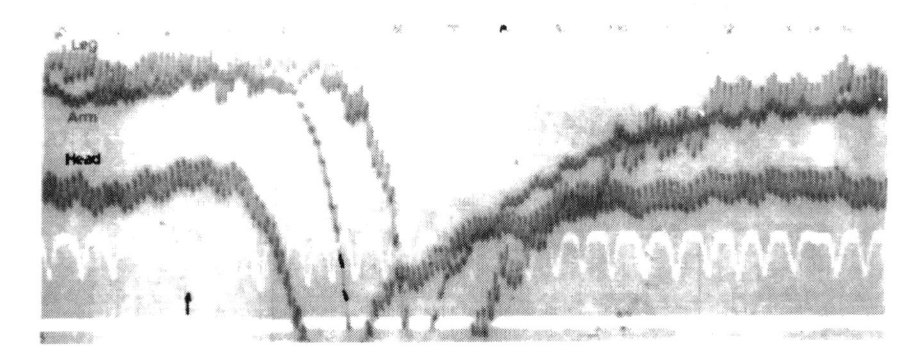

Figure 18.5. Record of S_{O_2} measured simultaneously with three Cyclops oximeters placed on the forehead, a forearm and a leg of a healthy volunteer.

A fall in S_{O_2} is provoked by having the subject take a few breaths of N_2 (arrow). The desaturated blood reaches the oximeters after different intervals depending on their distance from the lungs. Note the large heart-synchronous variations in the S_{O_2} signals (from ref. [454]).

applications, the pulsating part of the output signal was suppressed and/or disregarded. The much larger non-pulsating part of the signal was used for the measurement of S_{O_2}.

The relation between S_{O_2} and the heart-synchronous changes in the amount of light backscattered from the tissue can be understood with the aid of Figs 18.3 and 18.5. Figure 18.3 shows that the reflection of red light from the bloodless skin (RrE) decreases when blood with $S_{O_2} = 100\%$ enters the blood vessels ($R\text{r}_{100}$). Since, in the red part of the spectrum, the absorptivity of HHb is higher than that of O_2Hb (Fig. 8.4), this decrease will be greater when S_{O_2} is $< 100\%$. Consequently, the downward parts of the rapid pulsations in Fig. 18.5 correspond with an increase of arterial blood in the reflecting skin tissue, and these downward strokes will be greater the higher the fraction of HHb, hence the lower S_{O_2}, in the arterial blood. The pulsations in Fig. 18.5 can also be conceived as being, for a tiny part of the reflecting field (the part occupied by the varying arterial volume), analogous to the effect of the compression manoevre of Fig. 18.3 for the entire reflecting field.

A similar relationship holds for light transmission. Hence, the light intensity at the detector when the transilluminated volume is at its minimum can be conceived to be I_0 for the varying part of the transilluminated volume in the same way as I_0 was approximated for the entire volume (by squeezing out all blood) in the Wood oximeter (p. 246) The light received by the detector when the transilluminated volume is at its maximum is then analogous to I. Consequently, the variation in light transmission contains the same information as obtained by a conventional oximeter, but without the cumbersome preparatory manipulations, and with the advantage that the results of the measurement are not dependent on a single adjustment procedure before the start of the actual measurements.

Because the pulsating part of the light received by the detector is more prominent in the transmitted light than it is in light backscattered from the skin, most pulse oximeters are transmission oximeters. Reflection pulse oximeters have until now been developed only for special applications, such as measurements on the foetus during labour.

The photometer part of a pulse oximeter is not appreciably different from that of a conventional non-invasive *in vivo* oximeter. The essential difference is in the way the output signal of the photodetector is processed. The processing problems for which technical solutions had to be, and have been, found were quite formidable. Very small signals must be separated from much larger ones, amplified and processed so that an output is obtained that has a dependable relation with the oxygen saturation of the arterial blood.

Most pulse oximeters contain two light emitting diodes (LEDs) with maximum emission at 660 and 940 nm and a single photodetector sufficiently sensitive over this spectral range. The LEDs are switched on and off by a microprocessor. In some instruments three light levels are measured, the levels of the two LEDs and the level when both LEDs are off. Thus, the ambient light is measured and the photocurrent due to the ambient light is subtracted from the photocurrents of the red and infrared LED. The amplitude of the signals is kept at a sufficient level by automatically adjusting the light intensity of the LEDs and/or the amplification of the output of the photodetector.

Commonly a normalisation process is performed in which the varying part of the red and infrared photocurrents, resulting from expansion and relaxation of the arterial bed, is divided by the non-varying part, resulting from light that interacted with tissue components, with non-pulsatile arterial blood and with venous blood [260]. This gives the ratio of the relative changes of the intensities of the transmitted red and infrared light, $\Delta \ln[I(\mathrm{R})]/\Delta \ln[I(\mathrm{IR})]$. If Lambert–Beer's law is assumed to be valid, this ratio, which is henceforth designated R/IR, is determined by $\varepsilon_{\mathrm{HHb}}(\mathrm{R})$, $\varepsilon_{\mathrm{HHb}}(\mathrm{IR})$, $\varepsilon_{\mathrm{O_2Hb}}(\mathrm{R})$, $\varepsilon_{\mathrm{O_2Hb}}(\mathrm{IR})$ and $S_{\mathrm{O_2}}$ according to [135]:

$$R/IR = \Delta A(\mathrm{R})/\Delta A(\mathrm{IR}) = \frac{\varepsilon_{\mathrm{HHb}}(\mathrm{R}) \cdot [1 - S_{\mathrm{O_2}}] + \varepsilon_{\mathrm{O_2Hb}}(\mathrm{R}) \cdot S_{\mathrm{O_2}}}{\varepsilon_{\mathrm{HHb}}(\mathrm{IR}) \cdot [1 - S_{\mathrm{O_2}}] + \varepsilon_{\mathrm{O_2Hb}}(\mathrm{IR}) \cdot S_{\mathrm{O_2}}}. \tag{18.3}$$

When the values of HHb and O_2Hb at 660 and 940 nm (Table 8.4) are taken for the four εs, $S_{\mathrm{O_2}}$ is given by the following equation:

$$S_{\mathrm{O_2}} = \frac{\varepsilon_{\mathrm{HHb}}(660) - \varepsilon_{\mathrm{HHb}}(940) \cdot [R/IR]}{\varepsilon_{\mathrm{HHb}}(660) - \varepsilon_{\mathrm{O_2Hb}}(660) - [\varepsilon_{\mathrm{HHb}}(940) - \varepsilon_{\mathrm{O_2Hb}}(940)] \cdot (R/IR)}, \tag{18.4}$$

which is analogous to the equation that applies to spectrophotometric analysis of a two-component system, when neither of the wavelengths is isosbestic for the two components (equation 3.8).

However, equation 18.4 cannot simply be applied because the deviation from Lambert–Beer's law is considerable, as has already been shown for the Wood oximeter. The third of the basic problems of oximetry has definitely not been solved by the introduction of the pulse principle, as could hardly be expected. Therefore, pulse oximeters are calibrated empirically, usually by having healthy human subjects breathe gas mixtures with different low fractions of O_2, while arterial blood samples are collected and $S_{\mathrm{O_2}}$ of these samples is measured *in vitro*. For obvious reasons this procedure only yields the relationship between R/IR and $S_{\mathrm{O_2}}$ down to ~70%. The lower part of the calibration line has to be estimated by extrapolation or by means of animal experiments [292].

Around 1990, the pulse oximeter had evolved to a simple to use instrument that had only to be applied to the patient to obtain an $S_{\mathrm{O_2}}$ reading, usually in combination with an indication of the arterial pulsations and a heart rate reading. The pulse oximeter probes are mostly applied to a finger tip, but nose, ear and forehead are used too. In babies the probes are often applied to a hand or foot. A review published in 1989 mentions several pulse oximeters from different manufacturers, used for numerous clinical applications, especially in intraoperative, postoperative and critical care, with some 150 references to papers describing properties or applications of pulse oximeters [213]. Three years later another review appeared, summarising the literature since early 1989 with some 250 references [350]. It mentions more than 35 manufacturers and concludes that *pulse oximetry is now well-established — both inside and outside the operating room and intensive care unit — as a useful and sensitive detector of hypoxemia.* Soon reports appeared of comparative evaluations of the various types and brands of pulse oximeters [64, 65, 382].

The review articles [213, 260, 350, 412] mention several sources of error, among which poor perfusion and motion artefacts are the most prominent. When the arterial pulsations are less than ~ 0.2% of total light transmission, the signal becomes unsuitable for reliable calculation of S_{O_2}. Most instruments then display some kind of error signal. Patient motion may be a major source of error in awake patients; numerous methods of fixation have been devised to minimise this kind of interference. To this end, some instruments use synchronisation of the output signal of the photodetector with the R-wave of the electrocardiogram [260].

Because a pulse oximeter is a two-wavelength photometer for the analysis of the two-component system HHb/O_2Hb, the measurement can be interfered with by the incidental presence of other coloured substances, such as dyshaemoglobins, bilirubin or injected dyes. A reasonable estimate of errors in S_{O_2} as measured by a pulse oximeter operating at 660 and 940 nm that may ensue from the presence of COHb and Hi, as well as from the substitution of HbA by HbF can be made as follows [294, 480]. When in equation 18.4 R/IR is replaced by $A(660)/A(940)$, the corresponding equation for spectrophotometry *in vitro* is obtained (equation 3.8). Using data from Tables 8.4 and 8.11, $A(660)/A(940)$ can be calculated for any combination of HbA and HbF, in the form of any combination of HHb, O_2Hb, COHb and Hi. By substituting these values of $A(660)/A(940)$ into the equation, and taking for the constants the values of Table 8.4, the values of S_{O_2} that a pulse oximeter would give for these combinations of derivatives of HbA and HbF can be estimated. Some results are given in Table 18.1.

Table 18.1 shows that HbF has an appreciable influence on the values displayed by a 660/940 nm pulse oximeter only at very low S_{O_2}. At the level of 25% S_{O_2} and 100% HbF the underestimation may be some 5% S_{O_2} [294]. This deviation may be of clinical significance only during foetal monitoring. The error caused by the presence of COHb is insignificant at all levels of S_{O_2} and HbF. At high values of S_{O_2}, Hi causes a clear underestimation, while at 40% S_{O_2} there is a similar overestimation; in the 60–80% S_{O_2} range the error is hardly significant. Given the low frequency of high fractions of COHb and Hi in human blood [447] and the modest extent of the deviations, neither HbF

Table 18.1.
Influence of HbF, COHb and Hi on S_{O_2} by spectrophotometric measurement at 660 and 940 nm as a model of a pulse oximeter operating at the same wavelenghts

Haemoglobin	S_{O_2} (%)			
HbA	100	80	60	40
HbA/10% COHb	99.1	79.2	59.3	39.3
HbA/10% Hi	93.0	77.8	62.5	47.2
40% HbF	100.2	79.3	58.4	37.5
80% HbF	100.4	79.3	58.4	37.6
40% HbF/10% COHb	99.3	79.0	58.2	37.6
40% HbF/10% Hi	92.9	77.4	61.6	46.0
80% HbF/10% COHb	99.4	78.8	57.1	35.8
80% HbF/10% Hi	92.8	77.0	60.8	44.7

nor COHb and Hi constitute an appreciable source of error in pulse oximetry at the commonly used wavelengths. Dyshaemoglobins *per se* may of course be a serious cause of morbidity; any suspicion of their presence therefore calls for multicomponent analysis of haemoglobin in a blood sample.

In the last two decades, especially after pulse oximetry came into general use, much theoretical and experimental work has been done to solve also the first of the basic problems of oximetry, the absence of a satisfactory theory of the interaction of light with blood and tissue suitable to predict S_{O_2} from light transmission or light reflection. In most descriptions of pulse oximetry, however, a Lambert–Beer model has still been used to express R/IR as a function of S_{O_2}. Transmission through tissue is then compared with transmission through a cuvette containing layers of tissue, venous blood and arterial blood while the influence of scattering is neglected [432, 435, 437]. The relation of S_{O_2} and R/IR arrived at in this way (equations 18.3 and 18.4) deviates from the calibration curves that have been obtained empirically.

In early studies on reflectance pulse oximetry the limitation of the Lambert–Beer approach was already clearly shown by the fact that the ratio R/IR depends on the distance between the light sources and the detector at the skin surface, even when the numerator and the denominator are measured simultaneously [138]. The effect is obviously due to scattering of light in the tissue.

This observation was the starting point of an investigation of the influence of the optical properties of the tissue on reflectance pulse oximetry [135]. To support these investigations, a three-channel reflectance pulse oximeter was developed, containing red and infrared LEDs (660 and 940 nm) and signal processing electronics for the three silicon detectors, that are located at different distances from the light sources (Fig. 18.6).

The main reason for the limited value of the Lambert–Beer models of pulse oximetry is that scattering causes the light to follow paths of widely

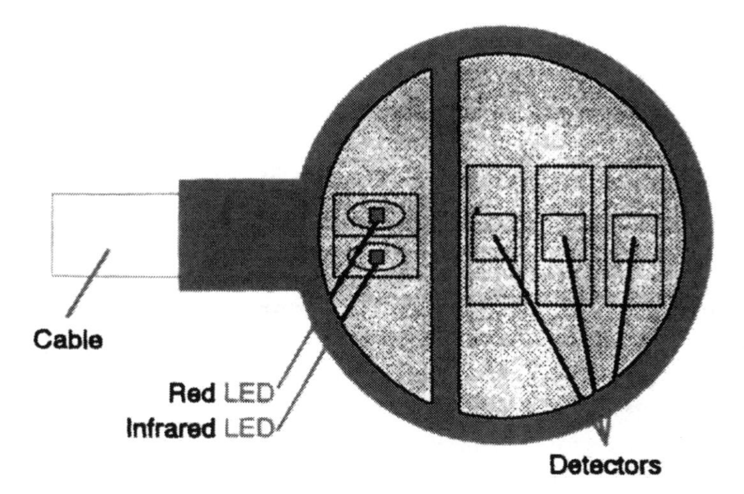

Figure 18.6. Diagram of sensor for reflectance pulse oximetry.
The sensor (diameter 18 mm) contains red and infrared light sources and three detectors at distances of 4.5–10 mm from the light sources (from ref. [135]).

different lengths on its way through the tissue from source to detector [142]. Firstly, the pathlengths are different for each of the applied wavelengths. Secondly, there is a distribution of lightpath lengths of the photons within the tissue at each of the wavelengths. The latter may be seen as a collection of unfolded lightpaths, each assigned to a beam in accordance with the pathlength the photon traverses through the tissue. We now assign imaginary cuvettes to each of the beams with the light that has passed through all these imaginary cuvettes collected by a common detector. The cuvettes are assumed to contain the absorbers that are present in the tissue and the blood. The analysis of the optical properties of such a complex system with a Lambert–Beer approach assuming a single lightpath length obviously leads to grave inaccuracies; a better understanding of (pulse) oximetry can only be attained when the influence of multiple scattering in the tissue is taken into account.

To this end, numerical models for reflectance pulse oximetry have been developed that are based on the simulation of light propagation in a semi-infinite turbid medium ('Monte Carlo simulation') [143]. Application of diffusion theory was avoided because it had been shown to be less suitable for this purpose [136, 137]. With the aim of modelling reflectance pulse oximetry, Monte Carlo results were first applied to *in vivo* measurements to obtain information on the optical properties of skin tissue [141]. Subsequently, Monte Carlo simulations were used to calculate calibration curves that proved to be dependent on the absorption and scattering coefficients of the tissue, including haemoglobin, especially at low values of S_{O_2} [142]. A similar dependence of the calibration curve on optical properties of the tissue was reported earlier by Schmitt [342], using the diffusion approximation. The observed influence of the optical properties of the tissue, including the non-pulsating part of the blood, is related to the fact that an increase in the absorption coefficient generally decreases the mean optical pathlength. This was observed by Benaron *et al.* [24] during measurements with phase-shift spectroscopy on the head of newborn infants (p. 274). Furthermore, the Monte Carlo models supported the observation that the value of R/IR depends on the distance between light sources and detector [143].

In **foetal pulse oximetry**, S_{O_2} values are encountered far below the range where empirical data are available from calibration curves obtained in adults. Because of the influence of tissue optical properties, calibration curves for the lower S_{O_2} range have to be based on experimental data, *e.g.* as obtained in the foetal lamb [85]. Several investigators have studied the influence of the optical properties of the tissue in foetal reflectance pulse oximetry. Johnson *et al.* [201] showed that changes in tissue properties after the formation of a caput succedaneum in the birth canal may result in an apparent decrease in S_{O_2} of 15%. Dassel *et al.* [83] have simulated the effect of pressure on the head of the foetus exerted by the birth canal by applying pressure on a ring around an oximeter on the head of foetal lambs. For applied circular pressures above the systolic blood pressure the value of R/IR increased. An increase measured by a detector that was 7 mm away from the light sources corresponded with an apparent fall in S_{O_2} of 7%. The signal obtained under these conditions originates from the cerebral circulation, since the cutaneous circulation was occluded.

Furthermore, the value of R/IR proved to depend on the location of the oximeter when it was applied, without pressure, to the head of human neonates [83, 84]. Differences in R/IR corresponding to an apparent change of 8 and 13% S_{O_2} were found for detectors at 6.9 and 9.4 mm from the light sources, respectively. These results have probably not only been due to the tissue properties immediately below the sensor, but also to the layered geometry of the tissue, because part of the signal is obtained from light that has penetrated through the thin neonatal skull. Moreover, the fractions of red and infrared light that have passed through the skull may have been different.

Another relevant factor in reflectance pulse oximetry is the influence of the venous blood [87]. Measurements on the forehead of normal adults revealed that application of a pressure >80 mm Hg significantly decreased the variability in R/IR, probably by forcing venous blood out of the tissue under the sensor. This observation would imply that under these circumstances elimination of venous blood improved the accuracy of the S_{O_2} measurement. Pressure on the reflectance oximeter when applied to the forehead of human neonates had little effect on R/IR, unless it exceeded 150 mm Hg [87]. It may be that then only pulsations from cerebral blood vessels contribute to the signal. During firm pressure, S_{O_2} calculated from R/IR decreased 9% for the detector at 7 mm from the light source, but only 2% for the detector at 10 mm from the light source. Consequently, when a reflectance pulse oximeter is applied to the foetal or neonatal head, R/IR measured at a larger distance from the light sources is mainly due to signals from the cerebral circulation and cannot be improved by pressure on the sensor.

Nijland [292] studied the influence of the wavelengths used in reflectance pulse oximetry with two different sensors taped to the shaved skin of piglets. Using 735 and 890 nm yielded, in the lower S_{O_2} range, definitely better results than when 660 and 890 nm were used. This may have been due to the elimination of the relatively high absorption of HHb at 660 nm when S_{O_2} is low. In a subsequent study in piglets, Nijland *et al.* [295] calibrated a new type of reflectance pulse oximeter (Nellcor), operating at 735 and 890 nm, for S_{O_2} in the range of 17–100%. The light source-to-detector distance was 14 mm. No differences were found between results obtained on either head or buttock of the animals.

These observations show that many factors influence the interaction of light with tissue elements on its way from source to detector. It appears that the effective mean ligthpath length is determined by the absorption and scattering coefficients and the geometry of the tissue, which may vary with time and may have different effects at different wavelengths. The combination with continuous determination of the mean lightpath length, as is now being developed in relation to NIRS (p. 274), therefore might considerably improve the performance of (reflection) pulse oximeters. This could bring nearer a solution of the remaining 'basic problem of oximetry' and constitute a big step in the direction of the ideal situation that S_{O_2} can be calculated from the signals of the measuring system without empirical calibration as is the case in spectrophotometry *in vitro*.

The rapid growth of the use of pulse oximetry in clinical medicine has not been associated with a similar growth in knowledge of the significance of S_{O_2} as measured by this technique and of its interpretation. Rodriguez *et al.* [328] evaluated, by means of a questionnaire, the knowledge of pulse oximetry of the paediatric house staff in five academic institutions in the New York City area and found it inadequate in a majority of the respondents. Such a knowledge deficit may be easily made up for by more emphasis on the subject in teaching. A more serious development is that a few grave misunderstandings have become entrenched in the literature [18, 19, 88, 131, 239, 312, 313, 398, 410], which, even in the technical literature concerning pulse oximetry [3, 412, 432], are not adequately discussed.

At the root of most of these misunderstandings is the decision of the designers of the IL282 haemoglobin photometer to display the O_2Hb fraction, F_{O_2Hb}, and to call it oxygen saturation, S_{O_2} (p. 231). The two quantities have been defined in Chapter 2, S_{O_2} by equations 2.7, 2.8, 2.9 and 2.14, and F_{O_2Hb} by equation 2.10. As the IL282 was the first commercially available multiwavelength haemoglobin photometer, it was widely used in testing the first generations of pulse oximeters. When one or more dyshaemoglobins were present in the blood, this necessarily led to different results, irrespective of the fact that the dyshaemoglobin, being a coloured substance, possibly also interfered with the measurement. That the observed difference was a matter of measuring different quantities was not immediately recognised. When this difference was hesitatingly recognised, it resulted in the new terms 'fractional saturation' for F_{O_2Hb}, and 'functional saturation' for S_{O_2}, and in discussions on the question which one was the physiologically relevant quantity.

The erroneous idea that the 'fractional saturation', *i.e.* F_{O_2Hb}, represents the O_2 concentration of the blood was a reason for considering this quantity as the physiologically relevant one. It should be realised, however, that the concentration of haemoglobin-bound O_2 is not given by F_{O_2Hb}, but by $F_{O_2Hb} \cdot ct_{Hb}$, and that, more importantly, F_{O_2Hb} has no unique relationship with p_{O_2} (Chapter 20). It is the S_{O_2} that is directly related to p_{O_2} through the oxygen dissociation curve (Fig. 20.2). This relationship may indeed be influenced by the presence of a dyshaemoglobin, *e.g.* COHb, but this influence can only be properly described by using F_{COHb} as a parameter influencing the relation between p_{O_2} and S_{O_2} (Fig. 20.12). The introduction of the terms 'fractional saturation' and 'functional saturation' has only perpetuated the misunderstanding [489].

The relative insensitivity of a pulse oximeter to the presence of a dyshaemoglobin such as COHb (Table 18.1) is a favourable property, not a deficiency. Statements like 'the oximeter measures the sum of COHb and O_2Hb' [18] or 'the oximeter interprets the presence of Hi as S_{O_2} 85%' [3] do not make sense. Nor is it a failure of the pulse oximeter that it does not detect hypercapnia and respiratory insufficiency [88]. The idea that the difference in oxygen affinity between HbA and HbF could have any influence on the photometric *measurement* of S_{O_2} [432], as well as the fact that it needed an editorial in a leading journal [184] to explain that the failure to detect hypercapnia by pulse oximetry is not a failure of *measurement* but a failure of *interpretation*, seems to demonstrate that the spectacular rise in monitoring technology has been accompanied by a considerable fall in knowledge and understanding of cardiorespiratory physiology and a fallacious notion of what the measuring instruments are intended for.

Another confusing factor has been created by the introduction of the symbol Sp_{O_2} for oxygen saturation measured by means of pulse oximetry [495]. It will not be intuitively grasped by medical students who have grown accustomed to the convention that S_{O_2} means oxygen saturation and p_{O_2} means oxygen tension. When the reason for using this symbol is explained to them, they will fail to appreciate that the pulse oximeter estimates the same quantity that is determined by analysing an arterial sample with the help of a haemoglobin photometer, or that is calculated from p_{O_2} by a blood gas analyser. The accuracy and reliability of these methods may be widely different, as are the techniques, but the measured quantity is the same: S_{O_2}.

NEAR INFRARED SPECTROSCOPY

Near infrared spectroscopy (NIRS) is a new branch of oximetry *in vivo*, based on the observation of Jöbsis [199] that in the spectral range of 700–1300 nm an appreciable amount of radiation can be transmitted through living tissue. This brought within reach the possibility to obtain, by optical methods, not only information on the oxygen supply to the tissue through measuring HHb and O_2Hb in the blood, but also on the presence of oxygen at the very site of consumption through measuring changes in the oxidation–reduction state of cytochrome aa_3 (cytochrome oxydase), the terminal enzyme in the respiratory chain that directly reacts with molecular oxygen. Cytochrome aa_3, in the oxidised state, has a weak absorption band in the near infrared with a broad maximum from 820–840 nm. Upon reduction of the enzyme, the absorbtion band disappears. Jöbsis *et al.* [200] demonstrated that it was indeed possible to obtain, by reflectance spectrophotometry, simultaneous information on the oxygenation state of haemoglobin and the oxidation–reduction state of cytochrome aa_3 in the exposed cerebral cortex of the cat.

Efforts were then made to develop practical instruments for noninvasive photometric multicomponent analysis in living tissues, with the brain of the newborn infant as the primary target. A noninvasive optical method for bedside observation of the cerebral oxygenation state in infants was first described by Brazy *et al.* [40]. They used a three-wavelength method with as light source three pulsed laser diodes at 775, 815 and 904 nm (1 kHz; 200 ns pulses). A fibre optic bundle conducted the light to a cylindrical disc, 1.8 cm in diameter, which was applied to the temple of the infant's head ('light delivering optode'). A similar optode at the opposite temple collected transmitted light that was then guided by a fibre optic bundle to a photomultiplier tube. The output signal of the photomultiplier was demodulated, amplified and logarithmised. The signals thus obtained were assumed to be linearly related to changes in concentration. Hence, changes in concentration of cytochrome aa_3, HHb and O_2Hb in the tissue could be recorded. Only relative changes were obtained because the lightpath length was unknown. The actual lightpath is, due to light scattering in the tissue, much longer than the distance between the optodes on the skull.

In the following decade the general design of instruments for multicomponent analysis in living tissues remained more or less the same. A technically more advanced instrument was described in 1988 by Cope and Delpy [76]. Four laser diodes at 778, 813, 867 and 904 nm were used and much attention was given to the production of pulses with very short rise and fall times and to a close control of pulse frequency and intensity. Skin reflection was monitored to check the coupling of the optic fibres to the skin continuously. The transmitted light was measured by means of a gallium arsenide photomultiplier kept at $\sim 5\,°C$ with Peltier cooling. This instrument was the prototype for a commercially available version, the Hamamatsu NIR1000.

That light scattering is an important factor influencing the results of the measurements was more readily appreciated in NIRS than in the development of conventional and pulse oximetry. The determination of the extended lightpath due to multiple scattering has been, from the beginning, in the centre of the research effort. In 1988 Delpy *et al.* [91] succeeded in measuring the optical pathlength in a scattering medium such as animal tissue. The pathlength was estimated from the time of flight of picosecond length light pulses. For the head of the rat they obtained a lightpath length of 5.3 ± 0.3 times the head diameter. This factor, with which the distance between the optodes must be multiplied to obtain the actual lightpath length was called **differential pathlength factor** (DPF) and led to a modification of Lambert–Beer's law for measurements in animal tissues:

$$A(\lambda) = \varepsilon(\lambda) \cdot b \cdot c \cdot l + A_t(\lambda), \qquad (18.5)$$

where l is the interoptode distance, b is the DPF, and $A_t(\lambda)$ represents light losses through scattering in the tissue expressed as absorbance ('optical density'; in most of the literature on near infrared spectroscopy OD is used instead of A). When $A_t(\lambda)$ is assumed to be constant over the observation period, a change in concentration (Δc) is then given by:

$$\Delta c = \Delta A(\lambda)/[\varepsilon(\lambda) \cdot b \cdot l]. \qquad (18.6)$$

This equation concerns a single chromophore. If more coloured substances are to be determined, the absorbance is measured at more wavelengths and Δc is calculated for each component analogous to established methods for MCA (p. 222).

For measuring the DPF by means of the time-of-flight method heavy and expensive equipment is needed. The DPF can, however, also be determined by phase-shift spectrophotometry [23, 24]. When intensity-modulated light travels through the tissue, a phase-shift is induced from which the distance travelled by the light can be calculated, if the modulation frequency and the tissue refractive index are known. Duncan *et al.* [99], using this method, made a large series of DPF measurements at four wavelengths; the results were not significantly different from those obtained with the time-of-flight method. There were, however, appreciable differences between various tissues and between male and female. There also was some dependence on wavelength, as shown by the values obtained across the head of (white) newborn infants: 5.41, 5.14, 5.02 and 4.64 at 690, 744, 807 and 832 nm, respectively.

Using intensity-modulated light for NIRS *in vivo* would allow the measurement of the DPF in addition to the absorbance [24, 25]. Such an approach would probably give more accurate and precise results, but the method has not yet been implemented in equipment suitable for practical application. All measurements in patients and experimental animals, up to now, have been made with continuous wave instruments.

In the first decade clinical application of NIRS concerned mainly the brain of the newborn infant because cerebral hypoxia is an important problem in neonatology and perhaps also because the available techniques were well-suited to this object of investigation [40, 105, 243, 245, 433, 438]. Edwards *et al.* [105] studied the effect of changes in arterial S_{O_2} and p_{CO_2} on the level of reduction of oxidised cytochrome oxidase. They used a NIR1000 photometer (Hamamatsu) and light at six wavelengths: 779.0, 802.5, 831.2, 848.7, 866.5 and 907.8 nm. Wyatt *et al.* [433] demonstrated that by inducing a known change in arterial S_{O_2} and measuring the effect on the ensuing changes in the cerebral O_2Hb and HHb concentrations, the cerebral blood volume can be determined. Livera *et al.* [245], using a slightly different measuring system with light at four wavelengths (775, 805, 845, and 904 nm) assessed the effects of hypoxaemia and bradycardia on the cerebral circulation in neonates; in this study changes in cerebral total Hb concentration were taken as a measure of changes in cerebral blood volume. Yoxall *et al.* [438] tried to estimate cerebral venous S_{O_2} by measuring the changes in cerebral O_2Hb and total Hb concentration following partial jugular vein occlusion. These investigations showed the considerable potential of NIRS for monitoring the oxygen status of the neonatal brain, especially when the method is combined with the determination of other quantities, such as arterial S_{O_2}, p_{O_2}, and p_{CO_2}, and when the response to various slight interventions is observed.

Further experiments showed the accessibility of the adult brain for similar measurements by NIRS [69, 71, 109], the suitability of the technique for studying blood flow and oxygen consumption in arm and leg muscles [72, 89, 106], including the assessment of imbalance between oxygen demand and oxygen delivery in the leg muscles of patients with peripheral arterial occlusive disease [222], and the value of NIRS in predicting, during an operation for cryptorchidism, the viability of the abdominal testis after spermatic blood vessel ligation [68, 70].

An improved continuous wave near infrared spectrophotometer has recently been developed by van der Sluijs *et al.* [362] in an effort to overcome the low temporal resolution of the available instruments and to improve the signal-to-noise ratio. A higher temporal resolution is especially needed for neurophysiological studies. Through the modular design, the new instrument is quite versatile; it can be equipped with up to eight laser diodes. The basic system contains one transmitter module that drives three semiconductor laser diodes emitting at 905, 845 and 770 nm. The peak power is 30 W. Through the low duty cycle (1 kHz; 50 ns pulses) the mean optical power of each laser is only 1.5 mW. The diodes are kept at a constant low temperature by Peltier cooling. The receiver module contains, as light detector, a large area avalanche photodiode, stabilised by temperature control. Transmitter and re-

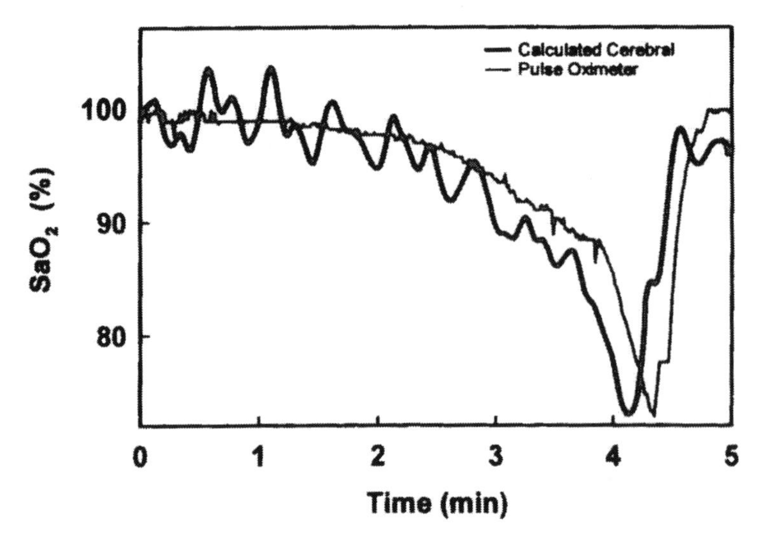

Figure 18.7. Arterial S_{O_2} of intracerebral arteries determined by NIRS during a short period of induced hypoxic hypoxaemia in a healthy adult human.

Interoptode distance 4.5 cm; sampling frequency 50 Hz. Note the good agreement with arterial S_{O_2} measured with a reflection pulse oximeter on the forehead. The time lag between the two S_{O_2} signals is due to the slower dynamic response of the pulse oximeter (from ref. [74]).

ceiver are connected by glass fibres to the object of investigation. A daylight cutoff filter protects the receiver from ambient light.

The low noise level of this instrument makes faster sampling possible. Using a sampling frequency of 10 Hz, cerebral oxygenation changes could be observed in the human motor cortex in response to a finger opposition task. The optodes were positioned over the left motor cortex area, with an interoptode distance of 3.5 cm. In a sequence and with a rate of 2 Hz, the thumb of the right hand had to oppose each of the other fingers. In all experimental subjects oxygenation changes were found in response to the finger opposition task [73].

When a sampling rate of 50 Hz is used, cerebral *arterial* S_{O_2} can be calculated from the HHb and O_2Hb pulse heights. Thus, the pulse oximeter principle is implemented in the NIRS instrument. Figure 18.7 shows the S_{O_2} signal from cerebral arteries in comparison with S_{O_2} as measured with an N200 pulse oximeter (Nellcor) on the forehead. The combination of the two methods in a single NIRS measuring system creates several new possibilities for noninvasive study of physiological phenomena and for monitoring vital functions in patients [74].

CHAPTER 19

THE OXYGEN BINDING CAPACITY OF HUMAN HAEMOGLOBIN

With the elucidation of the composition of human haemoglobin A and the calculation of the molecular mass of the haemoglobin tetramer (4 haem groups, 2 α and 2 β globin chains) to be 64458 g/mol [38, 39, 159], and with the knowledge that one haem group, *i.e.* one atom of iron, can reversibly bind one molecule of oxygen, one can calculate that 1 g of haemoglobin is able to bind 22 394/16 114.5 = 1.39 mL of oxygen (p. 15). As the reference HiCN method for measuring ct_{Hb} is based on the chemical composition of the haemoglobin molecule, this method would also be suitable for an easy determination of the oxygen capacity of blood, according to equations 2.4 and 2.5. However, all experimental determinations of β_{O_2} published before 1977 yielded values lower than 1.39 mL/g [55, 147, 180, 181, 385, 471, 484].

EARLY DETERMINATIONS OF THE OXYGEN BINDING CAPACITY OF HUMAN HAEMOGLOBIN

In 1894 Hüfner postulated that there was only one type of haemoglobin common to all mammals, even birds, with one molecular mass and one oxygen binding capacity [180]. From a series of experiments to determine the oxygen and the carbon monoxide binding capacity of bovine haemoglobin, he concluded that 1 g of haemoglobin was capable of binding 1.34 mL of oxygen, that the iron content of haemoglobin was 0.336%, and that haemoglobin had a relative molecular mass of 16 667 [180, 181]. For the next 60 odd years 1.34 mL/g remained the value quoted for the oxygen binding capacity of haemoglobin [467].

With the ability to accurately measure ct_{Hb} with the HiCN method, attention soon turned to the accurate experimental redetermination of β_{O_2}. In the course of redetermining $\varepsilon_{HiCN}(540)$, Zijlstra and van Kampen also determined the CO capacity of haemoglobin, finding a value of 1.35 mL/g [484]. In 1965 a series of 20 blood specimens from healthy, non-smoking individuals was studied by duplicate analysis of 1-mL samples using the Van Slyke–Neill manometric blood gas analysis technique [394] and correcting for physically dissolved O_2 with the solubility coefficients of Sendroy *et al.* [346]. A mean value of 1.311 mL/g, with a standard error of 0.0033 mL/g was found [471]. Other investigators did not fare much better: Theye found 1.30 mL/g [385], Burseaux *et al.* 1.31 mL/g [55], and Gregory 1.31 mL/g for β_{O_2} and 1.33 mL/g for β_{CO} [147].

These investigations all indicated, although blood specimens of non-smoking individuals were analysed, the presence of about 5.5% dyshaemoglobin in human blood; it also seemed that β_{CO} was slightly greater than β_{O_2}. Several hypotheses were put forward to explain the difference between the experimental and the theoretical values of β_{O_2}. (1) There actually might be about 5.5% inactive haemoglobin in the blood. (2) Part, or even all dyshaemoglobin might be produced during handling of the blood samples. (3) The usual gasometric analyses, especially the Van Slyke–Neill technique used in the investigations, might underestimate the amount of oxygen bound per unit amount of haemoglobin.

REFERENCE METHOD FOR TOTAL OXYGEN IN BLOOD

Methods other than the Van Slyke–Neill technique for measuring the total oxygen concentration, ct_{O_2}, of blood have been used, including gas chromatography, coulometry and polarography, but all show greater imprecision [100]. In organic microanalysis, a direct O_2 determination is used based on controlled conversion of O_2 via CO, into CO_2, with the titration of CO_2. The method is widely accepted and the reliability and accuracy of the results have been confirmed [373]. Dijkhuizen *et al.* [100, 104] adapted this method for blood gas analysis.

Haemoglobin-bound O_2 is set free by conversion of O_2Hb to Hi and the O_2 is stripped by a flow of pure N_2 as carrier gas. O_2 is converted to CO by contact with granular carbon at 1120°C. Next, CO is converted to CO_2 by contact with copper oxide, CuO, at 300°C and the CO_2 is titrated in a solution of barium chloride, $BaCl_2$, with NaOH of known titre. Conversion and titration take place according to the following reactions:

$$O_2 + 2C \rightarrow 2CO, \tag{19.1}$$

$$2CO + 2CuO \rightarrow 2CO_2 + 2Cu, \tag{19.2}$$

$$2CO_2 + 2H_2O + 2Ba^{2+} \rightarrow 2BaCO_3\downarrow + 4H^+, \tag{19.3}$$

$$4H^+ + 4OH^- \rightarrow 4H_2O. \tag{19.4}$$

A diagram of the measuring system is given in Fig. 19.1. The carrier gas, pure N_2, enters the system via a needle valve and a flow meter (Fig. 19.1, A) and passes through a drying tube with silica gel (B), a pressure relief valve (C), and a 10-cm-long reaction tube which contains copper turnings heated to 500°C by a furnace (D) to eliminate all traces of O_2. The gas then

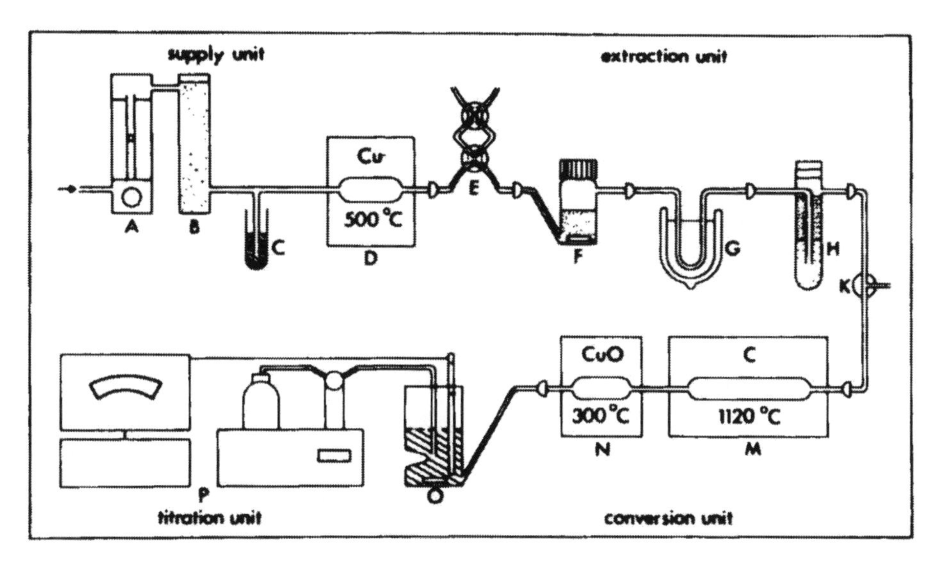

Figure 19.1. Diagram of the system for measuring the total oxygen concentration of blood. A, needle valve and gas flow meter; B, silica gel drying tube; C, pressure relief valve; D, furnace and tube containing copper turnings; E, stopcock assembly; F, extraction vessel with stirrer bar; G, liquid nitrogen trap to remove H_2O and CO_2; H, trap for residual CO_2; K, three-way stopcock to eliminate gas from the system; M, high-temperature furnace and tube containing granular carbon; N, medium temperature furnace and tube containing copper oxide; O, titration vessel; P, automatic titrator (from ref. [104]).

passes to an extraction vessel via a stopcock assembly (E). The stopcock assembly is composed of two interconnected 4-way stopcocks which form a loop, the volume of which has been determined accurately. The assembly serves to introduce (through a drying tube containing silica gel) a known volume of fresh air for checking the accuracy of the measuring system. The extraction vessel (F) has a volume of about 10 mL. The carrier gas enters the extraction vessel through a glass capillary with an inner diameter of 1 mm, inserted immediately above the vessel bottom in order to allow the gas to pass through the blood sample when the vessel has been filled. The top of the vessel is closed off by a 4-mm-thick silicon rubber membrane through which the blood sample can be injected; the membrane is held in place by a screw cap. The vessel includes a magnetic stirrer bar to mix the blood sample with $K_3Fe(CN)_6$ and to aid in stripping the blood. The extraction vessel is connected to two trapping units. The first (G) consists of a glass tube immersed in liquid nitrogen to trap H_2O and CO_2 released from the blood sample, the second (H) consists of a tube filled with soda asbestos and magnesium perchlorate, $Mg(ClO_4)_2$; it serves to remove the last traces of H_2O and CO_2.

The second trap is connected to the conversion unit by means of a 3-way stopcock (K). The conversion unit consists of a 60-cm-long quartz reaction tube filled with granular carbon and copper oxide, separated by quartz wool. Before filling the tube with carbon and copper oxide, the inside of the tube is thoroughly rinsed with hydrofluoric acid, HF, 40%; before heating the tube, the outside is cleaned with ethanol. The granular carbon is purged, before use, for 8 h at 1120°C under a constant flow of N_2 gas (0.5 L/min). After some 200 determinations the tube should be replaced because tube deterioration causes an increase in blank readings. The carbon-containing section of the reaction tube passes through a 20-cm-long furnace (M) and is heated to 1120°C; it is essential that this temperature is maintained over the entire furnace length. The CuO-containing section of the reaction tube is kept at 300°C by means of a 10-cm-long furnace (N).

The conversion unit is connected, via a glass capillary, to the 100-mL titration vessel (O), which is filled with 50 mL 0.5 mol/L $BaCl_2$ in a mixture of *t*-butanol and water (1 : 10, v/v). To ensure binding of all CO_2 and to prevent any CO_2 from escaping, an organic solvent (*tert*-butanol) is added to the $BaCl_2$ solution, a fast-rotating stirrer disperses the gas bubbles, and a constriction in the lower part of the titration vessel increases the contact time of the gas bubbles with the solution. The pH of the solution is adjusted to 10; the automatic titrator (P) maintains this value by adding a 0.05 mol/L NaOH solution in steps of 0.01–0.003 mL. The temperature of the titration vessel is thermostatically controlled; pH is measured with a combined glass/reference electrode because of the rotating magnetic field of the stirrer. To ensure a sufficiently fast electrode response, the 3 mol/L KCl solution of the liquid–liquid junction is renewed every 2 days of use; at the same time the glass membrane is cleansed by rinsing with HCl, 0.1 mol/L.

The tubes and vessels of the measuring system are made of glass, with solar radiation grade quartz for the reaction tubes in the furnaces. The components are interconnected by ground ball and socket joints. To keep the total internal volume as small as possible and to allow a short passage time, the inner diameter of the connecting tubes is 5 mm, that of the reaction tube 10 mm.

After filling and assembly, the measuring system is allowed to equilibrate for 48 h with the furnaces at operating temperature and a carrier gas flow of 0.5 L/h; a blank reading of 0.05–0.10 mL of 0.05 mol/L NaOH per hour should be obtained. Between series of determinations the furnaces should be kept on with the temperature of the carbon furnace adjusted to about 900°C, and a gas flow rate of 0.5 L/h should be maintained.

A measurement series then is comprised of the following steps:

(1) Place a small amount of $K_3Fe(CN)_6$ crystals and one drop of silicon antifoam in the extraction vessel and free the system from air with the 3-way stopcock K. (In practice, to speed up the measurement process, two extraction vessels may be used in tandem. After the first determination the first extraction vessel is emptied by suction, after the second determination both extraction vessels are replaced.)

(2) Allow ample time for the measuring system to reach steady state at the working temperature of the furnaces before the first determination is run.

(3) With stopcock K closed in all directions, immerse the H_2O/CO_2 trap (Fig. 19.1, G) in liquid nitrogen; open stopcock K in the direction of the conversion unit and take a blank reading over a period of about 30 min.

(4) Check proper function of the measuring system by introducing a known amount of dry outside air with the stopcock assembly (E). The measured amount of oxygen in the air sample must agree with the amount of oxygen calculated from stopcock volume, barometric pressure and temperature.

(5) Inject about 1 mL of blood into the extraction vessel through the rubber membrane; determine the exact amount of blood by weighing the sample syringe before and after the injection. Determine the density of the blood by also weighing a known volume, *e.g.* 0.5 or 1.0 mL delivered by a calibrated Ostwald pipette.

(6) Close stopcock K in all directions and haemolyse the blood by briefly immersing the extraction vessel in liquid nitrogen.

(7) Open stopcock K in the direction of the conversion unit, when, after thawing of the blood, gas begins to flow again through the extraction vessel; engage the magnetic stirrer. All O_2Hb is readily converted to Hi and the blood gases are stripped from the sample by the carrier gas.

(8) Allow about 50–60 min for all O_2 to reach the titration unit, then read the amount of NaOH delivered by the titrator. (This is made easier by using a printer.)

(9) Close stopcocks E and K, replace the extraction vessel(s) and the liquid nitrogen-cooled H_2O/CO_2 trap and refill the extraction vessel(s) with $K_3Fe(CN)_6$ and antifoam solution. Free the system from air with the 3-way stopcock K.

(10) Take a blank reading over approximately 30 min and inject the next blood sample.

The concentration of total oxygen in the blood sample is calculated as follows [104]. Equations 19.1–19.4 show that 1 mol O_2 corresponds with 4 mol OH^-. If a mL of x mol/L NaOH has been used in the titration and the blank reading is b mL, the sample contains $0.25 \times (a - b)x$ mmol O_2. If the volume of the sample is y mL, the total oxygen concentration of the blood, ct_{O_2}, is given by

$$ct_{O_2} = 0.25 \times 10^3 (a - b)x/y \text{ (mmol/L)}, \tag{19.5}$$

or

$$ct_{O_2} = 22.394 \times 0.25 \times 10^3 (a - b)x/y \text{ [mL(STPD)/L]}, \tag{19.6}$$

where 22.394 is the millimolar volume of oxygen in mL(STPD).

The accuracy and precision of the method were determined by introducing known volumes of outside air into the system by means of the stopcock assembly (Fig. 19.1, E). The volume of the stopcock assembly, as determined by weighing with mercury, was 1.576 mL at 22°C. A total of 36 outside air samples were introduced into the measuring system. A mean volume fraction of 0.2093, with a standard deviation of 0.0006 was found [104]. Compared to the value of 0.209476 reported in the literature [242] the inaccuracy is $< -0.1\%$; the coefficient of variation of the measurements was 0.29%. The coefficient of variation calculated from duplicate determinations of 62 blood specimens was 0.65%.

To determine the CO concentration of a blood specimen or the CO binding capacity of haemoglobin (β_{CO}) the same apparatus and procedure can be used with the carbon furnace bypassed or switched off [100, 102, 479]. The determination time is decreased to about 30 min.

REDETERMINATION OF THE OXYGEN BINDING CAPACITY OF HUMAN HAEMOGLOBIN

The principal purpose of the development of an accurate, absolute method for measuring the total oxygen concentration of blood, which may serve as a reference method for ct_{O_2}, was to further investigate the difference between the experimental and the theoretical values of the oxygen binding capacity, β_{O_2}, of human haemoglobin. An absolute method for measuring the concentration of a substance is a method that does not depend for calibration on samples with a known concentration of that substance. The chemical method described in the preceding section yields a value for ct_{O_2} that is traceable to the titre of an NaOH solution, which in turn depends on a weighed amount of an acidimetric standard, *e.g.* potassium hydrogen phthalate or benzoic acid. The samples of outside air are not for calibration, but are a check on the proper functioning of the method.

Since in most earlier determinations of β_{O_2}, the concentration of total oxygen, ct_{O_2}, had been measured with Van Slyke's technique [394], the obvious choice for a second method was this procedure. As a third method for measuring ct_{O_2} of blood an entirely different procedure was selected, based on the polarographic determination of p_{O_2} after adding a small volume of blood to a much larger volume of $K_3Fe(CN)_6$ solution [115]. In this solution lysis of the red cells occurs rapidly and the oxygen is released through the conversion of all O_2Hb to Hi. The fluid volume is large enough to dissolve all O_2 and the ensuing rise in p_{O_2} is measured. The three methods for the determination of ct_{O_2} in blood are schematically shown in Fig. 19.2.

For the determination of β_{O_2} using the three independent methods for measuring ct_{O_2}, the following procedure was used. Part of the blood specimens

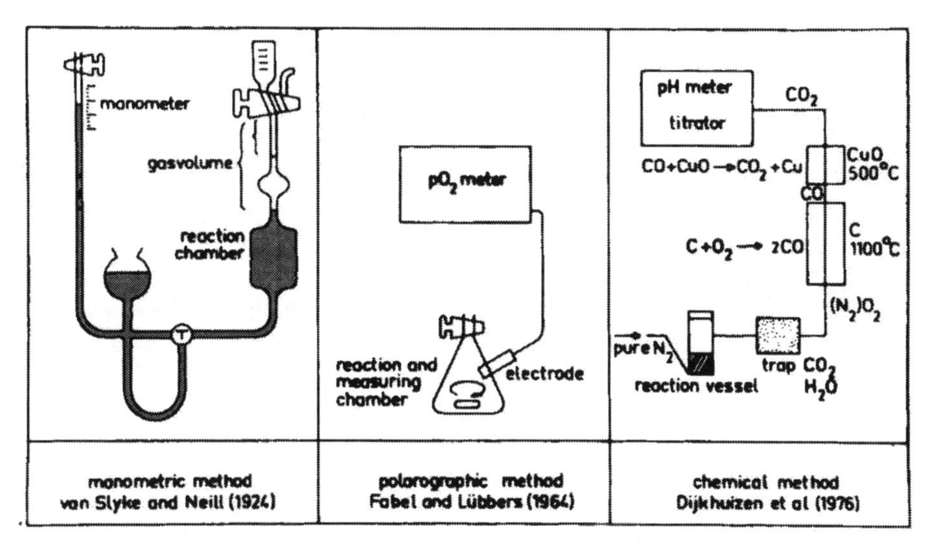

Figure 19.2. Diagram of three independent methods for measuring the total O_2 concentration of blood used for the determination of the oxygen capacity of haemoglobin (Hüfner's factor). The only common step is the release of O_2 through the conversion of O_2Hb into Hi with the help of $K_3Fe(CN)_6$ (from ref. [101]).

were diluted in their own plasma to a total haemoglobin concentration, ct_{Hb}, of approximately 110 g/L. About 50 mL was then equilibrated for at least 150 min at room temperature (~22°C) with humidified pure O_2 in an open, 300-mL, cylindrical glass tonometer, rotation speed 30 revolutions per min, gas flow rate 1.5 L/h. This time was considered to be amply sufficient for full oxygenation of all active haemoglobin, including conversion of any COHb to O_2Hb. For the various simultaneous replicate determinations samples were taken from the rotating tonometer for up to 5 h; ct_{Hb} was measured at regular intervals and the p_{O_2} was monitored with two oxygen electrodes.

For Van Slyke's method a Thomas–Van Slyke Magnetic Model manometric apparatus was used. One-mL samples were introduced with a calibrated Ostwald pipette. For the calculation of β_{O_2} of a blood specimen the mean of at least three determinations was used differing < 0.5% from each other.

The polarographic measurements were performed using an Eschweiler micro p_{O_2} analyser. For each determination 0.05 mL blood was introduced with a microliter syringe and a thin needle into a 2-mL reaction chamber containing 2 g/L $K_3Fe(CN)_6$ solution. For the calculation of β_{O_2} the mean of at least 10 analyses of each specimen was used.

In the chemical (titrimetric) method, 1.5-mL blood samples were injected with a glass syringe and an 8-gauge needle. The exact volume injected was determined by weighing the syringe as described on p. 281. For the calculation of β_{O_2} of a blood specimen the mean of three determinations differing < 0.5% from each other was used.

The amount of O_2 released per volume of blood was calculated and corrected for dissolved O_2 using for the Bunsen solubility coefficient of O_2 in blood a value of 0.0280 $mL \cdot mL^{-1} \cdot atm^{-1}$, according to Christoforides and Hedley-White [63] valid for $ct_{Hb} \approx 110$ g/L and ~22°C. This corresponds with a concentrational solubility coefficient of 0.0123 $mmol \cdot L^{-1} \cdot kPa^{-1}$. The total haemoglobin concentration was determined with the HiCN method, using the same Ostwald pipette as was used for the manometric method. $A_{HiCN}(540)$ was measured with an Optica CF4.

At the time of this investigation the methods for multicomponent analysis (p. 222) had not yet been developed and the dyshaemoglobin fractions were determined by separate determinations of COHb, Hi and SHb. The COHb fraction, F_{COHb}, was determined in the two-component system COHb/O_2Hb by the 562/540 nm method (p. 221), using an Optica CF4 and a lightpath length of 0.005 cm. F_{COHb} was determined after 15, 150 and 240 min from the start of tonometry. The haemiglobin fraction, F_{Hi}, was determined with the CN^-addition method (p. 218), carried out with an Optica CF4, measuring at 630 nm and using a lightpath length of 4.00 cm. The sample for this determination was taken after 150 min of tonometry.

The fraction of sulfhaemoglobin, F_{SHb}, was estimated by measuring at $A(620)$ with an Optica CF4 using a lightpath of 4.00 cm. Corrections were made for the known fractions of O_2Hb and Hi and a value of 18.1 $L \cdot mmol^{-1} \cdot cm^{-1}$ was used for the absorptivity of SHb at 620 nm [357]. In addition a complete absorption spectrum of the diluted blood was recorded with an Optica CF4DR, spectral range 450–700 nm, using a lightpath of 1.00 cm. The blood was taken from the tonometer after at least 150 min and was diluted 50-fold with a 0.15 mol/L phosphate buffer of pH = 7.2.

Two series of measurements of β_{O_2} have been made, each comprising 18 blood specimens of healthy adult donors. Table 19.1 summarises the result of these measurements. These data show on the one hand a smaller difference of the experimental values in comparison with the theoretical one (1.39 mL/g) than found in earlier investigations, but on the other hand still a gap between what is reasonably expected and what is actually measured. The good agreement in all determinations between the β_{O_2} values obtained with the three independent procedures for measuring ct_{O_2} and the high precision of the individual methods suggest that the blood gas measurement is not, and therefore in earlier investigations probably was not, an important factor in the causation of the low experimental values.

A possible cause of the low experimental values might be a too large correction of ct_{O_2} for dissolved O_2. Because of the high p_{O_2} in the tonometer gas, this correction is considerable. With $ct_{Hb} \approx 110$ g/L $\approx$ 6.8 mmol/L the maximum concentration of haemoglobin-bound O_2 is 6.8 mmol/L. The p_{O_2} in the open tonometer flushed with humidified pure O_2 at ~22°C was ~87 kPa [101]. With a concentrational solubility coefficient of 0.0123 $mmol \cdot L^{-1} \cdot kPa^{-1}$, $c_{O_2}(\text{free}) = 0.0123 \times 87 \approx 1.1$ mmol/L

Table 19.1.
Oxygen binding capacity (β_{O_2}) of adult human haemoglobin determined with three independent methods for the measurement of total oxygen concentration [101]

Method	Chemical	Manometric	Polarographic
Mean	1.365	1.371	1.360
SD	0.018	0.018	0.023
SEM	0.003	0.003	0.004
N	35	34	32

β_{O_2} in mL/g; *N* is number of specimens; *SD* is standard deviation; *SEM* is standard error of the mean.

(equation 2.3). Hence, $ct_{O_2} \approx 6.8 + 1.1 = 7.9$ mmol/L and the correction for c_{O_2}(free) is $\sim$ 14%.

This possible source of error was investigated by the following experiment: β_{O_2} of five specimens of human blood was measured in duplicate using the chemical method for the determination of ct_{O_2}. The first measurement was performed as usual, but in the second one the tonometry was continued for another 150 min with air instead of oxygen ($p_{O_2} \approx 20$ kPa). Using Student's paired t-test there proved to be no significant difference between the β_{O_2} values after equilibration with oxygen or with air. This makes a too high correction for freely dissolved oxygen rather improbable. Moreover, another careful study of the oxygen solubility in blood [332] yielded an even slightly higher solubility coefficient for O_2 in blood than that of Christoforides and Hedley-White [63] used by Dijkhuizen *et al.* [101].

For each specimen, the β_{O_2} values obtained with the chemical and the manometric method were taken together and the fraction of *active* haemoglobin was calculated by dividing the mean of the two values of β_{O_2} by 1.39 mL/g. From the measured dyshaemoglobin fractions it was subsequently calculated which part of the *inactive* haemoglobin could be accounted for by the presence of dyshaemoglobin. The remaining part of the inactive haemoglobin was called ***unidentified inactive*** haemoglobin. The results are presented in Table 19.2. Although there is on the average less than 1% unidentified inactive haemoglobin, it should be noted that the spread in this quantity is considerable. In the series of 36 specimens there were four with more than 2% unidentified inactive haemoglobin; the fractions were 2.8, 3.3, 3.9 and 4.3%. For each of these specimens the three methods for the measurement of ct_{O_2} yielded equally low values for β_{O_2}.

Table 19.2 also shows that the dyshaemoglobin fractions in the samples used in the determination of β_{O_2}, accounted for only 0.7% of the total haemoglobin concentration. F_{Hi} varied between 0 and 1.7% with a mean value of 0.45%. F_{COHb} was effectively removed from the blood by tonometry and was in most samples below the level of detection. Even in a blood specimen of a heavy smoker, which after 15 min of tonometry still contained

Table 19.2.
Oxygen binding capacity (β_{O_2}) of adult human haemoglobin by chemical and manometric methods, and fractions of dyshaemoglobin and unidentified inactive haemoglobin

	Range	Median	Mean	*SD*	*SEM*
β_{O_2} (mL/g)	1.322–1.387	1.372	1.368	0.017	0.003
Active Hb (%)	95.1–99.8	98.6	98.4	1.2	0.20
Dyshaemoglobin (%)	0–1.7	0.6	0.7	0.5	0.08
Unident inact Hb (%)	−1.0–4.3	0.8	0.9	1.2	0.20

For each specimen the mean value of β_{O_2} obtained with the chemical and the manometric method has been calculated first. Unident inact Hb is unidentified inactive haemoglobin; *SD* is standard deviation, *SEM* is standard error of the mean.

8% COHb, there was only 0.5% after 150 min. F_{SHb} was > 0.2% in only a single sample.

The presence of about 1% unidentified inactive haemoglobin in normal human blood has been confirmed by Shimizu *et al.* [352] for human, canine and murine blood. They used for the determination of ct_{O_2} the micromanometric method of Van Slyke and Plazin [395]. Hence, the much greater deviation from the theoretical value formerly found for β_{O_2} (p. 277) was most probably due to procedural sources of error. A possible explanation of the remaining difference might be that by the usual methods of tonometry no full saturation of all haemoglobin with oxygen is achieved. However, in light of the data given in Chapter 20 and in the literature [332] this is quite improbable, unless there is a tiny fraction of haemoglobin with decreased affinity.

The latter hypothesis is supported by observations of Van Slyke *et al.* [393] and of Gregory [147] who found β_{CO} slightly higher than β_{O_2}. Dijkhuizen [100] therefore measured β_{CO} in twelve specimens of human haemoglobin. Fresh blood from healthy donors, anticoagulated with heparin, was diluted with 150 mmol/L NaCl solution to $ct_{Hb} \approx 7$ mmol/L and haemolysed by ultrasonic treatment. Complete saturation was achieved by tonometry with CO for 120 min in darkness. The concentrations of total haemoglobin and the fractions of Hi and SHb were determined as in the determination of β_{O_2}. The total CO concentration of the samples was determined with the chemical method with slight modifications [102]. A Bunsen solubility coefficient of 0.0235 $\text{mL} \cdot \text{mL}^{-1} \cdot \text{atm}^{-1}$ at 22°C, corresponding with a concentrational solubility coefficient, α_{CO}, of 0.0104 $\text{mmol} \cdot \text{L}^{-1} \cdot \text{kPa}^{-1}$ was used. This solubility was calculated from the value found by Power [310] for blood at 37°C, using the temperature coefficient of α_{O_2} of Christoforides and Hedley-White [63] for calculating α_{CO} at 22°C.

For β_{CO} a mean value was found of 1.386 mL/g with a standard deviation of 0.007 mL/g ($N = 12$). This value proved to be significantly different from the corresponding value obtained for β_{O_2} ($p = 0.001$, Student's t-test

Table 19.3.
Oxygen and carbon monoxide binding capacity of a single blood sample of a healthy donor

	Mean (mL/g)	*SD*	*n*	Act Hb	Inact Hb
β_{O_2} (1)	1.349		6	97.1%	2.9%
β_{O_2} (2)	1.347	0.012	4	96.9%	3.1%
β_{CO}	1.379	0.009	4	99.2%	0.8%

For β_{O_2} (1), n is the mean of three manometric and three chemical determinations; for β_{O_2} (2) and β_{CO}, n is the mean of four chemical determinations.

for unpaired samples). $\beta_{CO} = 1.386$ mL/g corresponds with 99.7 % active haemoglobin and the difference from the theoretical value may be explained by a mean dyshaemoglobin fraction of 0.4%. Thus no unidentified inactive haemoglobin appeared to be present.

Subsequently, a second blood specimen was obtained from one of the 36 donors who had participated in the determination of β_{O_2}. The second specimen was used for four determinations of β_{O_2} and β_{CO}. The results are given in Table 19.3.

On the basis of these observations, the following hypothesis concerning the presence of unidentified inactive haemoglobin in human blood seems justified [100]. In a varying amount of haemoglobin there may be a slight modification in spatial structure which leaves the spectral properties unaffected and does not impede the conversion to HiCN, but hinders the binding of oxygen. A higher p_{O_2} is therefore necessary before full saturation is reached. Because the affinity of haemoglobin for CO is considerably greater, tonometry with CO results in the formation of 100% COHb. In this respect there may be a certain analogy with SHb, of which a CO derivative was found [207] long before it became known that SHb can bind O_2 when the gas pressure is elevated far above the atmospheric level [28, 59].

The inactive haemoglobin may be formed *in vivo*, the amount gradually increasing when the erythrocyte grows older. However, it may also be that some haemoglobin becomes changed during handling of the blood samples *in vitro*. The higher values of β_{O_2} found in a few experiments in which the haemoglobin was fully saturated *in vivo* through breathing oxygen [146, 261] point in this direction.

ROUTINE DETERMINATION OF THE OXYGEN CAPACITY OF HUMAN BLOOD

In clinical practice the unidentified inactive haemoglobin is hardly a relevant problem. In any case, it is only a tiny fraction of ct_{Hb}, and it may even be non-existent, if the hypothesis that it is formed *in vitro* is proven to be true. Therefore, no appreciable error is introduced when in the practical determination of the oxygen capacity and of the oxygen concentration of the blood this kind of dyshaemoglobin is disregarded. This, however, does not hold for the common dyshaemoglobins, COHb and Hi. These dyshaemoglobins may be present in clinically relevant amounts [447, 489].

The common multiwavelength haemoglobin photometers (p. 230) give reliable values for ct_{Hb}, F_{COHb} and F_{Hi}, so that the oxygen capacity can be calculated with equation 2.6, which gives B_{O_2} in mL/L when ct_{Hb} is in g/L.

$$B_{O_2} = \beta_{O_2} \cdot (ct_{Hb} - c_{COHb} - c_{Hi}). \tag{2.6}$$

The concentration of haemoglobin-bound oxygen, expressed in mL/L, is then given by

$$c_{O_2}(\mathrm{Hb}) = S_{O_2} \cdot B_{O_2} = S_{O_2} \cdot \beta_{O_2} \cdot (ct_{Hb} - c_{COHb} - c_{Hi}). \tag{19.7}$$

To obtain the concentration of total oxygen, p_{O_2} should also be measured and, as follows from equations 2.2 and 2.3, ct_{O_2}, expressed in mL/L, may then be calculated with equation 19.8.

$$ct_{O_2} = S_{O_2} \cdot \beta_{O_2} \cdot (ct_{Hb} - c_{COHb} - c_{Hi}) + 22.394 \times (\alpha_{O_2} \cdot p_{O_2}). \tag{19.8}$$

The factor 22.394 had to be added because α_{O_2} is expressed in $\mathrm{mmol \cdot L^{-1} \cdot kPa^{-1}}$, and p_{O_2} in kPa. To obtain the concentration of total oxygen in mmol/L, equation 19.9 can be used.

$$ct_{O_2} = S_{O_2} \cdot (ct_{Hb} - c_{COHb} - c_{Hi}) + (\alpha_{O_2} \cdot p_{O_2}), \tag{19.9}$$

in which ct_{Hb} is in mmol/L and p_{O_2} in kPa.

Sulfhaemoglobin has not been taken into account in these equations because it is seldom present in an amount that has an appreciable influence on the oxygen transport capability of the blood [442, 447]. When it has been measured, its fraction can be added as a third dyshaemoglobin to equations 2.6 and 19.7–19.9. There are no well-documented cases of life-threatening sulfhaemoglobinaemia, but measuring SHb may be useful to solve the diagnostic problem that SHb may pose through the severe cyanosis it may produce. Only 5 g/L ($F_{SHb} \approx 3\%$) is enough to cause readily detectable cyanosis [220].

CHAPTER 20

THE OXYGEN AFFINITY OF HUMAN HAEMOGLOBIN

The oxygen transport by the blood can be summarised in the equation:

$$V_{O_2} = Q \cdot (ct_{Hb} - c_{dysHb}) \cdot \beta_{O_2} \cdot (aS_{O_2} - vS_{O_2}), \qquad (20.1)$$

where V_{O_2} (mL/min) is oxygen consumption, Q (L/min) is blood flow rate, ct_{Hb} (g/L) is total haemoglobin concentration (Chapter 16), c_{dysHb} (g/L) is dyshaemoglobin concentration (Chapter 17), β_{O_2} is oxygen binding capacity (mL) per g haemoglobin (Chapter 19) and aS_{O_2} and vS_{O_2} are, respectively, arterial and venous oxygen saturation (Chapters 17 and 18). The equation is valid for organs and tissues as well as for the total body. In the latter case, the mixed venous S_{O_2}, *i.e.* S_{O_2} in the pulmonary artery, should be used for vS_{O_2}.

Equation 20.1, however, has no bearing on the diffusion of O_2 from the capillaries to the mitochondria in the oxygen-consuming cells. The driving force for the diffusion of O_2 is the p_{O_2} gradient from the capillaries to the cells. Since the oxygen affinity of cytochrome oxydase (p. 273) is very high, the p_{O_2} in active cells may be low; consequently there is hardly any O_2 stored in the tissue and the supply has to be continuous. Hence, the capillary p_{O_2} must be high enough to drive a continuous diffusive flow of O_2 to the cells. It is the **oxygen affinity of haemoglobin** in the erythrocytes, together with the arterial p_{O_2} and the active haemoglobin concentration ($ct_{Hb} - c_{dysHb}$), that determine the capillary p_{O_2}, and thus the p_{O_2} gradient available to drive oxygen to the cells.

This chapter describes how optical methods in combination with selective electrodes for measuring p_{O_2}, p_{CO_2} and pH may be used in the determination of the oxygen affinity of haemoglobin, and presents a set of data concerning the oxygen affinity of human blood, including the effects of changes in temperature, pH, p_{CO_2}, and of the concentration of 2,3-diphosphoglycerate (DPG) and dyshaemoglobins in the erythrocytes. All data are based on measurements with a measuring system in which all variables are determined directly and independently.

INSTRUMENTS AND PROCEDURES

To completely describe the oxygen affinity of the blood, an oxygen dissociation curve (ODC), *i.e.* a graph in which S_{O_2} is plotted as a function of p_{O_2}, should be made over the whole S_{O_2} range. However, for many clinical applications the determination of p_{50} (*i.e.* p_{O_2} when $S_{O_2} = 50\%$; p. 16) may suffice. This may be accomplished with a multiwavelength haemoglobin photometer and a bloodgas analyser [232, 414]. Using these instruments, it is even possible to establish the shape and position of the ODC by measuring p_{O_2} and S_{O_2} in a series of blood samples with widely different S_{O_2} [414].

Numerous methods for the determination of the complete ODC of haemoglobin have been developed [66, 442, 448]. Although many methods determine the ODC of haemoglobin solutions, it will be clear that the **oxygen affinity of whole blood** is the relevant factor in oxygen transport *in vivo* [21, 258]. In most methods measuring the ODC of whole blood, only part of the relevant quantities are measured directly; the ODC is then calculated from the measured data on the basis of plausible assumptions. Only in the method of Clerbaux *et al.* [67] and, with the exception of the pH read from a nomogram, also in the micromethod of Reeves [316], all relevant quantities are measured directly.

The data given in this chapter have been collected using a specially constructed measuring system that allows continuous, independent measurement

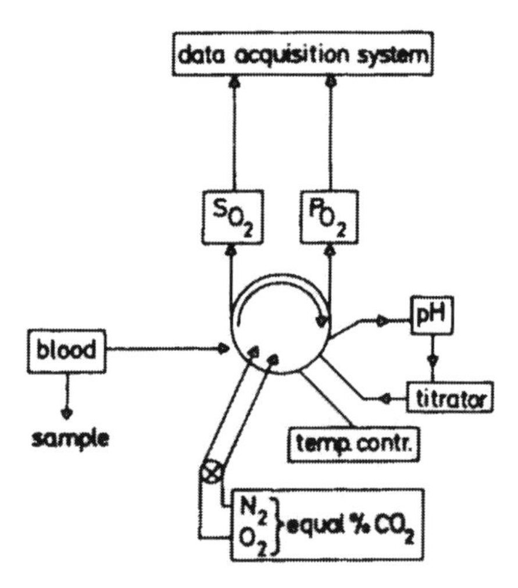

Figure 20.1. Diagram of measuring system for the determination of oxygen dissociation curves of whole blood.

Blood in the thermostatted central measuring chamber is kept in rapid motion by means of a magnetic stirrer. The composition of the gas mixture above the blood is controlled by an adjustable gas flow of which the p_{O_2} can be varied at constant p_{CO_2} (from ref. [448]).

of p_{O_2} and S_{O_2} at constant temperature, p_{CO_2} and pH [296, 448]. The central part of the measuring system is a double-walled stainless steel cylindrical measuring chamber with a magnetic stirrer on the bottom (Fig. 20.1). The temperature in the measuring chamber is kept constant (±0.02°C) through perfusion of the wall with water from a thermostatted waterbath, the temperature of which is checked with a calibrated certified mercury thermometer. A fibre optic oximeter catheter, a p_{O_2} electrode, a pH electrode and a burette outlet are connected to the measuring chamber through ports in the side wall. The measuring chamber is closed with a lid that contains holes through which preheated, water-saturated gases enter. Thus blood in the measuring chamber can be rapidly equilibrated with any gas mixture while S_{O_2}, p_{O_2} and pH are measured; the pH is controlled by an automated titrator. All data are stored in a data-acquisition system (Minc PDP-11/23, Digital).

The p_{O_2} of the blood was measured by a polarographic method using a fast-responding membrane-covered platinum electrode (Eak-1s, Eschweiler) [297]. For calibration of the electrode two exactly known p_{O_2} levels of 0 and 20 kPa were used. The latter calibration value was obtained by equilibrating water with air at 37°C, the zero value was obtained with pure nitrogen. The linearity of the p_{O_2} electrode was checked frequently by measuring at a third p_{O_2} level of about 4 kPa using an exactly known gas mixture made with gas-mixing pumps (Wösthoff). No deviations from linearity were found; the maximum error in blood p_{O_2} determined in this way is $\leqslant 0.2\%$.

S_{O_2} was measured by means of a fibre optic oximeter as described by Landsman *et al.* [235] (p. 260). Filter-photocell combinations were used that yield blood reflection measurements at 630 and 920 nm and the two signals were processed according to equation 20.2. Note that this differs from the procedure described in Chapter 18.

$$y = R(920)/[k \cdot R(920) + R(630)], \qquad (20.2)$$

where $R(630)$ and $R(920)$ represent the light reflection at 630 and 920 nm, respectively, and k is a factor that linearises the relationship between y and S_{O_2}. This factor must be determined experimentally and was 0.35 for the measuring system used in collecting the data presented in this chapter.

After an ODC had been recorded, a two-point calibration was carried out. The values of y measured at $p_{O_2} = 0$ were averaged and set to correspond to $S_{O_2} = 0$; the values of y measured at $p_{O_2} > 55$ kPa were averaged and set to correspond to $S_{O_2} = 100\%$. The performance of the oximeter was checked by adjusting the gas supply to the cuvette so that the S_{O_2} reading, displayed on a digital panel, became stable at 50%. A blood sample was then taken from the cuvette and analysed with an individually calibrated two-wavelength haemoglobin photometer (Hemoximeter OSM2, Radiometer) which, in turn, had been checked with the spectrophotometric 680/800 nm procedure described in Chapter 17 (p. 219; equation 17.2). If the result of this measurement deviated more than 1% S_{O_2} from the value obtained with the fibre optic oximeter, the ODC measurement was discarded.

In a subsequent version of the measuring system, the analogue processing of the light reflection signals in the fibre optic oximeter was replaced by a digital procedure. The signals from the photocells representing the light reflection at 630 and 920 nm were passed directly to the computer for digital processing and calibration with the aid of small samples taken at intervals from the cuvette and immediately measured with an OSM2 (Radiometer) that had been checked with a spectrophotometric procedure. The light reflection signals were converted into S_{O_2} values using four-point calibration and multiple regression [230].

The pH of the blood in the cuvette was measured continuously with a combined glass-reference electrode (7GR241, Electrofact; later: U402-M3, Ingold) connected to a pH meter (PHM27, Radiometer) frequently calibrated with phosphate buffers of pH = 6.841 and 7.383 at 37°C (NIST). The pH measurements were compared regularly with the results of a blood pH meter with a capillary electrode (G297/G2, Radiometer). This served to trace and correct for differences in liquid–liquid junction potential between buffers and blood. The least significant digit in the pH reading was 0.001.

The pH signal from the combined electrode in the cuvette was received by a titrator (TTT11, Radiometer) which controlled a burette (ABU12, Radiometer) connected to the cuvette; the burette was filled with 2 mol/L NaOH. This system serves to titrate the oxygen-linked protons that are released during oxygenation of haemoglobin. Hence, the blood pH can be kept constant at any desired value during the recording of the ODC. Moreover, the amount of protons released per mole of oxygen bound can be calculated from the total amount of NaOH added.

During the recording of an ODC, the O_2 fraction in the gas mixture flowing through the cuvette must change gradually, while the CO_2 fraction remains constant; however, it should be possible to adjust the CO_2 fraction to any desired value. This was first accomplished by means of two gas mixing pumps (Wösthoff) and three gas cylinders containing pure O_2, N_2, and CO_2. For an ODC to be made under standard conditions (T = 37°C, p_{CO_2} = 5.33 kPa and pH = 7.40) the CO_2 fraction should be 5.6%. The pumps are then set to provide CO_2/O_2 and CO_2/N_2 mixtures of 5.6/94.4 by volume. The mixtures are saturated with water at 37°C and the flow of the gas mixtures to the cuvette is controlled by needle valves. By changing the ratio of the flow rates of the gas mixtures, p_{O_2} can be varied from 0 to 80 kPa without any change in p_{CO_2}. Depending on the barometric pressure, the actual p_{CO_2} was 5.33 kPa with a maximum deviation of ±0.07 kPa.

In a later version of the measuring system, the gas mixing pumps were replaced by a gas mixing system containing four mass flow controllers (MFCs, Brinkhorst) [228]. With this system the O_2 fraction could be varied between 0 and 90%, the CO_2 fraction between 0 and 20%. An automated calibration procedure was carried out overnight using the computer-controlled data-acquisition system; the flow through each MFC was measured with a soap bubble meter, the O_2 concentration in the gas mixtures with an oxygen analyser (OM-11, Beckman), and pure N_2 and dry air were used as calibration gases [228]. The fractions of O_2 and CO_2 in the gas mixtures supplied to the cuvette were known within ±0.05%.

All blood specimens for the measurement of haemoglobin oxygen affinity were collected by venipuncture from healthy, adult human donors. The samples were anticoagulated with heparin (100 USP units/mL) and kept in ice: ct_{Hb} was measured with the HiCN method (p. 200) and the concentration of 2,3-diphosphoglycerate by means of a kit modification (No. 35 UV, Sigma) of the enzymatic method of Keitt [212]. Before and after recording a series of ODCs, each blood specimen was checked for the possible presence of COHb, Hi and SHb by means of the standard multiwavelength method (p. 227).

For the determination of an ODC ~8 mL of blood was placed in the cuvette. A few drops of a 10% antifoam silicon emulsion were added to prevent foaming during rapid stirring of the blood. This amount of silicon emulsion has no influence on the oxygen affinity of the blood nor on any of the measurements. Stirring with a speed of 2500 rpm caused no detectable haemolysis.

Before recording the ODC, the gas phase above the stirred blood was kept at $p_{O_2} = 0$ till the blood had a p_{O_2} of zero; this takes about 20 min. The pH of the blood was then adjusted to the desired value by adding either NaOH or HCl. The value was checked by taking a sample from the cuvette and measuring with the capillary blood pH meter. During recording the ODC the titrator kept the pH constant at the adjusted value.

After 100 data points had been sampled at $p_{O_2} = 0$, the p_{O_2} was gradually increased to a value > 55 kPa, while the p_{O_2} and S_{O_2} signals were sampled every 0.2 s and stored in the data-acquisition system. This takes about 8 min, thus each ODC is based on 2500 data points. When this level was reached another 100 data-points were sampled before data storage was stopped.

Error evaluation, taking into account the uncertainty in all measured quantities (p_{O_2}, S_{O_2}, pH, p_{CO_2} and temperature) revealed, for the first version of the measuring system, that the coefficient of variation of p_{O_2} at each value of S_{O_2} was 2.2, 3.1 and 2,1% for the lower, mid and upper part of the ODC, respectively [448]. The improvements realised in the later version of the measuring system, especially the modified calibration procedure for S_{O_2} measurement, decreased the overall imprecision considerably. Between 15 and 95% S_{O_2} the coefficient of variation of the corresponding p_{O_2} values was $\leqslant$ 2%: between 40 and 90% S_{O_2} this was ~1%. At very low and very high S_{O_2}, the imprecision in the measured p_{O_2} values increases rapidly. At the lower end of the S_{O_2} scale the coefficient of variation of p_{O_2} increases because the uncertainties in the measurements become large in comparison with the decreasing values of p_{O_2}. At the upper end of the S_{O_2} scale, minor errors in S_{O_2} are associated with large deviations in p_{O_2} due to the flat course of the ODC [230].

THE NORMAL OXYGEN DISSOCIATION CURVE OF HUMAN BLOOD

Using the instruments and techniques described in the preceding section, standard ODCs, *i.e.* ODCs recorded under **standard conditions**, have been recorded for haemoglobin in freshly drawn heparinised whole blood of 45 healthy adult volunteers. In the literature there is no general agreement as to the definition of standard conditions, *i.e.* as to which of the factors that modulate the oxygen affinity should be kept constant at a chosen 'normal' level during recording of the ODC. Usually only temperature, pH and p_{CO_2} are included, but some authors have proposed to also include the DPG concentration in the erythrocytes, as well as the COHb fraction. In our experience, such an extended definition of 'standard conditions' complicates rather than facilitates the proper description of the determinants of the oxygen affinity of the blood. Thus, standard conditions for the determination of an ODC, and of p_{50} (p_n), are defined as $T = 37°C$, $p_{CO_2} = 5.33$ kPa and pH = 7.40.

The mean ODC is shown in Fig. 20.2, while the numerical data in steps of 1% S_{O_2} with SD and SEM are presented in Table 20.1. Because these data

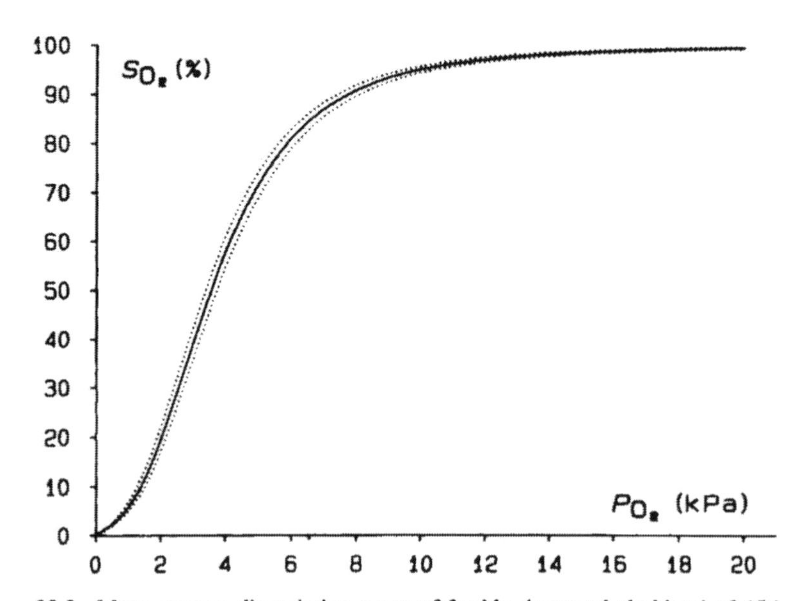

Figure 20.2. Mean oxygen dissociation curve of freshly drawn whole blood of 45 healthy human adults (solid line). The dotted lines indicate ± *SD*. pH = 7.40, $p_{CO_2} = 5.33$ kPa, $T = 37°C$.

Table 20.1.
Numerical values of the mean ODC of Fig. 20.2. For fresh human blood

S_{O_2} (%)	p_{O_2} (kPa)	*SD*	*SEM*
0	0.000	0.000	0.000
1	0.244	0.045	0.007
2	0.470	0.060	0.009
3	0.648	0.075	0.011
4	0.794	0.090	0.013
5	0.921	0.099	0.015
6	1.036	0.106	0.016
7	1.136	0.112	0.017
8	1.228	0.117	0.017
9	1.312	0.121	0.018
10	1.392	0.124	0.018
11	1.466	0.126	0.019
12	1.538	0.128	0.019
13	1.606	0.129	0.019
14	1.670	0.130	0.019
15	1.733	0.131	0.019
16	1.794	0.132	0.020
17	1.852	0.134	0.020
18	1.910	0.135	0.020
19	1.966	0.136	0.020
20	2.022	0.137	0.020
21	2.077	0.139	0.021
22	2.130	0.140	0.021
23	2.182	0.141	0.021
24	2.235	0.143	0.021
25	2.286	0.144	0.021
26	2.337	0.144	0.022
27	2.387	0.146	0.022
28	2.437	0.147	0.022
29	2.487	0.150	0.022
30	2.537	0.151	0.023
31	2.587	0.153	0.023
32	2.636	0.154	0.023
33	2.684	0.155	0.023
34	2.734	0.156	0.023
35	2.783	0.156	0.023
36	2.833	0.158	0.024
37	2.882	0.160	0.024
3B	2.931	0.161	0.024
39	2.981	0.163	0.024
40	3.031	0.164	0.024
41	3.082	0.165	0.025
42	3.132	0.167	0.025
43	3.184	0.168	0.025

Table 20.1.
(Continued)

S_{O_2} (%)	P_{O_2} (kPa)	*SD*	*SEM*
44	3.235	0.169	0.025
45	3.287	0.172	0.026
46	3.340	0.173	0.026
47	3.392	0.174	0.026
48	3.445	0.176	0.026
49	3.499	0.177	0.026
50	3.554	0.178	0.027
51	3.609	0.179	0.027
52	3.664	0.181	0.027
53	3.720	0.183	0.027
54	3.778	0.186	0.028
55	3.836	0.188	0.028
56	3.895	0.189	0.028
57	3.955	0.190	0.028
58	4.016	0.192	0.029
59	4.078	0.194	0.029
60	4.142	0.196	0.029
61	4.206	0.198	0.029
62	4.273	0.201	0.030
63	4.341	0.203	0.030
64	4.411	0.205	0.031
65	4.482	0.207	0.031
66	4.555	0.209	0.031
67	4.630	0.212	0.032
68	4.709	0.215	0.032
69	4.788	0.218	0.032
70	4.869	0.221	0.033
71	4.954	0.224	0.033
72	5.041	0.227	0.034
73	5.131	0.230	0.034
74	5.226	0.233	0.035
75	5.325	0.237	0.035
76	5.428	0.241	0.036
77	5.536	0.246	0.037
78	5.649	0.250	0.037
79	5.768	0.254	0.038
80	5.892	0.259	0.039
81	6.024	0.264	0.039
82	6.163	0.271	0.040
83	6.313	0.277	0.041
84	6.476	0.284	0.042
85	6.650	0.291	0.043
86	6.840	0.301	0.045
87	7.046	0.311	0.046
88	7.272	0.322	0.048

Table 20.1.
(Continued)

S_{O_2} (%)	P_{O_2} (kPa)	*SD*	*SEM*
89	7.525	0.336	0.050
90	7.810	0.355	0.053
90.1	7.838	0.357	0.053
90.2	7.868	0.359	0.054
90.3	7.900	0.361	0.054
90.4	7.932	0.362	0.054
90.5	7.964	0.364	0.054
90.6	7.997	0.367	0.055
90.7	8.030	0.369	0.055
90.8	8.065	0.370	0.055
90.9	8.099	0.372	0.055
91	8.133	0.375	0.056
91.1	8.168	0.376	0.056
91.2	8.205	0.379	0.056
91.3	8.241	0.381	0.057
91.4	8.277	0.383	0.057
91.5	8.315	0.385	0.057
91.6	8.353	0.387	0.058
91.7	8.392	0.391	0.058
91.8	8.431	0.392	0.058
91.9	8.471	0.393	0.059
92	8.510	0.396	0.059
92.1	8.551	0.400	0.060
92.2	8.593	0.401	0.060
92.3	8.635	0.404	0.060
92.4	8.677	0.407	0.061
92.5	8.720	0.409	0.061
92.6	8.767	0.411	0.061
92.7	8.813	0.414	0.062
92.8	8.860	0.416	0.062
92.9	8.906	0.419	0.062
93	8.958	0.423	0.063
93.1	9.007	0.428	0.064
93.2	9.058	0.432	0.064
93.3	9.109	0.436	0.065
93.4	9.160	0.439	0.065
93.5	9.213	0.443	0.066
93.6	9.267	0.447	0.067
93.7	9.323	0.453	0.067
93.8	9.381	0.457	0.068
93.9	9.440	0.461	0.069
94	9.502	0.466	0.069
94.1	9.567	0.475	0.071
94.2	9.631	0.479	0.071
94.3	9.697	0.487	0.073

Table 20.1.
(Continued)

S_{O_2} (%)	p_{O_2} (kPa)	*SD*	*SEM*
94.4	9.770	0.495	0.074
94.5	9.843	0.504	0.075
94.6	9.913	0.511	0.076
94.7	9.987	0.518	0.077
94.8	10.067	0.526	0.078
94.9	10.144	0.533	0.079
95	10.223	0.542	0.081
95.1	10.305	0.549	0.082
95.2	10.390	0.558	0.083
95.3	10.475	0.564	0.084
95.4	10.565	0.574	0.086
95.5	10.656	0.582	0.087
95.6	10.754	0.592	0.088
95.7	10.856	0.602	0.090
95.8	10.961	0.610	0.091
95.9	11.065	0.620	0.092
96	11.180	0.631	0.094
96.1	11.294	0.645	0.096
96.2	11.418	0.659	0.098
96.3	11.538	0.672	0.100
96.4	11.674	0.685	0.102
96.5	11.804	0.701	0.105
96.6	11.944	0.722	0.108
96.7	12.090	0.741	0.110
96.8	12.249	0.760	0.113
96.9	12.414	0.782	0.117
97	12.585	0.803	0.120
97.1	12.762	0.831	0.124
97.2	12.949	0.857	0.128
97.3	13.150	0.889	0.133
97.4	13.359	0.915	0.136
97.5	13.587	0.949	0.142
97.6	13.827	0.985	0.147
97.7	14.080	1.020	0.152
97.8	14.351	1.052	0.157
97.9	14.651	1.098	0.164
98	14.973	1.138	0.170
98.1	15.311	1.184	0.176
98.2	15.671	1.233	0.184
98.3	16.087	1.296	0.193
98.4	16.536	1.360	0.203
98.5	17.034	1.438	0.214
98.6	17.603	1.541	0.230
98.7	18.262	1.687	0.251
98.8	18.966	1.804	0.269

Table 20.1.
(Continued)

S_{O_2} (%)	p_{O_2} (kPa)	*SD*	*SEM*
98.9	19.813	1.938	0.289
99	20.781	2.143	0.319
99.1	21.931	2.363	0.352
99.2	23.232	2.609	0.389
99.3	24.850	2.969	0.443
99.4	26.844	3.311	0.494
99.5	29.407	3.690	0.550
99.6	32.684	4.515	0.673
99.7	36.946	5.198	0.775
99.8	42.591	5.790	0.863
99.9	46.175	5.777	0.861
100	48.300	5.935	0.885

pH = 7.40, p_{CO_2} = 5.33 kPa, T = 37°C, N = 45. *SD* is standard deviation; *SEM* is standard error of the mean.

[231] have not been published elsewhere, Table 20.2 presents a comparison with data from the literature. The ODC from Severinghaus is based on the principal investigations that at the time had appeared in the literature. The literature data were critically reviewed, then used to construct a standard ODC that was incorporated in the well-known blood gas calculator [347]. Although this ODC was constructed before the strong influence of DPG on the oxygen affinity of blood had been discovered [26, 61], it has long been used as representative for the normal ODC of human blood. Table 20.2 shows that there is indeed little difference between Severinghaus' ODC and those from later investigations where the influence of DPG was taken into account. The ODC of Fig. 20.2 and Table 20.1 is based on the measurement of fresh human blood of which the DPG concentration in the erythrocytes was measured and found to vary considerably between the specimens (Fig. 20.10).

Instead of an ODC (S_{O_2} *vs* p_{O_2}) to express the oxygen affinity of haemoglobin, a graph can be used of $\log[S_{O_2}/(1 - S_{O_2})]$, or logit S_{O_2}, *vs* $\log p_{O_2}$. This presentation was introduced in 1910 by Hill [158]. Figure 20.3 shows a **Hill plot** corresponding with the mean ODC of Fig. 20.2. $\mathrm{Log}[S_{O_2}/(1-S_{O_2})] = 0$ corresponds to S_{O_2} = 50%. The slope of this graph is called the Hill coefficient (*n*) [126]. When Hill proposed this way of expressing the oxygen affinity of haemoglobin, he did not ascribe any physical meaning to it. Wyman, however, has shown that Hill's *n* can be conceived as an index of the cooperativity of the four subunits of the haemoglobin molecule [434]. When all four oxygen binding sites are identical and independent, there is no cooperativity and the value of *n* is 1. When there is maximum cooperativity, *n* is 4, *i.e.* equal to the number of subunits. Figure 20.4 shows the Hill

Table 20.2.
Comparison of p_{O_2} values corresponding to fixed values of S_{O_2} from ODCs of various origin

S_{O_2} %	p_{O_2} (Ref. [347])		p_{O_2} (Refs [6, 325])		p_{O_2} (Ref. [383])		p_{O_2} (Table 20.1)	
	kPa	mmHg	kPa	mmHg	kPa	mmHg	kPa	mmHg
2	0.45	3.4	—	—	—	—	0.47	3.5
4	0.76	5.7	—	—	—	—	0.79	6.0
6	1.00	7.5	—	—	—	—	1.04	7.8
10	1.37	10.3	1.41	10.6	1.37	10.27	1.39	10.4
20	2.05	15.4	2.05	15.4	1.98	14.88	2.02	15.2
30	2.56	19.2	2.60	19.5	2.51	18.82	2.54	19.0
40	3.04	22.8	3.12	23.4	3.00	22.52	3.03	22.7
50	3.55	26.6	3.65	27.4	3.55	26.59	3.55	26.7
60	4.16	31.2	4.25	31.9	4.14	31.05	4.14	31.1
70	4.92	36.9	5.00	37.5	4.86	36.45	4.87	36.5
80	5.93	44.5	6.01	45.1	5.89	44.19	5.89	44.2
90	7.71	57.8	7.76	58.2	7.77	58.24	7.81	58.6
94	9.25	69.4	9.15	68.6	—	—	9.50	71.3
96	10.80	81.0	10.31	77.3	—	—	11.18	83.9
98	14.80	111.0	12.31	92.3	—	—	14.97	112.3

Ref. [347]: pH = 7.40, T = 37°C; Refs [6, 325]: N = 73, pH = 7.40, T = 37°C, p_{CO_2} = 40 mm Hg, DPG = 5 mmol/L, F_{COHb} = 1%; Ref. [383]: N = 25, pH = 7.40, T = 37°C, p_{CO_2} = 40 mm Hg; Table 20.1: N = 45, pH = 7.40, T = 37°C, p_{CO_2} = 40 mm Hg. 1 kPa = 7.5 mm Hg.

coefficient plotted as a function of S_{O_2}. Between 20 and 80% S_{O_2} there is little variation in n: the Hill plot is a nearly straight line in this S_{O_2} range. When S_{O_2} = 50%, n = 2.60. At very high and very low values of S_{O_2} n approximates 1.

The various **modulators** of the oxygen affinity influence the ***position*** of the ODC, some also the ***shape***. A shift in position may be characterised with the help of p_{50}. A change in the Hill coefficient may be used as a sign of a change in shape of the ODC. The normal modulators that have a major influence on the oxygen affinity of the blood are **temperature**, **pH** and **DPG**. Although it is the pH in the erythrocytes that counts in this respect, it is, for practical reasons, the plasma pH that is measured and is kept at 7.40 in the standard conditions. The influence of p_{CO_2} is mainly exerted through its effect on the pH, but there is a discernible influence of the p_{CO_2} *per se*. Of the common dyshaemoglobins **COHb** has the strongest effect on the oxygen affinity; a similar effect is exerted by **Hi**. Quantitative data on the effects of the important modulators are presented in the following sections.

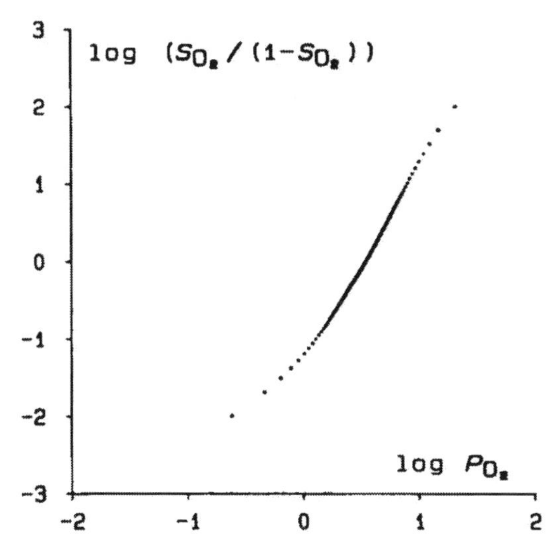

Figure 20.3. Hill plot of the oxygen dissociation curve of Fig. 20.2 and Table 20.1.

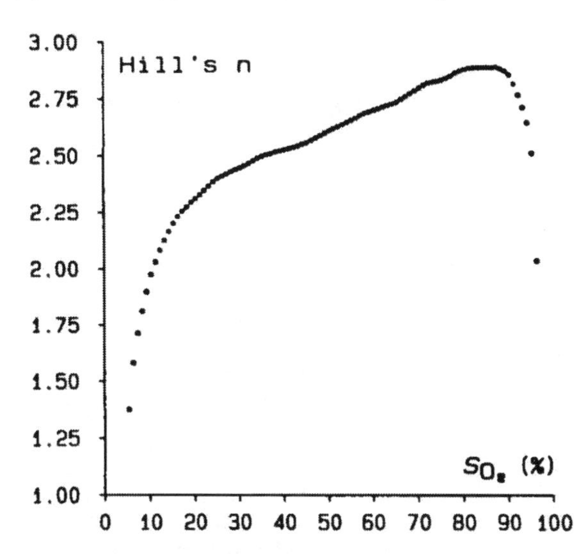

Figure 20.4. Hill coefficient plotted as a function of S_{O_2}.
Each value of n was calculated over an S_{O_2} range of 8%; the value for $S_{O_2} = 5\%$ was thus calculated over the range of 1 to 9% S_{O_2}.

EFFECT OF TEMPERATURE

The effect of temperature on the oxygen affinity of haemoglobin has been recognised for more than a century [27, 179]. An increase in temperature decreases the oxygen affinity of blood. Figure 20.5 shows five ODCs of a single blood specimen determined at five different temperatures: 22, 27, 32, 37 and 42°C, at pH = 7.40 and $p_{CO_2} = 5.33$ kPa. Similar groups of ODCs

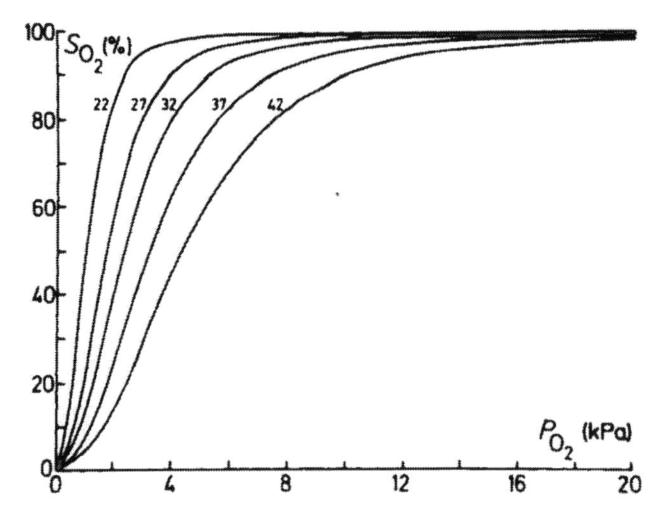

Figure 20.5. Oxygen dissociation curves from blood of a single donor at five different temperatures (°C); pH = 7.40 and p_{CO_2} = 5.33 kPa. These curves are not the result of a fitting procedure but are composed of actually measured data points (from ref. [442]).

Table 20.3.
p_{O_2} at fixed values of S_{O_2} at five different temperatures, with pH = 7.40 and p_{CO_2} = 5.33 kPa

S_{O_2}	$p_{O_2}(22)$	$p_{O_2}(27)$	$p_{O_2}(32)$	$p_{O_2}(37)$	$p_{O_2}(42)$
10	0.53 ± 0.06	0.73 ± 0.05	0.96 ± 0.05	1.28 ± 0.10	1.72 ± 0.13
20	0.78 ± 0.08	1.06 ± 0.06	1.43 ± 0.06	1.90 ± 0.13	2.45 ± 0.13
30	0.98 ± 0.08	1.38 ± 0.07	1.82 ± 0.06	2.42 ± 0.15	3.03 ± 0.12
40	1.18 ± 0.10	1.66 ± 0.08	2.20 ± 0.09	2.93 ± 0.17	3.65 ± 0.15
50	1.39 ± 0.12	1.97 ± 0.09	2.61 ± 0.09	3.46 ± 0.19	4.31 ± 0.18
60	1.63 ± 0.13	2.32 ± 0.11	3.06 ± 0.11	4.02 ± 0.22	5.04 ± 0.23
70	1.92 ± 0.15	2.73 ± 0.12	3.61 ± 0.14	4.75 ± 0.26	5.95 ± 0.26
80	2.33 ± 0.16	3.31 ± 0.15	4.39 ± 0.15	5.77 ± 0.30	7.23 ± 0.36
90	3.03 ± 0.21	4.34 ± 0.19	5.82 ± 0.22	7.82 ± 0.37	9.87 ± 0.47
95	3.86 ± 0.31	5.51 ± 0.28	7.56 ± 0.33	10.21 ± 0.51	13.05 ± 0.76

S_{O_2} in %, p_{O_2} in kPa with temperature in °C in parentheses. The p_{O_2} values are the mean ± *SD* of six specimens. The mean DPG concentration in the erythrocytes was 4.55 mmol/L; mean c_{DPG}/c_{Hb_4} = 0.87 mol/mol. N = 6. In each sample F_{COHb} and F_{Hi} were < 1%, F_{SHb} was not detectable.

were determined of the blood of 6 healthy non-smoking donors [450]. The temperature coefficient, $TC = \partial \log p_{O_2}/\partial T$, calculated from these curves proved to be almost independent of S_{O_2}: for S_{O_2} = 30–80% the mean *TC* was 0.024 ± 0.002 K^{-1} (*SEM*); for S_{O_2} = 10, 20 and 90% a mean *TC* of 0.025 ± 0.002 K^{-1} was found; and for S_{O_2} = 95% it was 0.025 ± 0.004 K^{-1}.

Table 20.3 gives the mean p_{O_2} values at each temperature of the six blood specimens at fixed values of S_{O_2}. The invariance of the temperature

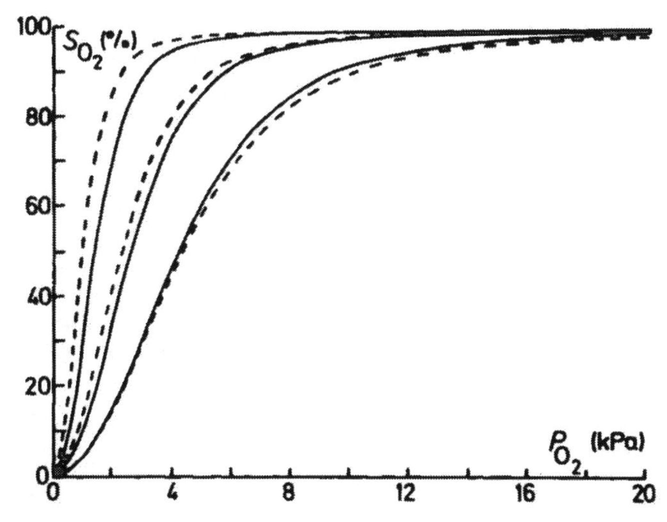

Figure 20.6. Oxygen dissociation curves at three different temperatures at $p_{CO_2} = 5.33$ kPa and pH = 7.40 (solid lines) and with pN − pH = 0.6 (dashed lines) and temperatures from left to right 22, 32 and 42°C (from ref. [488]).

coefficient with S_{O_2} implies that the shape of the ODC does not change with a change in temperature, *i.e.* that the five ODCs of Fig. 20.5 can be made to coincide with a single scaling factor [450]. This is corroborated by the constancy of the Hill coefficient over the temperature range studied. Hill's n was determined as the slope of the linear regression of logit S_{O_2} *vs* log p_{O_2} for all data points between S_{O_2} = 20 and 80%. The mean values (± *SEM*) were 2.52±0.037, 2.47±0.020, 2.46±0.041, 2.46±0.045 and 2.53±0.065 for 22, 27, 32, 37 and 42°C, respectively.

In vivo, a change in temperature is accompanied by a change in pH. As described by Rahn *et al.* [311], most animals keep their extracellular fluid at a constant alkalinity relative to neutrality. Hence, when the temperature falls the blood pH rises in parallel with the rise in the pH of neutrality, neutrality being the state where the activities of H^+ and OH^- are equal (pH = pOH = pN). The change in pH invokes a proton Bohr effect (p. 304), which enhances the effect of the fall in temperature *per se*. In Fig. 20.6 three of the ODCs of Fig. 20.5 have been redrawn — those at 22, 32 and 42°C (solid lines). In recording these ODCs, the pH was kept at 7.40 at each temperature. The dashed lines show what happens when *in vivo* the extracellular pH changes with the temperature. Now pH − pN is 0.6 at each temperature [311]. When a fall in temperature from 37 to 22°C causes pN to increase from 6.80 to 7.04, there is a concomitant rise in pH from 7.40 to 7.64. Therefore, the dashed line corresponding to 22°C has been calculated for pH = 7.64. Accordingly, the lines corresponding to 32 and 42°C have been calculated for pH = 7.48 and 7.33, respectively. The proton Bohr factors used in these calculations [496] were −0.505, −0.427 and −0.365, for 22, 32 and 42°C, respectively.

The procedure is illustrated by the following example in which the change in p_{O_2} at 50% S_{O_2} is calculated when the temperature falls from 37 to 22°C. The change in log p_{O_2} due to the change in temperature and the concomitant change in pH at constant base excess is given by

$$\begin{aligned} d\log p_{O_2} &= (\partial \log p_{O_2}/\partial T)dT + (\partial \log p_{O_2}/\partial pH)dpH \\ &= 0.024 \times (22 - 37) - 0.505 \times (7.64 - 7.40) \\ &= -0.48. \end{aligned} \quad (20.3)$$

At $S_{O_2} = 50\%$ and 37°C, $p_{O_2} = 3.55$ kPa (Table 20.1) and thus log $p_{O_2}(37) = 0.55$; log $p_{O_2}(22)$ then is $0.55-0.48 = 0.07$. Hence, p_{O_2} at 50% S_{O_2} and 22°C is $10^{0.07} = 1.17$ kPa.

From the experimentally determined change in plasma pH with a change in body temperature [54, 323], $dpH/dT = -0.015\ K^{-1}$, and the proton Bohr factor, $\Phi_H = \partial \log p_{O_2}/\partial pH = -0.43$ (Table 20.5), an overall (*in vivo*) temperature coefficient can be calculated:

$$\begin{aligned} d\log p_{O_2}/dT &= \partial \log p_{O_2}/\partial T + (\partial \log p_{O_2}/\partial pH) \cdot (dpH/dT) \\ &= 0.024 + (-0.43) \times (-0.015) = 0.0305\ K^{-1}. \end{aligned} \quad (20.4)$$

Note that in this calculation the temperature-dependence of the proton Bohr factor has been neglected. For a comparison of the data presented in this section with those from the literature see refs [442] and [450]; for a further discussion of the effect of the temperature on the oxygen affinity of the blood see refs [126, 442, 450].

THE BOHR AND HALDANE EFFECTS

The effect of CO_2 on the oxygen affinity of whole blood was first described by Bohr *et al.* [33]. Generally, the oxygen affinity decreases with an increase in p_{CO_2}, but CO_2 interacts with haemoglobin in different ways [21, 330, 331]. It binds to the globin moiety of haemoglobin forming carbamate compounds as well as exerts an indirect effect through its influence on the pH. For the quantitative description of these effects the following terminology has been proposed [230].

The influence of pH on the oxygen affinity of the blood at constant p_{CO_2} is called **proton Bohr effect** and is expressed by the **proton Bohr factor**: $\Phi_H = [\partial \log p_{O_2}/\partial pH]_{p_{CO_2}}$.

The influence of CO_2 on the oxygen affinity of the blood at constant pH is called **carbamate Bohr effect** and is expressed by the **carbamate Bohr factor**: $\Phi_C = [\partial \log p_{O_2}/\partial \log p_{CO_2}]_{pH}$.

The influence of CO_2 on the oxygen affinity of blood at constant base excess (BE) is called **total Bohr effect** and is expressed by the **total Bohr factor**: $\Phi_{HC} = [d \log p_{O_2}/dpH]_{BE}$.

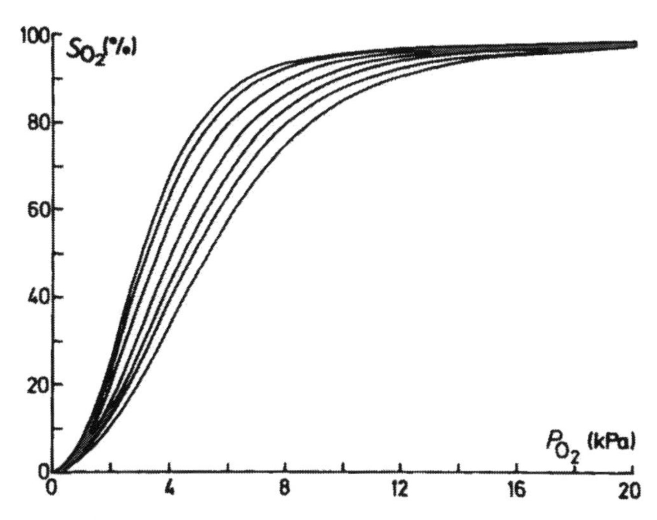

Figure 20.7. Oxygen dissociation curves of human whole blood at seven different pH values. The pH varies from left to right from 7.6 to 7.0 in steps of 0.1; p_{CO_2} = 5.33 kPa and $T = 37\,°C$. The ODCs are shown as recorded (from ref. [442]).

Table 20.4.
Quantitative data concerning the blood specimens used for the determination of the Bohr factors given in Table 20.5

Quantity (unit)	Women	Men
ct_{Hb} (mmol/L)	7.77 ± 0.32	9.15 ± 0.12
Haematocrit (%)	38.0 ± 1.3	44.0 ± 1.0
c_{DPG}/c_{Hb_4} (mol/mol)	0.94 ± 0.07	0.93 ± 0.05
$d\log p_{CO_2}/dpH$ ($S_{O_2} = 0$)	−1.41 ± 0.05	−1.43 ± 0.08
$d\log p_{CO_2}/dpH$ ($S_{O_2} = 1$)	−1.41 ± 0.06	−1.42 ± 0.07

Values are means ± *SD*. $N = 12$; Hb denotes the haemoglobin monomer, Hb_4 the tetramer.

The relationship between the Bohr factors is given by the equation:

$$\Phi_{HC} = \Phi_H + \Phi_C \cdot (d\log p_{CO_2}/dpH), \tag{20.5}$$

where $d\log p_{CO_2}/dpH$ is the slope of the $\log p_{CO_2}$−pH equilibrium curve [354]. Because of the mutual effects of p_{CO_2} and pH on Φ_H and Φ_C as shown in Tables 20.7 and 20.8, the relationship is more accurately described by adding a second order term to equation 20.5.

$$\Phi_{HC} = \Phi_H + \Phi_C \cdot (d\log p_{CO_2}/dpH) + (\partial\Phi_H/\partial \log p_{CO_2}) \cdot \Delta \log p_{CO_2}. \tag{20.6}$$

In these equations, and throughout this chapter, the partial derivative sign, ∂, implies that all other variables are kept constant.

Figure 20.7 shows the proton Bohr effect by means of seven ODCs of a single specimen of anticoagulated fresh human whole blood, made at pH values from 7.0 to 7.6. To obtain accurate quantitative values for the three Bohr factors and to study the S_{O_2}-dependence of these factors, ODCs were recorded at different values of pH and p_{O_2} in the blood of 12 donors (6 female, 6 male). Table 20.4 gives additional data of the 12 blood specimens used in the measurement of the Bohr factors. The slope of the pH–log p_{CO_2} equilibrium curve, $d \log p_{CO_2}/dpH$, was calculated from the p_{CO_2} and pH values measured at the beginning and the end of each ODC, for $S_{O_2} = 0$ and 1 (0 and 100%). In all calculations a mean value of 1.415 was used for $d \log p_{CO_2}/dpH$. F_{COHb}, F_{Hi} and F_{SHb} in all specimens were well below the upper limit of the reference range determined in more than 4000 specimens [447].

For the calculation of the Bohr factors, the $p_{O_2}-S_{O_2}$ data pairs of each ODC were transformed according to the following rules [230]. In the range of $S_{O_2} = 0-90\%$, the measured p_{O_2} values of the S_{O_2} interval of $x - 0.5\%$ to $x + 0.5\%$ were averaged to find the p_{O_2} corresponding to $S_{O_2} = x\%$, x being a whole number; for $S_{O_2} = 90-100\%$, p_{O_2} values from the S_{O_2} interval of $x - 0.1\%$ to $x + 0.1\%$ were used. The slopes of the log p_{O_2} *vs* pH (or log p_{CO_2}) lines were then calculated for each value of S_{O_2}. Φ_{HC} was measured as well as calculated with the help of equation 20.5. The Bohr factors were first calculated for the specimens of female and male donors separately. Because no significant differences were found ($p \gg 0.2$; unpaired t-test), the data were combined.

The Bohr factors are presented in Table 20.5. There is a reasonable agreement between Φ_{HC} as determined directly and as calculated from Φ_H and Φ_C using equation 20.5. However, when the effect of p_{CO_2} on Φ_H (Table 20.7) and the effect of pH on Φ_C (Table 20.8) are taken into account by using equation 20.6, a better agreement is indeed obtained. For Φ_{HC} at $S_{O_2} = 20$, 50 and 80% values of −0.500, −0.495 and −0.474 are then obtained instead of −0.509, −0.504 and −0.481, while the directly measured values are −0.500, −0.488 and −0.466.

For $S_{O_2} > 30\%$ the proton Bohr effect is hardly dependent on S_{O_2}, but at $S_{O_2} = 10\%$, Φ_H is 18% higher (less negative) than at $S_{O_2} = 50\%$. The difference is statistically significant ($p < 0.001$; paired t-test). This decline of the absolute value of Φ_H in the lower S_{O_2} range means that, when the pH decreases, the shift to the right of the ODC is less in the lower part of the curve, *i.e.* while shifting to the right the ODC becomes less steep. Table 20.6 shows that the Hill coefficient indeed tends to decrease with decreasing pH. The carbamate Bohr effect, on the other hand, increases over the whole range when S_{O_2} falls: Φ_C increases from ~ 0.03 when $S_{O_2} > 90\%$ to ~ 0.10 when $S_{O_2} = 10\%$. This means that the ODC becomes steeper, corroborated by the increase in Hill's n on increasing p_{CO_2} (Table 20.6). Although to a lesser degree, the same holds for the total Bohr effect.

Blood specimens of 12 donors (11 female, 1 male) were used for the determination of the effect of p_{CO_2} on Φ_H and the effect of pH on Φ_C. The

Table 20.5.
Bohr factors of human whole blood at $T = 37\,°C$ and various values of S_{O_2}

S_{O_2} (%)	Φ_H	Φ_C	Φ_{HC}	Φ_{HC}(calc)
10	−0.350 ± 0.017	0.097 ± 0.009	−0.486 ± 0.020	−0.487 ± 0.016
15	−0.374 ± 0.014	0.091 ± 0.007	−0.495 ± 0.016	−0.502 ± 0.013
20	−0.391 ± 0.013	0.084 ± 0.006	−0.500 ± 0.014	−0.509 ± 0.011
25	−0.401 ± 0.012	0.078 ± 0.007	−0.502 ± 0.012	−0.510 ± 0.011
30	−0.410 ± 0.011	0.071 ± 0.007	−0.501 ± 0.010	−0.510 ± 0.010
35	−0.413 ± 0.011	0.067 ± 0.006	−0.498 ± 0.009	−0.508 ± 0.010
40	−0.419 ± 0.011	0.062 ± 0.006	−0.494 ± 0.008	−0.507 ± 0.010
45	−0.426 ± 0.010	0.057 ± 0.006	−0.491 ± 0.007	−0.507 ± 0.009
50	−0.428 ± 0.010	0.054 ± 0.006	−0.488 ± 0.007	−0.504 ± 0.009
55	−0.430 ± 0.010	0.050 ± 0.006	−0.484 ± 0.007	−0.500 ± 0.009
60	−0.431 ± 0.010	0.046 ± 0.005	−0.483 ± 0.007	−0.496 ± 0.009
65	−0.433 ± 0.009	0.042 ± 0.005	−0.479 ± 0.006	−0.493 ± 0.009
70	−0.434 ± 0.009	0.040 ± 0.004	−0.474 ± 0.007	−0.490 ± 0.008
75	−0.433 ± 0.009	0.039 ± 0.004	−0.470 ± 0.006	−0.488 ± 0.008
80	−0.429 ± 0.009	0.037 ± 0.005	−0.466 ± 0.006	−0.481 ± 0.008
85	−0.425 ± 0.009	0.035 ± 0.005	−0.462 ± 0.006	−0.475 ± 0.008
90	−0.423 ± 0.009	0.035 ± 0.005	−0.456 ± 0.006	−0.472 ± 0.008
95	−0.413 ± 0.010	0.032 ± 0.006	−0.444 ± 0.007	−0.459 ± 0.012

Φ_H = proton Bohr factor; Φ_C = carbamate Bohr factor; Φ_{HC} = total Bohr factor; Φ_{HC}(calc) = total Bohr factor, calculated with equation 20.4. Values are means ± *SEM*; $N = 12$.

Table 20.6.
Hill's n at various values of pH and p_{CO_2}

Proton Bohr effect ($p_{CO_2} = 5.33$ kPa)		Carbamate Bohr effect (pH = 7.40)		Total Bohr effect (BE = 0)	
pH	n	p_{CO_2}	n	p_{CO_2}	n
7.6	2.69 ± 0.019	2	2.54 ± 0.020	2	2.59 ± 0.021
7.4	2.63 ± 0.026	6	2.67 ± 0.020	6	2.69 ± 0.024
7.2	2.62 ± 0.023	10	2.75 ± 0.017	10	2.70 ± 0.023
7.0	2.54 ± 0.029	14	2.79 ± 0.017	14	2.73 ± 0.023

Values are means ± *SEM*; BE is base excess. $T = 37°C$. $N = 12$.

influence of p_{CO_2} on Φ_H is shown in Table 20.7. There is an appreciable fall in the absolute value of Φ_H with increasing p_{CO_2}. Table 20.8 shows the strong influence of pH on Φ_C; with decreasing pH the carbamate Bohr effect diminishes. At pH < 7.2, Φ_C approaches zero showing that at low pH haemoglobin hardly binds CO_2. For a further discussion of the results and a comparison with those in the literature see ref. [230]; for the underlying chemical interactions see refs [21, 126, 330, 331].

Table 20.7.
Proton Bohr factor at $T = 37°C$ and various values of S_{O_2} and p_{CO_2}

S_{O_2}	Φ_H ($p_{CO_2} = 2$ kPa)	Φ_H ($p_{CO_2} = 6$ kPa)	Φ_H ($p_{CO_2} = 10$ kPa)	Φ_H ($p_{CO_2} = 14$ kPa)
10	−0.45 ± 0.014	−0.33 ± 0.008	−0.31 ± 0.009	−0.25 ± 0.012
20	−0.47 ± 0.011	−0.36 ± 0.008	−0.33 ± 0.008	−0.27 ± 0.010
30	−0.48 ± 0.010	−0.37 ± 0.006	−0.34 ± 0.007	−0.29 ± 0.010
40	−0.49 ± 0.010	−0.39 ± 0.006	−0.35 ± 0.006	−0.30 ± 0.010
50	−0.49 ± 0.009	−0.40 ± 0.006	−0.36 ± 0.006	−0.31 ± 0.010
60	−0.49 ± 0.009	−0.40 ± 0.006	−0.36 ± 0.007	−0.32 ± 0.009
70	−0.48 ± 0.009	−0.40 ± 0.006	−0.37 ± 0.007	−0.32 ± 0.009
80	−0.47 ± 0.009	−0.40 ± 0.006	−0.37 ± 0.007	−0.33 ± 0.009
90	−0.46 ± 0.008	−0.40 ± 0.008	−0.37 ± 0.006	−0.33 ± 0.011

Φ_H = proton Bohr factor. Values are means ± *SEM*; $N = 12$.

Table 20.8.
Carbamate Bohr factor at $T = 37°C$ and various values of S_{O_2} and pH

S_{O_2} (%)	Φ_C (pH = 7.00)	Φ_C (pH = 7.20)	Φ_C (pH = 7.40)
10	0.033 ± 0.013	0.035 ± 0.006	0.095 ± 0.009
20	0.024 ± 0.011	0.030 ± 0.004	0.083 ± 0.006
30	0.016 ± 0.009	0.022 ± 0.005	0.072 ± 0.006
40	0.010 ± 0.009	0.016 ± 0.005	0.062 ± 0.005
50	0.006 ± 0.009	0.013 ± 0.005	0.054 ± 0.005
60	0.005 ± 0.009	0.009 ± 0.006	0.047 ± 0.004
70	0.005 ± 0.009	0.008 ± 0.005	0.041 ± 0.004
80	0.005 ± 0.009	0.007 ± 0.005	0.038 ± 0.004
90	0.011 ± 0.009	0.008 ± 0.006	0.034 ± 0.004

S_{O_2} (%)	Φ_C (pH = 7.55)	Φ_C (pH = 7.70)	Φ_C (pH = 7.80)
10	0.106 ± 0.013	0.157 ± 0.019	0.193 ± 0.018
20	0.109 ± 0.010	0.154 ± 0.017	0.184 ± 0.015
30	0.095 ± 0.009	0.146 ± 0.015	0.176 ± 0.014
40	0.087 ± 0.009	0.137 ± 0.014	0.167 ± 0.014
50	0.078 ± 0.010	0.127 ± 0.014	0.157 ± 0.015
60	0.068 ± 0.010	0.118 ± 0.013	0.147 ± 0.015
70	0.062 ± 0.009	0.108 ± 0.012	0.137 ± 0.015
80	0.057 ± 0.009	0.102 ± 0.011	0.129 ± 0.016
90	0.049 ± 0.009	0.093 ± 0.012	0.120 ± 0.016

Φ_C = carbamate Bohr factor. Values are means ± *SEM*; $N = 18$ at pH = 7.40; $N = 6$ at other pH values.

The counterpart of the strong influence of pH on the oxygen affinity of the blood is the release of protons (H^+ ions) on oxygenation. This phenomenon, which is called the **Haldane effect**, was first observed by Christiansen *et*

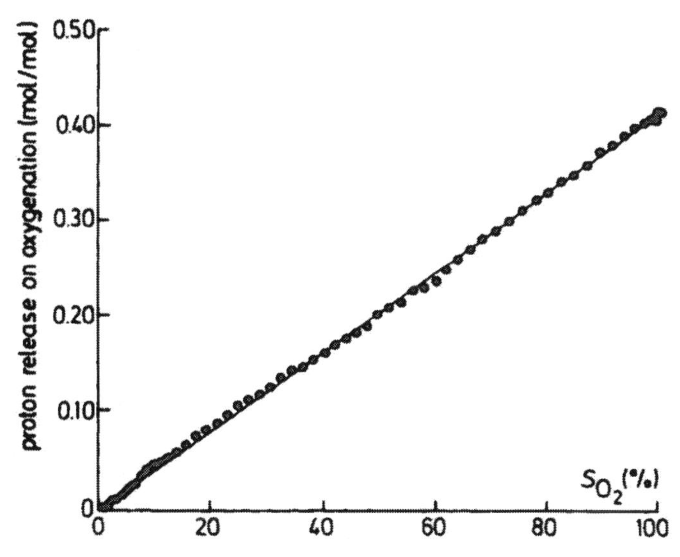

Figure 20.8. Amount of protons released per mol active Hb ($= ct_{Hb} - c_{COHb} - c_{Hi} - c_{SHb}$) plotted *vs* S_{O_2}.
Standard conditions: pH = 7.40, p_{CO_2} = 5.33 kPa, T = 37°C.

al. [62], who described in 1914 that deoxygenated blood binds considerably more CO_2 than blood that is almost saturated with oxygen. The effect proved to be caused by a decrease in carbamate formation as well as a release of protons when S_{O_2} is increased. Hence, a **carbamate Haldane effect** and a **proton Haldane effect** were defined as the thermodynamic counterparts of the carbamate and proton Bohr effects [356, 390].

During the recording of an ODC, the H^+ ions set free when HHb is converted to O_2Hb are neutralised by the addition of NaOH to keep the pH constant (p. 292). The amount of NaOH solution added in this procedure may be used to calculate the proton Haldane effect, *i.e.* the amount of H^+ released from HbH^+ per mole O_2 bound (or per mole O_2Hb formed). The calculation is based on the equation

$$\Delta c_{HbH^+}/\Delta c_{O_2Hb} = V_a \cdot c_{NaOH}/V_b \cdot ct_{Hb} \cdot (1 - F_{COHb} - F_{Hi} - F_{SHb}), \quad (20.7)$$

where V_a is the volume of NaOH solution added in the titration, c_{NaOH} is the concentration of the NaOH solution and V_b is the blood volume in the cuvette; ct_{Hb} is measured by the HiCN method (p. 200) and F_{COHb}, F_{Hi} and F_{SHb} by the standard multiwavelength method (p. 227). Figure 20.8 shows the liberation of H^+ from HHb during oxygenation from S_{O_2} = 0 to 100%. The value at S_{O_2} = 100% equals the absolute value of the proton Haldane factor.

Theoretically, the proton Haldane factor is numerically equal to the proton Bohr factor:

$$\partial c_{HbH^+}/\partial c_{O_2Hb} = \partial \log p_{O_2}/\partial pH. \quad (20.8)$$

There has been some controversy in the literature as to the experimental equality of the proton Bohr and Haldane factors [356, 390]. However, when taking the S_{O_2} dependence of Φ_H into account by integrating Φ_H over the entire S_{O_2} range a good agreement with the proton Haldane factor as determined with equation 20.7 is obtained.

In the measurement of the ODCs of the six specimens of human blood for the determination of the temperature coefficient (Table 20.3) the proton Haldane factor was also determined. By linear regression the relationship between proton Haldane factor and temperature proved to be $\Delta c_{HbH^+}/\Delta c_{O_2Hb} = 0.006 \times T - 0.644$, where T is the temperature in °C. This gives a value of −0.42 for the proton Haldane factor at 37°C, which is in excellent agreement with the value of −0.416 found by numerical integration of the Φ_H data of Table 20.5.

EFFECT OF 2,3-DIPHOSPHOGLYCERATE

Although 2,3-diphosphoglycerate (DPG) had long been known to be present in high concentration in human erythrocytes and its formation and breakdown had been studied extensively [314], its effect on the oxygen affinity of blood was not described until 1967. The influence of DPG on the haemoglobin oxygen affinity, almost simultaneously discovered by Benesch and Benesch [26] and Chanutin and Curnisch [61], is due to its preferential binding to HHb. The affinity of DPG for HHb is nearly 100 times that for O_2Hb. The preferential binding only occurs with Hb_4, to which it is bound between the β-chains in a molar ratio of 1 : 1.

When in a solution of haemoglobin the ratio c_{DPG}/c_{Hb_4} is increased above 1, no further decrease in oxygen affinity occurs. In erythrocytes however, a c_{DPG}/c_{Hb_4} ratio > 1 leads to a further fall in oxygen affinity. Since DPG is a poly-anion that does not pass through the red cell membrane, it lowers the intracellular pH through its effect on the Gibbs–Donnan distribution across the cell membrane. This further decreases the oxygen affinity of haemoglobin through the proton Bohr effect [98].

Figure 20.9 shows the strong influence of DPG on the position and shape of the ODC of human blood [231]. The left ODC is similar to the one found for human blood stored at 4°C for about two weeks. Almost all DPG has then disappeared from the erythrocytes. The right ODC is displaced far to the right, resulting, to a considerable extent, from the proton Bohr effect due to the lowered intra-erythrocytic pH caused by artificially increasing c_{DPG} in the red cells to a supernormal level [442].

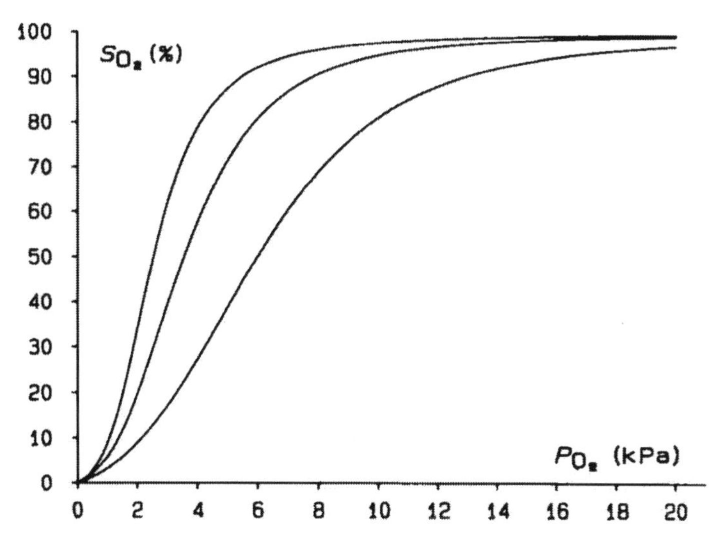

Figure 20.9. Effect of the DPG concentration in the erythrocytes on the ODC of human whole blood.

The central curve is the mean ODC of Fig. 20.2; the left curve is from blood with hardly any DPG; the right one from blood with about three times the normal DPG concentration. pH = 7.40, p_{CO_2} = 5.33 kPa, T = 37°C.

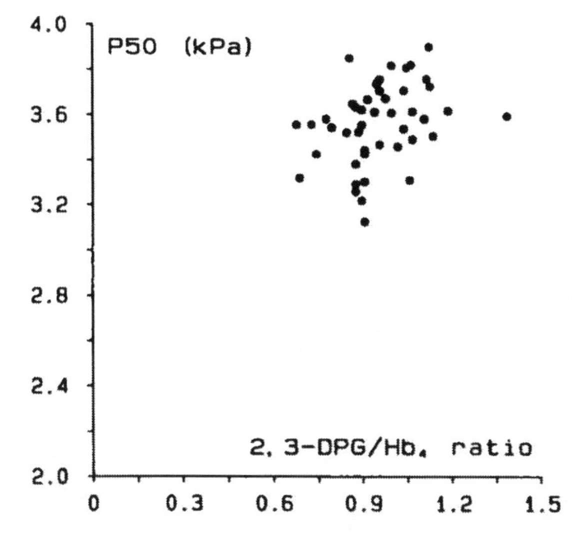

Figure 20.10. p_{50} of the ODCs underlying the mean ODC of Fig. 20.2 plotted *vs* the corresponding c_{DPG}/c_{Hb_4} ratios. pH = 7.40, p_{CO_2} = 5.33 kPa, T = 37°C. N = 45.

The c_{DPG}/c_{Hb_4} ratio has been determined in all 45 blood specimens that underlie the mean ODC of Fig. 20.2. The p_{50} values were then plotted against the c_{DPG}/c_{Hb_4} ratios (Fig. 20.10). It appears that the variation in c_{DPG}/c_{Hb_4} is from 0.7 to 1.4, while the corresponding spread in p_{50} is only from 3.12

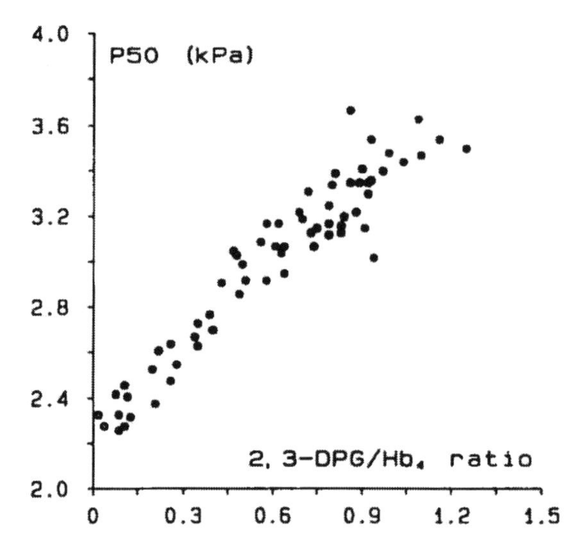

Figure 20.11. p_{50} plotted *vs* c_{DPG}/c_{Hb_4} of six specimens of human whole blood with falling erythrocytic DPG concentration. pH = 7.40, p_{CO_2} = 5.33 kPa, T = 37°C. The regression lines of the individual specimens are presented in Table 20.9.

Table 20.9.
Relationship between p_{50} and c_{DPG}/c_{Hb_4} of six specimens of human whole blood with falling erythrocytic DPG concentration

Regression equation	*n*	*r*
$p_{50} = 2.484 + 0.927 \cdot c_{DPG}/c_{Hb_4}$	13	0.963
$p_{50} = 2.189 + 1.240 \cdot c_{DPG}/c_{Hb_4}$	10	0.985
$p_{50} = 2.314 + 1.365 \cdot c_{DPG}/c_{Hb_4}$	12	0.983
$p_{50} = 2.239 + 1.249 \cdot c_{DPG}/c_{Hb_4}$	11	0.984
$p_{50} = 2.307 + 0.969 \cdot c_{DPG}/c_{Hb_4}$	10	0.965
$p_{50} = 2.254 + 1.187 \cdot c_{DPG}/c_{Hb_4}$	11	0.956

pH = 7.40, p_{CO_2} = 5.33 kPa, T = 37°C.

to 3.89 kPa. There is only a weak correlation between the two quantities ($r = 0.328$). This is in contrast with the relation between p_{50} and c_{DPG}/c_{Hb_4}, shown in Fig. 20.11, resulting from serial determinations of p_{50} of blood with declining c_{DPG}/c_{Hb_4} ratio: p_{50} and c_{DPG}/c_{Hb_4} were measured in six blood specimens in which the DPG concentration in the erythrocytes was allowed to decline to almost zero by storage at 4°C for several days. A closer correlation between p_{50} and c_{DPG}/c_{Hb_4} is now found. Table 20.9 shows the relationship between p_{50} and c_{DPG}/c_{Hb_4} for the individual specimens [231].

The results of Fig. 20.11 and Table 20.9 obviously approximate the effect of DPG on the oxygen affinity of haemoglobin much better then the data of Fig. 20.10. The spread in p_{50} in the group of 45 human blood specimens (mean ± 2 *SD*; Table 20.1) is from 3.2 to 3.9 kPa. It may be that the concomitant rather wide variation in c_{DPG}/c_{Hb_4} indicates that under physiological conditions DPG contributes to keep p_{50} within a rather narrow range by compensating for the variation caused by other modulators of oxygen affinity. It may well be that one blood specimen needs a c_{DPG}/c_{Hb_4} ratio of 0.7 to keep its p_{50} at 3.6 kPa, while another needs double this amount of DPG for the same value of p_{50}. Factors not taken into account are variations in the pH difference between erythrocyte fluid and plasma and the effects of the Cl^- activity in the red cells [21, 123]. This interpretation is no more than an extension of the established role of DPG to compensate for disturbances in the oxygen transport properties of the blood under pathological conditions [126].

To calculate a DPG factor for quantitatively expressing the effect of DPG on the oxygen affinity of the blood, it seems worthwhile to separate the allosteric effect on haemoglobin from the effect on the pH of the erythrocyte fluid. Therefore, $\log p_{O_2}$ was related to c_{DPG}/c_{Hb_4} for the data of the six specimens of Fig. 20.11 and Table 20.9 with $c_{DPG}/c_{Hb_4} \leqslant 0.7$ ($n = 37$). This yields $\partial \log p_{O_2}/\partial(c_{DPG}/c_{Hb_4}) = 0.225$ for $S_{O_2} = 50\%$, pH $= 7.40$, $p_{CO_2} = 5.33$ kPa and $T = 37°C$.

EFFECT OF DYSHAEMOGLOBINS

Dyshaemoglobins influence the oxygen capacity as well as the oxygen affinity of the blood. The effect on the capacity has been discussed in Chapter 19. The effect on the oxygen affinity, especially the considerable increase in affinity caused by the presence of COHb, may contribute significantly to the disturbing effect of dyshaemoglobin formation on blood oxygen transport. An *increase* in oxygen affinity impedes the release of oxygen from the blood in the tissue capillaries and diminishes the p_{O_2} gradient that drives the diffusive oxygen flow into the cells. However, because of the flat course of the ODC at $p_{O_2} > 10$ kPa, a (moderate) *decrease* in oxygen affinity has little influence on the oxygen uptake in the lungs. Hi has a similar effect as COHb, though less strong. SHb is not important in this respect because it does not occur in toxic concentrations [489]. This section describes how ODCs can be measured in the presence of COHb and Hi with some results given.

To study the **effect of CO on the oxygen affinity** seven specimens of human blood were collected by venipuncture from healthy adults, and anticoagulated with heparin. About 35 mL was immediately transferred to a revolving glass tonometer (Fig. 6.1) and exposed to a humidified gas mixture of 94.4% N_2 and 5.6% CO_2. After 30 min, when S_{O_2} was near zero, 10 mL of the blood was removed from the tonometer and stored in a gastight syringe. The rest of the blood was then exposed, for a few minutes, to a humidified gas mixture of 5% CO and 95% N_2. After flushing the tonometer with the N_2/CO_2 mixture to remove all CO from the gas phase, a blood sample was

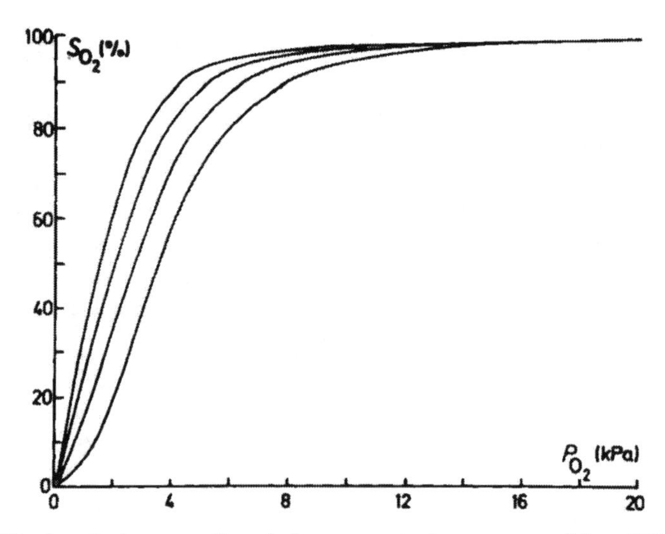

Figure 20.12. Standard oxygen dissociation curves in the presence of four different COHb fractions.

Blood from a single donor, ODCs directly recorded. From left to right: F_{COHb} = 50.8, 32.5, 14.6 and < 1% (from ref. [449]).

Table 20.10.
Hill's n at various values of F_{COHb}

F_{COHb} (%)	n
$\leqslant 1$	2.56 ± 0.015
~ 14	2.05 ± 0.045
~ 30	1.81 ± 0.024
~ 52	1.63 ± 0.018

Values are means ± SEM. pH = 7.40, $p_{CO_2} = 5.33$ kPa and $T = 37$°C. $N = 4$.

taken from the tonometer and F_{COHb} was determined. If F_{COHb} was below the intended level, the blood was again exposed to the CO/N_2 mixture; if F_{COHb} was above this level some deoxygenated CO-free blood from the syringe was added. The procedure was repeated until a series of blood samples with properly spaced F_{COHb} levels between 10 and 60% had been obtained.

ODCs were determined of the COHb-containing blood with the first version of the measuring system described previously (p. 290). In the calibration and performance check of the fibre optic oximeter, due corrections were made for the presence of COHb. F_{COHb}, F_{Hi}, F_{SHb}, ct_{Hb} and c_{DPG} were determined as described previously (p. 293). The binding of CO to haemoglobin proved to be strong enough for full oxygenation and subsequent deoxygenation to S_{O_2} = 50% to be carried out without a discernible change in F_{COHb}.

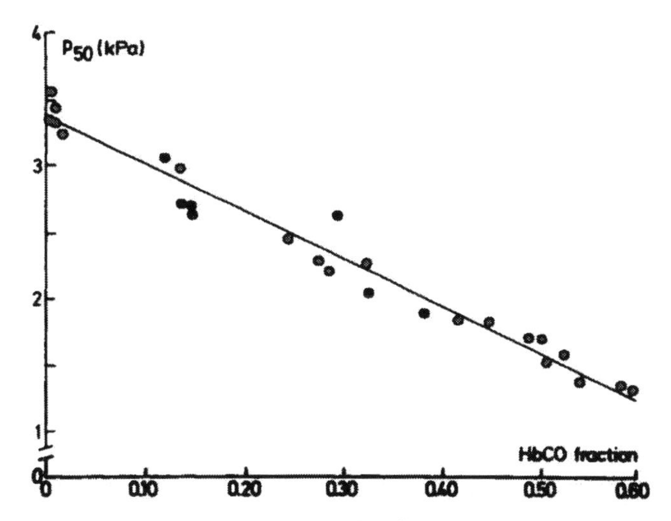

Figure 20.13. Standard p_{50} of 26 blood samples plotted *vs* F_{COHb} for seven specimens of human blood.
Regression equation: $p_{50} = -3.6 \cdot F_{COHb} + 3.4$; $r = -0.98$ (from ref. [449]).

Figure 20.12 shows, for blood of a single donor, four ODCs in the presence of four different COHb fractions under otherwise standard conditions. The considerable shift of the ODC to the left shows the increasing oxygen affinity with increasing F_{COHb}. To quantify this relationship, standard p_{50} of the remaining active haemoglobin was determined for 26 samples prepared from all seven specimens of human blood. Figure 20.13 shows these p_{50} values plotted against F_{COHb}. The Hill coefficient was calculated for the ODCs of four blood specimens and averaged for the four levels of F_{COHb}. Table 20.10 shows a fall in Hill's n, reflecting the change in shape of the ODC on increasing F_{COHb}.

A COHb factor ($\partial \log p_{O_2}/\partial F_{COHb}$) can be calculated from the experimental relation between p_{50} and F_{COHb} given in Fig. 20.13. For $S_{O_2} = 50\%$, pH $= 7.40$, $p_{CO_2} = 5.33$ kPa and $T = 37\,°C$:

$$p_{O_2} = -3.6F_{COHb} + 3.4, \tag{20.9}$$

$$dp_{O_2}/dF_{COHb} = -3.6,$$

hence,

$$\begin{aligned} \partial \log p_{O_2}/\partial F_{COHb} &= (\partial \log p_{O_2}/\partial p_{O_2}) \cdot (\partial p_{O_2}/\partial F_{COHb}) \\ &= -3.6[1/(p_{O_2} \cdot \ln 10)] \\ \partial \log p_{O_2}/\partial F_{COHb} &= -1.56/(-3.6F_{COHb} + 3.4). \end{aligned} \tag{20.10}$$

For $F_{COHb} = 0.1, 0.2, 0.3, 0.4$ and 0.5, $\partial \log p_{O_2}/\partial F_{COHb}$ thus equals $-0.51, -0.58, -0.67, -0.80$ and -0.98, respectively.

Table 20.11.
Proton Bohr factor at various values of S_{O_2} and F_{COHb}

S_{O_2} (%)	Proton Bohr factor (Φ_H)			
	$F_{COHb} \leqslant 1\%$	$F_{COHb} = 14 \pm 1\%$	$F_{COHb} = 30 \pm 2\%$	$F_{COHb} = 52 \pm 2\%$
10	-0.32 ± 0.06	-0.34 ± 0.11	$-0.45 \pm 0.09^*$	$-0.51 \pm 0.04^*$
20	-0.39 ± 0.02	-0.37 ± 0.06	-0.46 ± 0.08	$-0.56 \pm 0.05^*$
30	-0.42 ± 0.01	-0.40 ± 0.04	-0.48 ± 0.07	$-0.54 \pm 0.07^*$
40	-0.43 ± 0.01	-0.43 ± 0.06	-0.49 ± 0.07	$-0.53 \pm 0.06^*$
50	-0.44 ± 0.02	-0.45 ± 0.06	-0.50 ± 0.07	$-0.54 \pm 0.07^*$
60	-0.45 ± 0.02	-0.46 ± 0.06	-0.50 ± 0.06	$-0.55 \pm 0.06^*$
70	-0.44 ± 0.02	-0.47 ± 0.06	-0.49 ± 0.06	$-0.56 \pm 0.06^*$
80	-0.44 ± 0.02	-0.48 ± 0.06	-0.49 ± 0.05	$-0.56 \pm 0.06^*$
90	-0.43 ± 0.02	-0.48 ± 0.08	-0.49 ± 0.06	$-0.56 \pm 0.06^*$
95	-0.40 ± 0.03	-0.47 ± 0.08	-0.46 ± 0.06	$-0.52 \pm 0.08^*$

Values are means $\pm$ *SD*. $c_{DPG}/c_{Hb_4} = 0.86 \pm 0.07$ mol/mol (*SD*). $N = 4$.
*= significantly different from corresponding value of COHb-free blood (paired *t*-test, $p < 0.05$).

The proton Bohr factor was determined for five specimens of human blood at three COHb levels: F_{COHb} = 14, 30 and 52%. At each value of F_{COHb}, ODCs were measured at three different pH values: pH = 7.1, 7.3 and 7.5. Each sample contained < 1% Hi, SHb was not detectable. Table 20.11 shows that the absolute value of the proton Bohr factor is significantly increased at F_{COHb} = 52% and that the dependence on S_{O_2} is less than in COHb-free blood. For a further discussion of the results presented and a comparison with those in the literature see ref. [449]; for the underlying chemical interactions see refs [50] and [126].

It has been shown that partial oxidation of haemoglobin iron, *i.e.* the conversion of part of the haemoglobin into Hi, has a similar effect on the oxygen affinity as the conversion of part of the haemoglobin into COHb [111, 399]. There is a shift to the left and a change in shape of the ODC.

To quantify the **effect of Hi on the oxygen affinity** of the blood, a procedure was developed similar to the one used for COHb [231]. Hi formation in the erythrocytes was induced by addition of 4-DMAP. In six specimens of fresh human blood from healthy donors three levels of Hi (F_{Hi} ≈ 15, 30 and 50%) were prepared. When the blood had been deoxygenated in the cuvette of the ODC measuring system, a sample was taken to measure F_{Hi}. F_{COHb}, F_{Hi}, F_{SHb}, ct_{Hb} and c_{DPG} were determined as described previously (p. 293).

ODCs were recorded at the three levels of F_{Hi} and pH = 7.1, 7.3 and 7.5. There was a clear shift to the left of the ODCs with a tendency to a steeper course on increasing F_{Hi}. Similar to Fig. 20.13 (COHb), p_{50} was plotted *vs* F_{Hi} and, by linear regression, an equation analogous to equation 20.9 was obtained:

$$p_{O_2} = -2.9F_{Hi} + 3.5, \tag{20.11}$$

$$N = 6; n = 74; r = -0.89.$$

Analogous to equation 20.10, the Hi factor follows from equation 20.11:

$$\partial \log p_{O_2}/\partial F_{Hi} = -1.26/(-2.9F_{Hi} + 3.5). \tag{20.12}$$

Although the spread in these measurements is considerable, there is a reasonable agreement with the results of Versmold *et al.* [399] obtained for haemoglobin solutions. For the F_{Hi} range of 0–55% they found $\log p_{O_2} = -0.402F_{Hi} + 1.41$ for $S_{O_2} = 50\%$, pH = 7.15, $T = 37°C$ and $c_{DPG} = 5$ mmol/L. This gives $\partial \log p_{O_2}/\partial F_{Hi} = -0.40$, while equation 20.8 gives for $F_{Hi} = 25\%$: $\partial \log p_{O_2}/\partial F_{Hi} = -0.45$ ($c_{DPG}/c_{Hb_4} = 0.93 \pm 0.09$ mol/mol; *SD*).

The change in shape of the ODCs is reflected in a decrease of the Hill coefficient on increasing F_{Hi}: $n = 2.56, 2.26, 2.10$ and 1.77 for $F_{Hi} \leqslant 1$, ~15, ~30, ~50%, respectively. For the proton Bohr factor, Φ_H, a mean value of −0.43 was found with no dependence on F_{Hi}. This agrees with the findings of Enoki *et al.* [111].

These data have been measured with greater inaccuracy and imprecision than the other results given in this chapter. It is worthwhile to mention the difficulties encountered in recording ODCs in the presence of Hi. (1) The strong absorption of Hi in the 'oximeter band', the region around 630 nm, made it necessary to increase the sensitivity of the oximeter; in spite of an appropriate change in signal processing this decreased the signal-to-noise ratio. (2) The powerful reduction systems in the erythrocytes tend to reduce the Hi concentration during the measurements carried out at 37°C; this interferes with the correction of the oximeter output for the presence of Hi in addition to the fact that the ODC is not measured at constant F_{Hi}. (3) Considerable pH differences are necessary to measure the proton Bohr factor accurately; however, the absorptivity of Hi varies with the pH of the solution, which again influences the oximeter signals.

CONCLUSIONS

The results presented in this chapter demonstrate that the functional properties of blood as an O_2 transporter can best be studied with the help of a measuring system that determines all relevant quantities concerning the oxygen affinity of haemoglobin directly and independently. Optical methods in combination with ion-selective electrodes are at present most suited, and probably are amenable to further development as a research tool as well as an instrument for clinical application.

The good agreement between the directly measured values of the total Bohr factor, Φ_{HC}, with those calculated from the proton and carbamate Bohr factors, Φ_H and Φ_C, measured separately may inspire confidence in these methods. Together with the demonstrated high precision of the measuring system, this may also justify the use of the data presented, especially those concerning the temperature coefficient, the Bohr factors and the effect of COHb formation, in calculations concerning the oxygen transport by the blood, *e.g.* Siggaard-Andersen's oxygen status algorithm [358, 359].

The preceding sections contain all data necessary to convert an ODC or p_{50} (or p_n), to standard conditions or any other combination of values for pH, p_{CO_2} and T in the pathophysiological range. The change in p_{O_2} for a fixed value of S_{O_2} due to any change in pH, p_{CO_2} and T is given by

$$\mathrm{d}\log p_{O_2} = \Phi_H \cdot \mathrm{dpH} + \Phi_C \cdot \mathrm{d}\log p_{CO_2} + (\partial \log p_{O_2}/\partial T) \cdot \mathrm{d}T. \quad (20.13)$$

The use of this equation is illustrated by the following example. An arterial blood sample from a febrile patient with an acute infectious respiratory disease (rectal temperature 40.1 °C) was analysed at 37 °C with the following result: S_{O_2} = 92%, pH = 7.18, p_{CO_2} = 8.15 kPa, p_{O_2} = 11.6 kPa. For conversion of p_{92} to standard conditions, Φ_H (p_{CO_2} = 8.15 kPa; S_{O_2} = 92%) is obtained, by interpolation, from Table 20.7: −0.384, and Φ_C (pH = 7.18; S_{O_2} = 92%) from Table 20.8: 0.0083. Equation 20.13 then gives:

$$\mathrm{d}\log p_{O_2} = -0.384(7.40 - 7.18) + 0.008(\log 5.33 - \log 8.15) = -0.0860,$$

$$\log p_{92}(\mathrm{std}) = \log p_{92} + \mathrm{d}\log p_{O_2} = 1.0645 - 0.0860 = 0.9785,$$

$$p_{92}(\mathrm{std}) = 10^{0.9785} = 9.52 \text{ kPa}.$$

Table 20.1 gives for S_{O_2} = 92% a p_{O_2} value of 8.51 kPa, showing that the relatively high p_{O_2} is mainly the result of the Bohr effect. Probably there is a slightly increased DPG concentration in the erythrocytes.

For conversion to 'patient conditions' (*in vivo* conditions) pH and p_{CO_2} are first converted to 40.1 °C, using as temperature coefficients [54] $\Delta\mathrm{pH}/\Delta T = -0.015\ \mathrm{K}^{-1}$ and $\Delta \log p_{CO_2}/\Delta T = 0.021\ \mathrm{K}^{-1}$; this gives pH = 7.134 and p_{CO_2} = 9.47 kPa. The corresponding values for Φ_H and Φ_C are obtained, by interpolation, from Tables 20.7 and 20.8: −0.374 and 0.009, respectively. As temperature coefficient 0.025 K^{-1} should be used (p. 302). Equation 20.13 then gives:

$$\mathrm{d}\log p_{O_2} = -0.374(7.134 - 7.18) + 0.009(\log 9.47 - \log 8.15) + 0.025(40.1 - 37) = 0.0953,$$

$$\log p_{92}(\textit{in vivo}) = \log p_{O_2} + \mathrm{d}\log p_{O_2} = 1.0645 + 0.0953 = 1.1598,$$

$$p_{92}(\textit{in vivo}) = 10^{1.1598} = 14.4 \text{ kPa}.$$

The high p_{O_2} resulting from the body temperature, the Bohr effect and the displacement of standard ODC to the right promotes the oxygen diffusion from the blood to the oxygen consuming cells in the tissues. This helps to meet the increased demand resulting from the high temperature.

The Hill equation [232] (p. 299) may be used to estimate p_{50}:

$$\log p_{50} = \log p_{O_2} - (\text{logit } S_{O_2})/n. \tag{20.14}$$

This equation is used to calculate p_{50}(std) from p_{92}(std), using for n a value of 2.8 (Fig. 20.4):

$$\log p_{50}(\text{std}) = \log 9.52 - \{\log[0.92/(1 - 0.92)]\}/2.8 = 0.5998,$$

$$p_{50}(\text{std}) = 3.98 \text{ kPa}.$$

In comparison with $p_{50}(\text{std}) = 3.55$ kPa from Table 20.1 this again shows the displacement to the right of the patient's ODC. It should be noted that this displacement causes an error in S_{O_2} as calculated by a blood-gas analyser from p_{O_2}, pH and p_{CO_2}. The analyser probably gives a value near 94%, as can be read from Table 20.1 for $p_{O_2} = 9.52$ kPa. The difference from the measured S_{O_2} of 92% is in this case of only minor clinical importance, but more serious errors can be made [489]; S_{O_2} should preferably be measured.

Another example concerns the estimation of oxygen capacity, B_{O_2}, and oxygen affinity, expressed as p_{50}(std) for the blood of a hypothermic patient with acute dyshaemoglobinaemia of unknown origin. Rectal temperature was 31.8°C. A venous blood sample was analysed with the following result: ct_{Hb} (HiCN method) = 143 g/L, F_{COHb} and F_{Hi} (MCA) = 34% and 5%, respectively. B_{O_2} follows from equation 2.6:

$$B_{O_2} = \beta_{O_2} \cdot (ct_{Hb} - c_{COHb} - c_{Hi}) = 1.39(143 - 0.34 \times 143 - 0.05 \times 143),$$

$$B_{O_2} = 121.2 \text{ mL/L}.$$

Because the blood specimen is from a young adult who was presumably in good health before the present acute affliction, it may be assumed that p_{50}(std) was 3.55 kPa (Table 20.1). The presence of dyshaemoglobin and the fall in temperature have both increased the oxygen affinity of the blood. The change in $\log p_{O_2}$ resulting from the dyshaemoglobins is given by

$$\mathrm{d}\log p_{O_2} = (\partial \log p_{O_2}/\partial F_{COHb}) \cdot \mathrm{d}F_{COHb} + (\partial \log p_{O_2}/\partial F_{Hi}) \cdot \mathrm{d}F_{Hi}. \tag{20.15}$$

Values for the COHb and Hi factors are obtained by substituting the measured COHb and Hi fractions in equations 20.10 and 20.12. Hence,

$$\mathrm{d}\log p_{O_2} = -0.717 \times 0.34 - 0.367 \times 0.05 = -0.262.$$

The change in $\log p_{O_2}$ from the fall in temperature should also be taken into account. For venous blood a temperature coefficient of 0.024 K^{-1} is appropriate (p. 302). Consequently,

$$\mathrm{d}\log p_{O_2} = -0.262 + 0.024 \times (32 - 37) = -0.262 - 0.120 = -0.382.$$

The present value of p_{50}(std), designated p_{50}(std)*, then follows from

$$\log p_{50}(\mathrm{std})^* = \log 3.55 - 0.382 = 0.5502 - 0.382 = 0.1681,$$

$$p_{50}^* = 10^{0.1681} = 1.47\ \mathrm{kPa}.$$

It should be noted that these calculations are only fair approximations, although good enough for most practical applications. The calculations are not exact because there is a mutual influence between most of the affinity modulators [126, 161, 162, 496]. An important instance has already been shown by the interdependence of Φ_H and Φ_C (Tables 20.7 and 20.8 and equation 20.6), but there are many more. Thus, the proton Bohr factor is dependent on the temperature; Φ_H varies from −0.34 to −0.51 over the temperature range of 45–20°C. In the calculation of the dashed lines of Fig. 20.6 this has been taken into account. The complement of the temperature-dependence of the proton Bohr factor is the pH-dependence of the temperature coefficient [442, 496]. The corresponding temperature-dependence of the proton Haldane factor has already been mentioned on p. 310.

Changes in the c_{DPG}/c_{Hb_4} ratio have a considerable effect on the proton Bohr factor, especially in the supernormal range [496]. An increase of c_{DPG}/c_{Hb_4} from 0.95 to 3.66 has been found to be associated with a fall in Φ_H of −0.40 to 0.19; for $c_{DPG}/c_{Hb_4} < 0.9$ Φ_H proved to be hardly dependent on the c_{DPG}/c_{Hb_4} ratio. The influence of F_{COHb} on the proton Bohr factor has been shown in Table 20.11. This implies that the effect of F_{COHb} on the oxygen affinity will be dependent on the pH.

Finally, some notes on the terminology used in this chapter. The names chosen for the three Bohr effects have been justified in ref. [230] and should need no further clarification. As to the use of the terms 'factor' and 'coefficient' for quantities such as Φ_H and Φ_C, the literature is quite chaotic. These terms have been used indifferently, even in one and the same paper [162], and appear interchangeable. In an attempt to give an orderly description of the modulators of the oxygen affinity of haemoglobin in a previous paper [230], the term 'factor' was used for dimensionless quantities and the term 'coefficient' when a unit had to be included: hence, 'Bohr factor' and 'temperature coefficient'. In this book we have continued to adhere to this principle. This rule, however, also could not be applied consequently because it appeared unwise to rename the time-honoured (dimensionless) Hill 'coefficient'. Perhaps it would have been better to have followed Garby and Meldon who used the term 'coefficient' throughout their book [126]. In retrospect, looking beyond the description of the oxygen affinity, it appears that the whole attempt may have been folly. The oxygen binding capacity of haemoglobin can be expressed either as mol/mol or as mL/g: thus, Hüfner's 'factor' or Hüfner's 'coefficient'?

Jede Konsequenz führt zum Teufel.

REFERENCES

1. Aarnoudse JG, Oeseburg B, Kwant G, Huisjes HJ, Zijlstra WG: Continuous measurement of scalp tissue oxygen tension and carotid arterial oxygen tension in the fetal lamb. Biol Neonate **38**, 49–60, 1980.

2. Aarnoudse JG, Oeseburg B, Kwant G, Zwart A, Zijlstra WG, Huisjes HJ: Influence of variations in pH and P_{CO_2} on scalp tissue oxygen tension and carotid arterial oxygen tension in the fetal lamb. Biol Neonate **40**, 252–263, 1981.

3. Alexander CM, Teller LE, Gross JB: Principles of pulse oximetry: theoretical and practical considerations. Anesth Analg **68**, 368–376, 1989.

4. American Society for Testing Materials: Manual on recommended practices in spectrophotometry. Philadelphia, 1966 (ASTM).

5. American Society for Testing Materials: Manual on recommended practices in spectrophotometry, 3rd edition, Philadelphia, 1969 (ASTM).

6. Arturson G, Garby L, Robert M, Zaar B: The oxygen dissociation curve of normal human blood with special reference to the influence of physiological effector ligands. Scan J Clin Lab Invest **34**, 9–13, 1974.

7. van Assendelft OW: Spectrophotometry of haemoglobin derivatives. Ph D thesis, University of Groningen/Monograph, Van Gorcum & Comp, Assen, The Netherlands, and CC Thomas, Springfield IL, USA, 1970.

8. van Assendelft OW, Buursma A, Zijlstra WG: Stability of haemiglobincyanide standards. J Clin Pathol **49**, 275–277, 1996.

9. van Assendelft OW, Buursma A, Holtz AH, van Kampen EJ, Zijlstra WG: Quality control in haemoglobinometry with special reference to the stability of international haemiglobincyanide reference solutions. Clin Chim Acta **70**, 161–169, 1976.

10. van Assendelft OW, Holtz AH, van Kampen EJ, Zijlstra WG: Control data of international haemiglobincyanide reference solutions. Clin Chim Acta **18**, 78–81, 1967.

11. van Assendelft OW, van Kampen EJ, Zijlstra WG: Standardisation of haemoglobinometry: the use of haemiglobinazide instead of haemiglobincyanide for the determination of haemoglobin. Proc Kon Ned Ak Wet C **72**, 249–253, 1969.

12. van Assendelft OW, Zijlstra WG: The formation of haemiglobin using nitrites. Clin Chim Acta **11**, 571–577, 1965.

13. van Assendelft OW, Zijlstra WG: Extinction coefficients for use in equations for the spectrophotometric analysis of haemoglobin mixtures. Anal Biochem **69**, 43–48, 1975.

14. van Assendelft OW, Zijlstra WG: Observations on the alkaline haematin/detergent complex proposed for measuring haemoglobin concentration. J Clin Chem Clin Biochem **27**, 191–195, 1989.

15. van Assendelft OW, Zijlstra WG, van Kampen EJ, Holtz AH: Stability of haemiglobincyanide reference solutions. Clin Chim Acta **13**, 521–524, 1966.

16. Astrup P, Severinghaus JW: The history of blood gases, acids and bases. Copenhagen (Munksgaard) 1986, pp 117–119.

17. Barbour HG, Hamilton WF: The falling drop method for determining specific gravity. J Biol Chem **69**, 625–640, 1926.

18. Barker SJ, Tremper KK: The effect of carbon monoxide inhalation on pulse oximetry and transcutaneous P_{O_2}. Anesthesiol **66**, 677–679, 1987.

19. Barker SJ, Tremper KK, Hyatt J: Effects of methemoglobinemia on pulse oximetry and mixed venous oximetry. Anesthesiol **70**, 112–117, 1989.

20. Baud FJ, Barriot P, Toffis V, Riou B, Vicaut E, Lecarpentier Y, Bourdon R, Astier A, Bismuth C: Elevated blood cyanide concentration in victims of smoke inhalation. New Engl J Med **325**, 1761–1766, 1991.

21. Baumann R, Bartels H, Bauer C: Blood gas transport. In: Handbook of Physiology, Section 3: The respiratory System, Vol IV Gas exchange, edited by AP Fishman *et al.*, Bethesda 1987 (American Physiological Society), pp 147–172.

22. Baumann R, Bauer C, Rathschlag-Schaefer AM: Causes of the postnatal decrease of blood oxygen affinity in lambs. Resp Physiol **15**, 151–158, 1972.

23. Benaron DA, Gwiazdowski S, Kurth CD, Steven J, Delivoria-Papadopoulos M, Chance B: Optical path length of 754 nm and 816 nm light emitted into the head of infants. Ann Int Conf IEEE Eng Med Biol Soc **12**, 1117–1119, 1990.

24. Benaron DA, Kurth CD, Steven JM, Delivoria-Papadopoulos M, Chance B: Transcranial optical pathlength in infants by near-infrared phase-shift spectroscopy. J Clin Monit **11**, 109–117, 1995.

25. Benaron DA, Kurth CD, Steven J, Wagerle LC, Chance B, Delivoria-Papadopoulos M: Non-invasive estimation of cerebral oxygenation and oxygen consumption using phase-shift spectrophotometry. Ann Int Conf IEEE Eng Med Biol Soc **12**, 2004–2006, 1990.

26. Benesch R, Benesch RE: The effect of organic phosphates from the human erythrocyte on the allosteric properties of hemoglobin. Biochem Biophys Res Commun **26**, 162–167, 1967.

27. Bert P: La pression barométrique. Recherches de physiologie expérimentale. Paris 1878 (Masson), p 691.

28. Berzofsky JA, Peisach J, Blumberg WE: Sulfheme proteins. II. The reversible oxygenation of ferrous sulfmyoglobin. J Biol Chem **246**, 7366–7372, 1971.

29. Beutler E, West C: Simplified determination of carboxyhemoglobin. Clin Chem **30**, 871–874, 1984.

30. Bjure J, Nilsson NJ: Spectrophotometric determination of oxygen saturation of hemoglobin in the presence of carboxyhemoglobin. Scand J Clin Lab Invest **17**, 491–500, 1965.

31. Bland JM, Altman DG: Statistical methods for assessing agreement between two methods of clinical measurement. Lancet 307–310, 1986(I).

32. Bleeker WK, van der Plas J, Agterberg J, Rigter G, Bakker JC: Prolonged vascular retention of a hemoglobin solution modified by cross-linking with 2-nor-2-formylpyridoxal 5′-phosphate. J Lab Clin Med **108**, 448–455, 1986.

33. Bohr C, Hasselbalch KA, Krogh A: Über einen in biologischer Beziehung wichtigen Einfluß, den die Kohlensäurespannung des Blutes auf dessen Sauerstoffbindung übt. Skand Arch Physiol **16**, 402–412, 1904.

34. de Boroviczeny ChG: Foundation of Standardizing Committee of the European Society of Haematology. Bibl Haematol **18**, 108–109, 1964.

35. de Boroviczeny ChG: Decision of the Standardizing Committee of the European Society of Haematology concernig haemoglobinometry. Bibl Haematol **18**, 110–111, 1964.

36. de Boroviczeny ChG: Recommendations and requirements for haemoglobinometry in human blood. Bibl Haematol **21**, 213–216, 1965.

37. Bossina KK, Mook GA, Zijlstra WG: Direct reflection oximetry in routine cardiac catheterisation. Circulation **22**, 908–912, 1960.

38. Braunitzer G: The molecular weight of haemoglobin. Bibl. Haematol **18**, 59–60, 1964.

39. Braunitzer G, Gehring-Müller R, Hilschmann N, Hilse K, Rudolf V, Wittmann-Liebold B: Die Konstitution des normalen adulten Humanhämoglobins. Hoppe Seylers Z Physiol Chem **325**, 283–288, 1961.

40. Brazy JE, Lewis DV, Mitnick MH, Jöbsis van der Vliet FF: Noninvasive monitoring of cerebral oxygenation in preterm infants: preliminary observations. Pediatrics **75**, 217–225, 1985.

41. Bridges N, Parvin RM, van Assendelft OW: Evaluation of a new system for hemoglobin measurement. Am Clin Prod Rev **6**, 22–25, 1987.

42. Brinkman R, Cost WS, Koopmans RK, Zijlstra WG: Continuous observation of the percentage oxygen saturation in capillary blood in patients. Arch Chir Neerl **1**, 184–191, 1949.

43. Brinkman R, Zijlstra WG: Determination and continuous registration of the percentage oxygen saturation in clinical conditions. Arch Chir Neerl **1**, 177–183, 1949.

44. Brinkman R, Zijlstra WG, Jansonius NJ: Quantitative evaluation of the rate of rouleaux formation of erythrocytes by measuring light reflection ('Syllectometry'). Proc Kon Ned Ak Wet C **66**, 236–248, 1963.

45. Brinkman R, Zijlstra WG, Koopmans RK: A method for continuous observation of the percentage oxygen saturation in patients. Arch Chir Neerl **1**, 333–341, 1950.

46. Broden PN: Concerns for the use and disposal of cyanide-containing reagents in the hematology laboratory. Sysmex J Intern **2**, 156–161, 1992.

47. Brown LJ: A new instrument for the simultaneous measurement of total hemoglobin, % oxyhemoglobin, % carboxyhemoglobin, % methemoglobin, and oxygen content in whole blood. IEEE Trans Biomed Engin **27**, 132–138, 1980.

48. Brown EG, Krouskop RW, McDonnel FE, Monge CC, Winslow RM: A technique to continuously measure arteriovenous oxygen content difference and P_{50} *in vivo*. J Appl Physiol **58**, 1383–1389, 1985.

49. Bull BS, Houwen B, Koepke JA, Simson E, van Assendelft OW: Reference and selected procedures for the quantitative determination of hemoglobin in blood — Second Edition; Approved Standard. NCCLS Document H15-A2, Vol 14, No 6, 1994.

50. Bunn HF, Forget BG: Hemoglobin: molecular, genetic and clinical aspects. Philadelphia 1986 (Saunders).

51. Burker K: Ein neues Hämoglobinometer. Pflügers Arch **203**, 274, 1924.

52. Burkhard O, Barnikol WKR: Dependence of visible spectrum [$\varepsilon(\lambda)$] of fully oxygenated hemoglobin on concentration of hemoglobin. J Appl Physiol **52**, 124–130, 1982.

53. Burnett RW: Standard for spectrophotometry. In: Handbook of Clinical Chemistry, Vol II, edited by M Werner, Boca Raton FL 1985 (CRC Press), pp 25–29.

54. Burnett RW, Covington AK, Fogh-Andersen N, Külpmann WR, Maas AHJ, Müller-Plathe O, van Kessel AL, Wimberley PD, Zijlstra WG, Siggaard-Andersen O, Weisberg HF: Approved IFCC recommendations on definitions of quantities and conventions related to blood gases and pH. Eur J Clin Chem Clin Biochem **33**, 399–404, 1995.

55. Burseaux E, Dubos C, Poyart CF: Pouvoir oxyphorique et P_{50} du sang humain. Bull Physio-pathol Res **7**, 729–742, 1971.

56. Cannan RK: Proposal for distribution of a hemoglobin standard. Science **122**, 59–60, 1955.

57. Cannan RK: Proposal for a certified standard for use in hemoglobinometry. Second and final report. Clin Chem **4**, 246–251, 1958.

58. Cannan RK (for the Division of Medical Sciences of the National Academy Sciences — National Research Council): Proposal for adoption of an international method and standard solution for hemoglobinometry, specifications for preparation of the standard solution, and notification of availability of a reference standard solution. Blood **26**, 104–107, 1965.

59. Carrico RJ, Blumberg WE, Peisach J: The reversible binding of oxygen to sulfhemoglobin. J Biol Chem **253**, 7212–7215, 1978.

60. Cary HH, Beckman AO: A quartz photoelectric spectrophotometer. J Opt Soc Am **31**, 682–689, 1941.

61. Chanutin A, Curnish RR: Effect of organic and inorganic phosphates on the oxygen equilibrium of human erythrocytes. Arch Biochem Biophys **121**, 96–102, 1967.

62. Christiansen J, Douglas CG, Haldane JS: The absorption and dissociation of carbon dioxide by human blood. J Physiol **48**, 244–271, 1914.

63. Christoforides C, Hedley-White J: Effect of temperature and hemoglobin concentration on solubility of O_2 in blood. J Appl Physiol **27**, 592–596, 1969.

64. Clayton DG, Webb RK, Ralston AC, Duthie D, Runciman WB: A comparison of the performance of 20 pulse oximeters under conditions of poor perfusion. Anaesthesia **46**, 3–10, 1991.

65. Clayton DG, Webb RK, Ralston AC, Duthie D, Runciman WB: Pulse oximeter probes. A comparison between finger, nose, ear and forehead probes under conditions of poor perfusion. Anaesthesia **46**, 260–265, 1991.

66. Clerbaux Th: Méthodes pour déterminer la courbe de dissociation de l'hémoglobine. Bull Europ Physiopath Resp **12**, 487–505, 1976.

67. Clerbaux Th, Fesler R, Bourgeois J: A dynamic method for continuous recording of the whole blood oxyhemoglobin dissociation curve at constant temperature, pH and P_{CO_2}. J Med Lab Technol **30**, 1–9, 1973.

68. Colier WNJM: Near infrared spectroscopy: toy or tool? An investigation on the clinical applicability of near infrared spectroscopy. Ph D thesis, University of Nijmegen, Nijmegen, The Netherlands, 1995.

69. Colier WNJM, Binkhorst RA, Hopman MTE, Oeseburg B: Cerebral and circulatory haemodynamics before vasovagal syncope induced by orthostatic stress. Clin Physiol **17**, 83–94, 1997.

70. Colier WNJM, Froeling FMJA, de Vries JDM, Oeseburg B: Measurement of the blood supply to the abdominal testis by means of near infrared spectroscopy. Eur Urol **27**, 160–166, 1995.

71. Colier WNJM, van Haaren NJCW, van de Ven MJT, Folgering HTM, Oeseburg B: Age dependency of cerebral oxygenation assessed with near infrared spectroscopy. J Biomed Optics **2**, 162–170, 1997.

72. Colier WNJM, Meeuwsen IBAE, Degens H, Oeseburg B: Determination of oxygen consumption in muscle during exercise using near infrared spectroscopy. Acta Anaesthesiol Scand **39**, suppl 107, 151–155, 1995.

73. Colier WNJM, Quaresima V, Barattelli G, Cavallari P, van der Sluijs M, Ferrari M: Detailed evidence of cerebral hemoglobin oxygenation changes in response to motor cortical activation revealed by continuous wave spectrophotometer with 10 Hz temporal resolution. Spie **2979**, 390–396, 1997.

74. Colier WNJ, de Vries J, van der Zee H, van der Sluijs M, Houston RJF, Oeseburg B: Simultaneous, direct and continuous measurement of cerebral hemodynamics and cerebral arterial saturation using near infrared spectroscopy. Postgraduate Assembly Meeting, New York, December 1998, P-9157.

75. Comroe JH, Botelho S: The unreliability of cyanosis in the recognition of arterial anoxemia. Am J Med Sc **214**, 1–6, 1947.

76. Cope M, Delpy DT: System for long-term measurement of cerebral blood and tissue oxygenation on newborn infants by near infra-red transillumination. Med Biol Eng Comp **26**, 289–294, 1988.

77. Copeland BE, King J, Willis C: The National Bureau of Standards carbon yellow filter as a monitor for spectrophotometric performance. Am J Clin Pathol **49**, 459–466, 1968.

78. Cordone L, Cupane A, Leone M, Vitrano E: Optical absorption spectra of deoxy- and oxyhemoglobin in the temperature range 300–20 K. Biophys Chem **24**, 259–275, 1986.

79. Dalinghaus M: Chronic hypoxemia in lambs with experimental cyanotic heart disease. Ph D thesis, University of Groningen, Groningen, The Netherlands, 1994.

80. Dalinghaus M, Gratama JWC, Koers JH, Gerding AM, Zijlstra WG, Kuipers JRG: Left ventricular oxygen and substrate uptake in chronically hypoxemic lambs. Pediatr Res **34**, 471–477, 1993.

81. Dalinghaus M, Gratama JWC, Zijlstra WG, Kuipers JRG: Cardiovascular adjustments to acute hypoxemia superimposed on chronic hypoxemia in lambs. Am J Physiol **268**, H974–H979, 1995.

82. Dalziel K, O'Brien JRP: Side reactions in the de-oxygenation of dilute oxyhaemoglobin solutions by sodium dithionite. Biochem J **67**, 119–124, 1957.

83. Dassel ACM: Experimental studies on reflection pulse oximetry: specific aspects of *intrapartum* monitoring. Ph D thesis, University of Groningen, Groningen, The Netherlands, 1997.

84. Dassel ACM, Graaff R, Aardema M, Zijlstra WG, Aarnoudse JG: Effect of location of the sensor on reflectance pulse oximetry. Brit J Obst Gynaecol **104**, 910–916, 1997.

85. Dassel ACM, Graaff R, Aarnoudse JG, Elstrodt JM, Heida P, Koelink MH, de Mul FFM, Greve J: Reflectance pulse oximetry in fetal lambs. Pediatr Res **31**, 266–269, 1992.

86. Dassel ACM, Graaff R, Meijer A, Zijlstra WG, Aarnoudse JG: Reflectance pulse oximetry at the forehead of newborns: influence of varying pressure on the probe. J Clin Monit **12**, 421–428, 1996.

87. Dassel ACM, Graaff R, Sikkema M, Meijer A, Zijlstra WG, Aarnoudse JG: Reflectance pulse oximetry at the forehead improves by pressure on the probe. J Clin Monit **11**, 237–244, 1995.

88. Davidson JAH, Hosie HE: Limitations of pulse oximetry: respiratory insufficiency — a failure of detection. Brit Med J **307**, 372–373, 1993.

89. De Blasi RA, Ferrari M, Natali A, Conti G, Mega A, Gasparetto A: Noninvasive measurement of forearm blood flow and oxygen consumption by near-infrared spectroscopy. J Appl Physiol **76**, 1388–1393, 1994.

90. Delhaas T, Mook GA, Zijlstra WG: Respiration and measurement of cardiac output by thermodilution and by central and peripheral dye dilution. J Appl Physiol **73**, 1047–1051, 1992.

91. Delpy DT, Cope M, van der Zee P, Arridge S, Wray S, Wyatt J: Estimation of optical pathlength through tissue from direct time of flight measurement. Phys Med Biol **33**, 1433–1442, 1988.

92. Dennis RC, Valeri CR: Measuring percent oxygen saturation of hemoglobin, percent carboxyhemoglobin and methemoglobin, and concentrations of total hemoglobin and oxygen in blood of man, dog and baboon. Clin Chem **26**, 1304–1308, 1980.

93. Drabkin DL: Photometry and spectrophotometry. In: Medical Physics, edited by O Glasser, Chicago 1944 (Yearbook Publishers), pp 967–1008.

94. Drabkin DL: Spectrophotometric studies. XIV. The crystallographic and optical properties of the hemoglobin of man in comparison with those of other species. J Biol Chem **164**, 703–723, 1946.

95. Drabkin DL, Austin JH: Spectrophotometric studies. I. Spectrophotometric constants for common hemoglobin derivatives in human, dog, and rabbit blood. J Biol Chem **98**, 719–733, 1932.

96. Drabkin DL, Austin JH: Spectrophotometric studies. II. Preparations from washed blood cells; nitric oxide hemoglobin and sulfhemoglobin. J Biol Chem **112**, 51–65, 1935.

97. Drabkin DL, Austin JH: Spectrophotometric studies. V. A technique for the analysis of undiluted blood and concentrated hemoglobin solutions. J Biol Chem **112**, 105–115, 1935.

98. Duhm J: Dual effect of 2,3-diphosphoglycerate on the Bohr effects of human haemoglobin. Pflügers Arch **363**, 55–60, 1976.

99. Duncan A, Meek JH, Clemence M, Elwell CE, Tyszczuk L, Cope M, Delpy DT: Optical pathlength measurement on adult head, calf and forearm and the head of the newborn infant using phase resolved optical spectroscopy. Phys Med Biol **40**, 295–304, 1995.

100. Dijkhuizen P: The oxygen capacity of human blood. Ph D thesis, University of Groningen, Groningen, The Netherlands, 1977.

101. Dijkhuizen P, Buursma A, Fongers TME, Gerding AM, Oeseburg B, Zijlstra WG: The oxygen binding capacity of human haemoglobin. Hüfner's factor redetermined. Pflügers Arch **369**, 223–231, 1977.

102. Dijkhuizen P, Buursma A, Gerding AM, van Kampen EJ, Zijlstra WG: Carboxyhaemoglobin. Spectrophotometric determination tested and calibrated using a new reference method for measuring carbon monoxide in blood. Clin Chim Acta **80**, 95–104, 1977.

103. Dijkhuizen P, Buursma A, Gerding AM, Zijlstra WG: Sulfhaemoglobin. Absorption spectrum, millimolar extinction coefficient at $\lambda = 620$ nm, and interference with the determination of haemiglobin and of haemiglobincyanide. Clin Chim Acta **78**, 479–487, 1977.

104. Dijkhuizen P, Kwant G, Zijlstra WG: A new reference method for the determination of the oxygen content of blood. Clin Chim Acta **68**, 79–85, 1976.

105. Edwards AD, Brown GC, Cope M, Wyatt JS, McCormick DC, Roth SC, Delpy DT, Reynolds EOR: Quantification of concentration changes in neonatal human cerebral oxidized cytochrome oxidase. J Appl Physiol **71**, 1907–1913, 1991.

106. Edwards AD, Richardson C, van der Zee P, Elwell C, Wyatt JS, Cope M, Delpy DT, Reynolds EOR: Measurement of hemoglobin flow and blood flow by near-infrared spectroscopy. J Appl Physiol **75**, 1884–1889, 1993.

107. Ehrmeyer S, Burnett RW, Chatburn RL, Clausen JL, Durst RA, Fallon KD, Moran RF, van Kessel AL, Wandrup J: Fractional oxyhemoglobin, oxygen content and saturation, and related quantities in blood: terminology, measurement, and reporting; approved guideline. NCCLS document C25-A, Vol 17, No 3, 1997.

108. Elam JO, Neville JF, Sleater W, Elam WN: Sources of error in oximetry. Ann Surg **130**, 755–773, 1949.

109. Elwell CE, Cope M, Edwards AD, Wyatt JS, Delpy DT, Reynolds EOR: Quantification of adult cerebral hemodynamics by near infrared spectroscopy. J Appl Physiol **77**, 2753–2760, 1994.

110. Emanuel R, Norman J: Evaluation of a dynamic method for calibration of dye dilution curves. Brit Heart J **25**, 308–312, 1963.

111. Enoki Y, Tokui H, Tyuma I: Oxygen equilibria of partially oxidized hemoglobin. Resp Physiol **7**, 300–309, 1969.

112. Enson Y, Briscoe WA, Polanyi ML, Cournand A: *In vivo* studies with an intravascular and intracardiac reflection oximeter. J Appl Physiol **17**, 552–558, 1962.

113. Evans GO, Smith DEC: Preliminary studies with an SLS method for haemoglobin determination in three species. Sysmex J Intern **3**, 88–90, 1993.

114. Evelyn KA, Malloy HT: Microdetermination of oxyhemoglobin, methemoglobin, and sulfhemoglobin in a single sample of blood. J Biol Chem **126**, 655–662, 1938.

115. Fabel H, Lübbers DW: Eine schnelle Mikromethode zur serienmäßigen Bestimmung der O_2-Konzentration im Blut. Pflügers Arch **281**, 32–33, 1964.

116. Falholt W: Blood oxygen saturation determined spectrophotometrically. Scand J Clin Lab Invest **15**, 67, 1963.

117. Fogh-Andersen N, Siggaard-Andersen O, Lundsgaard FC, Wimberley PD: Diode-array spectrophotometry for simultaneous measurement of hemoglobin pigments. Clin Chim Acta **166**, 283–289, 1987.

118. Fogh-Andersen N, Siggaard-Andersen O, Lundsgaard FC, Wimberley PD: Spectrophotometric determination of hemoglobin pigments in neonatal blood. Clin Chim Acta **166**, 291–296, 1987.

119. Fox IJ, Brooker LGS, Heseltine DW, Essex HE, Wood EH: A tricarbocyanine dye for continuous recording of dilution curves in whole blood independent of variations in blood oxygen saturation. Mayo Clinic Proc **32**, 478–484, 1957.

120. Fox IJ, Wood EH: Indocyanine green: physical and physiologic properties. Proc Mayo Clin **35**, 732–744, 1960.

121. Fox IJ, Wood EH: Circulatory system: Methods; blood flow measurements by dye dilution technics. In: Medical Physics, Vol 3, edited by O Glasser, Chicago 1960 (Yearbook Publishers), pp 155–163.

122. Fox IJ, Wood EH: Circulatory system: Methods; indicator dilution technics in study of normal and abnormal circulation. In: Medical Physics, Vol 3, edited by O Glasser, Chicago 1960 (Yearbook Publishers), pp 163–178.

123. Fronticelli C: A possible new mechanism of oxygen affinity modulation in mammalian haemoglobins. Biophys Chem **37**, 141–146, 1990.

124. Fujiwara C, Hamaguchi Y, Toda S, Hayashi M: The reagent SULFOLYSER™ for hemoglobin measurement by hematology analyzers. Sysmex J **14**, 212–219, 1990.

125. Gamble WJ, Hugenholtz PG, Monroe RG, Polanyi M, Nadas AS: The use of fiberoptics in clinical cardiac catheterization. I. Intracardiac oximetry. Circulation **31**, 328–343, 1965.

126. Garby L, Meldon J: The respiratory functions of blood. New York 1977 (Plenum Press).

127. Gaspar-Rosas A, Thurston GB: Erythrocyte aggregate rheology by transmitted and reflected light. Biorheology **25**, 471–487, 1988.

128. Geubelle F: Sémi-micromethode spectrophotométrique de détermination du rapport oxyhémoglobine/hémoglobine dans le sang. Clin Chim Acta **1**, 225, 1956.

129. Goldie EAC: Device for the continuous indication of oxygen saturation of circulating blood in man. J Sc Instr **19**, 23–25, 1942.

130. Gong AK: Near-patient measurements of methemoglobin, oxygen saturation, and total hemoglobin: Evaluation of a new instrument for adult and neonatal intensive care. Crit Care Med **23**, 193–201, 1995.

131. González A, Gómez-Arnau J, Pensado A: Carboxyhemoglobin and pulse oximetry. Anesthesiol **73**, 573, 1990.

132. Gordy E, Drabkin DL: Spectrophotometric studies XVI. Determination of the oxygen saturation of blood by a simplified technique, applicable to standard equipment. J Biol Chem **227**, 285–299, 1957.

133. Gould SA, Moore EE, Moore FA, Haenel JB, Burch JM, Seghal H, Seghal L, DeWoskin R, Moss GS: Clinical utility of human polymerized hemoglobin as a blood substitute after acute trauma and urgent surgery. J Trauma **43**, 325–332, 1997.

134. Gowers W: An apparatus for the clinical estimation of haemoglobin. Trans Clin Soc London **12**, 64–67, 1879.

135. Graaff R: Tissue optics applied to reflectance pulse oximetry. Ph D thesis, University of Groningen, Groningen, The Netherlands, 1993.

136. Graaff R, Aarnoudse JG, de Mul FFM, Jentink HW: Light propagation parameters for anisotropically scattering media based on a rigorous solution of the transport equation. Appl Opt **26**, 2273–2279, 1989.

137. Graaff R, Aarnoudse JG, de Mul FFM, Jentink HW: Similarity relations for anisotropic scattering in absorbing media. Opt Eng **32**, 244–252, 1993.

138. Graaff R, Aarnoudse JG, Zijlstra WG, Heida P, de Mul FFM, Koelink MH, Greve J: Reflection pulse oximetry depends on source–detector distance. Intensive Care Med **16**, S99, 1990.

139. Graaff R, Buursma A, Zijlstra WG: The influence of sample refractive index and multiple reflections in spectrophotometry. To be published.

140. Graaff R, Buursma A, Zijlstra WG: The influence of the type of spectrophotometer on apparent hemolysate absorptivity. To be published.

141. Graaff R, Dassel ACM, Koelink MH, de Mul FFM, Aarnoudse JG, Zijlstra WG: Optical properties of human dermis *in vitro* and *in vivo*. Appl Opt **32**, 435–447, 1993.

142. Graaff R, Dassel ACM, Zijlstra WG, de Mul FFM, Aarnoudse JG: How tissue optics influences reflectance pulse oximetry. In: Oxygen transport to tissue, edited by C Ince *et al.*, New York 1996 (Plenum Press), pp 117–132.

143. Graaff R, Koelink MH, de Mul FFM, Zijlstra WG, Dassel ACM, Aarnoudse JG: Condensed Monte Carlo simulations for the description of light transport. Appl Opt **32**, 426–434, 1993.

144. Gratama JWC, Dalinghaus M, Meuzelaar JJ, Gerding AM, Koers JH, Zijlstra WG, Kuipers JRG: Blood volume and body fluid compartments in lambs with aortopulmonary left-to-right shunts. J Clin Invest **90**, 1745–1752, 1992.

145. Gratama JWC, Meuzelaar JJ, Dalinghaus M, Koers JH, Werre AJ, Zijlstra WG, Kuipers JRG: Maximal exercise capacity and oxygen consumption of lambs with an aortopulmonary left-to-right shunt. J Appl Physiol **69**, 1479–1485, 1990.

146. Gregory IC: The *in vivo* oxygen capacity of haemoglobin in man. J Physiol **219**, 31, 1971.

147. Gregory IC: The oxygen and carbon monoxide capacities of foetal and adult blood. J Physiol **236**, 625–634, 1974.

148. Grenvik Å: Errors of the dye dilution method compared to the direct Fick method in determination of cardiac output in man. Scand J Clin Lab Invest **18**, 486–492, 1966.

149. Groom D, Wood EH, Burchel HB, Parker RL: The application of an oximeter for whole blood to diagnostic cardiac catheterisation. Proc Staff Meet Mayo Clin **23**, 601–609, 1948.

150. Haldane J: The colorimetric determination of haemoglobin. J Physiol **26**, 497–504, 1901.

151. Handelé MJ, Zijlstra WG: Spectrophotometric determination of methemoglobin in human blood. Proc Kon Ned Akad Wet C **58**, 652–658, 1955.

152. Haveman R: Die Bestimmung von Hämoglobin mit dem lichtelektrischen Kolorimeter. Klin Wschr **19**, 503, 1940.

153. Heeres SG: The influence of plasma substitutes on the suspension stability and viscosity of blood. Ph D thesis, University of Groningen, Groningen, The Netherlands, 1963.

154. Heeres SG, Zijlstra WG: The influence of plasma substitutes on the suspension stability of human blood. Bibl Haemat **29**, 1167–1171, 1968.

155. Helledie NR, Rolfe P: Near infra red spectroscopy: pH sensitivity of the absorbances of dilute haemoglobin solutions and suspended erythrocytes at four wavelengths. Ann Int Conf IEEE Eng Med Biol Soc **12**, 1540–1541, 1990.

156. Hellung-Larsen P, Kjeldsen K, Mellemgaard K, Astrup P: Photometric determination of oxyhemoglobin saturation in the presence of carbon monoxide hemoglobin, especially at low oxygen tensions. Scand J Clin Lab Invest **18**, 443–449, 1966.

157. Henri HL: Molecular spectrophotometry. In: Handbook of Clinical Chemistry, edited by M Werner, Boca Raton FL 1985 (Plenum Press), pp 3–23.

158. Hill AV: The possible effects of aggregation of the molecules of haemoglobin on its dissociation curve. J Physiol **40**, 4–7, 1910.

159. Hill RJ, Konigsberg W, Guidotti G, Craig LC: The structure of human hemoglobin. I. The separation of the α and β chains and their amino acid composition. J Biol Chem **237**, 1549–1554, 1962.

160. Hillis LD, Firth BG, Winniford MD: Analysis of factors affecting the variability of Fick versus indicator dilution measurements of cardiac output. Am J Cardiol **56**, 764–768, 1985.

161. Hlastala MP, Woodson RD: Saturation dependences of the Bohr effect: interactions among H^+, CO_2, and DPG. J Appl Physiol **38**, 1126–1131, 1975.

162. Hlastala MP, Woodson RD, Wranne B: Influence of temperature on hemoglobin ligand interaction in whole blood. J Appl Physiol **43**, 545–550, 1977.

163. Hoek W, Kamphuis N, Gast R: Hemiglobincyanide: molar lineic absorbance and stability constant. J Clin Chem Clin Biochem **19**, 1209–1210, 1981.

164. van der Hoeven MAHBM, Maertzdorf WJ, Blanco CE: Feasibility and accuracy of a fiberoptic catheter for measurement of venous oxygen saturation in newborn infants. Acta Paediatr **84**, 122–127, 1995.

165. van der Hoeven MAHBM, Maertzdorf WJ, Blanco CE: Mixed venous oxygen saturation and biochemical parameters of hypoxia during progressive hypoxemia in 10- to 14-day-old piglets. Pediatr Res **42**, 878–884, 1997.

166. Holmgren A, Pernow B: Spectrophotometric measurement of oxygen saturation of blood in the determination of cardiac output. A comparison with the Van Slyke method. Scand J Clin Lab Invest **11**, 143–149, 1959.

167. Holtz AH: Some experience with a cyanhaemiglobin solution. Bibl Haemat **21**, 75–78, 1965.

168. Hoogeveen YL, Zock JP, Rispens P, Zijlstra WG: Influence of changes in cardiac output on the acid–base status of arterial and mixed venous blood. Pflügers Arch **410**, 257–262, 1987.

169. ten Hoor F: Determination of cardiac output with dye dilution methods. Ph D thesis, University of Groningen, Groningen, The Netherlands, 1969.

170. ten Hoor F, Mook GA: A linear reflection densitometer. Cardiovasc Res **3**, 373–380, 1969.

171. ten Hoor F, Mook GA, Sparling CM, Zijlstra WG: A dynamic calibration procedure for dye dilution curves. Acta Physiol Pharmacol Neerl **10**, 280–282, 1962.

172. ten Hoor F, Rispens P, Zijlstra WG: Determination of cardiac output in the dog by dye dilution and by the direct Fick procedure. Proc Kon Ned Ak Wet **C77**, 453–459, 1974.

173. Hoppe-Seyler F: Über das Verhalten des Blutfarbstoffes im Spectrum des Sonnenlichtes. Arch pathol Anat u Physiol **23**, 446–449, 1862.

174. Horecker BL: The absorption spectrum of hemoglobin and its derivatives in the visible and near-infrared regions. J Biol Chem **148**, 173–183, 1943.

175. Horecker BL, Brackett FS: A rapid spectrophotometric method for the determination of methaemoglobin and carboxyhaemoglobin in blood. J Biol Chem **152**, 669–677, 1944.

176. Houwen B, Hsiao-Liao N, Kubose-Perry KK, Mast BJ, Westengard JC, Bull BS, van Assendelft OW, Zijlstra WG: The reference method for hemoglobin measurement revisited. Lab Hematol **2**, 86–93, 1996.

177. Huch R, Huch A, Tuchschmid P, Zijlstra WG, Zwart A: Carboxyhaemoglobin concentration in fetal cord blood. Pediatrics **71**, 461–462, 1983.

178. Hüfner G: Über ein neues Spektrophotometer. Z phys Chemie **3**, 562–571, 1889.

179. Hüfner G: Über die Tension des Sauerstoffes im Blute und in Oxyhämoglobinlösungen. Z phys Chemie **13**, 285–291, 1889.

180. Hüfner G: Neue Versuche zur Bestimmung der Sauerstoffcapacität des Blutfarbstoffs. Arch Anat Physiol, Physiol Abt, 130–176, 1894.

181. Hüfner G: Noch ein Mal die Frage nach der Sauerstoffcapacität des Blutfarbstoffes. Arch Anat Physiol, Physiol Abt, 217–224, 1903.

182. Hugenholtz PG, Gamble WJ, Monroe RG, Polanyi M: The use of fiberoptics in clinical cardiac catheterization. I. *In vivo* dye-dilution curves. Circulation **31**, 344–355, 1965.

183. Hughes HK: Suggested nomenclature in applied spectroscopy. Report No 6 of the Joint Committee on Nomenclature in Applied Spectroscopy. Anal Chem **24**, 1349–1354, 1952.

184. Hutton P, Clutton-Brocke T: The benefits and pitfalls of pulse oximetry. Brit Med J **307**, 457–458, 1993.

185. Huisman A: Haemorheological changes during pregnancy. Ph D thesis, University of Groningen, Groningen, The Netherlands, 1986.

186. Huisman A, Aarnoudse JG, Krans M, Huisjes HJ, Fidler V, Zijlstra WG: Red cell aggregation during normal pregnancy. Brit J Haematol **68**, 121–124, 1988.

187. ICSH Expert Panel on Cytometry: The assignment of values to fresh blood used for calibrating automated cell counters. Clin Lab Haemat **9**, 387–393, 1987.

188. ICSH Expert Panel on Cytometry: Recommendations of the International Council for Standardisation in Haematology for ethylenediaminetetraacetic acid anticoagulation of blood for blood cell counting and sizing. Am J Clin Pathol **100**, 371–372, 1993.

189. International Committee for Standardisation in Haematology (ICSH): Recommendations and requirements for haemoglobinometry in human blood. J Clin Pathol **18**, 353, 1965.

190. International Committee for Standardisation in Haematology (ICSH): Recommendations for haemoglobinometry in human blood. Brit J Haematol **13** (Suppl), 71–75, 1967.

191. International Committee for Standardisation in Haematology (ICSH): Recommendations for reference method for haemoglobinometry in human blood (ICSH Standard EP 6/2: 1977) and specifications for international haemiglobincyanide reference preparation (ICSH Standard EP 6/3: 1977). J Clin Pathol **31**, 139–143, 1978.

192. International Committee for Standardisation in Haematology (ICSH): Recommendations for reference method for haemoglobinometry in human blood (ICSH standard 1986) and specifications for international haemiglobincyanide reference preparation (3rd edn). Clin Lab Haemat **9**, 73–79, 1987.

193. International Committee for Standardisation in Haematology (ICSH): Rules and operating procedures. ICSH Secretariat, Leuven, Belgium, 1991.

194. International Council for Standardisation in Haematology (ICSH): Recommendations for reference method for haemoglobinometry in human blood (ICSH standard 1995) and specifications for international haemiglobincyanide reference preparation (4th edn). J Clin Pathol **49**, 271–274, 1996.

195. Itano HA: Molar extinction coefficients of cyanmethemoglobin (hemiglobincyanide) at 540 and 281 nm. In: Modern Concepts in Hematology, edited by G Izak and SM Lewis, New York 1972 (Academic Press), pp 26–28.

196. Jansen AP, van Kampen EJ, Steigstra H, van der Ploeg PHW, Zwart A: Simultaneous spectrophotometric calibration of wavelength and absorbance in an interlaboratory survey using holmium oxide (Ho_2O_3) in perchloric acid as reference, compared with *p*-nitrophenol and cobaltous sulphate solutions. J Clin Chem Clin Biochem **24**, 141–146, 1986.

197. Jansonius NJ: Quantitative studies on rouleaux formation of erythrocytes by reflection measurement (syllectometry). Ph D thesis, University of Groningen, Groningen, The Netherlands, 1959.

198. Jansonius NJ, Zijlstra WG: Various factors influencing rouleaux formation of erythrocytes studied with the aid of syllectometry. Proc Kon Ned Ak Wet **C68**, 121–127, 1965.

199. Jöbsis FF: Noninvasive, infrared monitoring of cerebral and myocardial oxygen sufficiency and circulatory parameters. Science **198**, 1264–1267, 1977.

200. Jöbsis FF, Keizer JH, Lamanna JC, Rosenthal M: Reflectance spectrophotometry of cytychrome aa_3 *in vivo*. J Appl Physiol **43**, 858–872, 1977.

201. Johnson N, Johnson VA, Bannister J, Lilford RJ: The effect of caput succedaneum on oxygen saturation measurements. Brit J Obst Gynaecol **97**, 493–498, 1990.

202. Jonxis JHP, Boeve JHW: A spectrophotometric determination of oxygen saturation in small amounts of blood. Acta Med Scand **155**, 157–160, 1956.

203. Jonxis JHP, Visser HKA: Determination of low percentages of fetal haemoglobin in blood of normal children. Am J Dis Child **92**, 588–591, 1956.

204. Kapany NS, Silbertrust N: Fibre optics spectrophotometer for *in vivo* oximetry. Nature **204**, 138–142, 1964.

205. van Kampen EJ, van Assendelft OW: Quality control and hematology. In: Quality control in clinical chemistry, edited by G Anido *et al.*, Berlin 1975 (Walter de Gruyter), pp 325–333.

206. van Kampen EJ, Klouwen H: Spectrophotometric determination of carboxyhemoglobin. Rec Trav Chim Pays Bas **73**, 119, 1954.

207. van Kampen EJ, Volger HC, Zijlstra WG: Spectrophotometric determination of carboxyhemoglobin in human blood and the application of this method to the estimation of carbonmonoxide in gas mixtures. Proc Kon Ned Akad Wet C **57**, 320–331, 1954.

208. van Kampen EJ, Zijlstra WG: Standardization of hemoglobinometry. II. The hemiglobincyanide method. Clin Chim Acta **6**, 538–544, 1961.

209. van Kampen EJ, Zijlstra WG: A simple hemoglobin photometer to be used in standardised hemoglobinometry. Clin Chim Acta **7**, 147–148, 1962.

210. van Kampen EJ, Zijlstra WG: Spectrophotometry of hemoglobin and hemoglobin derivatives. Adv Clin Chem **23**, 199–257, 1983.

211. van Kampen EJ, Zijlstra WG, van Assendelft OW, Reinkingh WA: Determination of hemoglobin and its derivatives. Adv Clin Chem **8**, 141–187, 1965.

212. Keitt AS: Reduced nicotinamide adenine dinuleotide linked analysis of 2,3-DPG: spectrophotometric and fluorometric procedures. J Lab Clin Med **77**, 470–475, 1972.

213. Kelleher JF: Pulse oximetry. J Clin Monit **5**, 37–62, 1989.

214. Kinsman JM, Moore JW, Hamilton WF: Studies on the circulation. I. Injection method: physical and mathematical considerations. Am J Physiol **89**, 322ff., 1929.

215. Kirkwood TBL: Predicting the stability of biological standards and products. Biometrics **33**, 736–742, 1977.

216. Kirkwood TBL: Design and analysis of accelerated degradation tests for the stability of biological standards. III. Principles of design. J Biol Standardiz **12**, 215–224, 1984.

217. Klempt HW, Schmidt E, Bender F, Most E, Hewing R: Ein Fiberoptiksystem zur kontinuierlichen Messung der O_2-Sättigung und zur Bestimmung des Herzzeitvolumens mit der Farbstoffverdünnungstechnik. Z Kardiol **66**, 257–264, 1977.

218. Klose HJ, Volger E, Brechtelsbauer H, Heinich L, Schmid-Schönbein H: Microrheology and light transmission of blood. I. The photometric effects of red cell aggregation and red cell orientation. Pflügers Arch **333**, 126–139, 1972.

219. Klungsøyr L, Stöa KF: Spectrophotometric determination of hemoglobin oxygen saturation. Rec Trav Chim Pays Bas **74**, 571, 1955.

220. Kneezel LD, Kitchens CS: Phenacetin-induced sulfhemoglobinemia: report of a case and review of the literature. Johns Hopkins Med J **139**, 175–177, 1976.

221. Koch R, Saunier C: Mesure de la saturation oxyhémoglobinée du sang par la méthode de Van Slyke et par l'hémoréflecteur de Brinkman. Étude comparative. Rev Franç d'études clin biol **3**, 793–794, 1958.

222. Kooijman HM, Hopman MTE, Colier WNJM, van der Vliet JA, Oeseburg B: Near infrared spectroscopy for noninvasive assessment of claudication. J Surg Res **72**, 1–7, 1997.

223. Korner PI: Some factors influencing the dispersion of indicator substances in the mammalian circulation. Progr Biophys **11**, 111–176, 1961.

224. Kramer K: Bestimmung des Sauerstoffgehaltes und der Hämoglobinkonzentration in Hämoglobinlösungen und hämolysiertem Blut auf lichtelektrischem Wege. Z Biol **95**, 126–134, 1934.

225. Kramer K: Ein Verfahren zur fortlaufenden Messung des Sauerstoffgehaltes im strömenden Blute an uneröffneten Gefäßen. Z Biol **96**, 61–75, 1935.

226. Krauss XH, Verdouw PD, Hugenholtz PG, Nauta J: On-line monitoring of mixed venous oxygen saturation after cardiothoracic surgery. Thorax **30**, 636–643, 1975.

227. Kulig K: Cyanide antidotes and fire toxicology. New Engl J Med **325**, 1801–1802, 1991.

228. Kwant G, Oeseburg B: Gas mixing for biomedical application using mass flow controllers. Med Biol Eng Comput **27**, 634–636, 1989.

229. Kwant G, Oeseburg B, Zwart A, Zijlstra WG: Calibration of a practical haemoglobinometer. Clin Lab Haemat **9**, 387–393, 1987.

230. Kwant G, Oeseburg B, Zwart A, Zijlstra WG: Human whole-blood O_2 affinity: effect of CO_2. J Appl Physiol **64**, 2400–2409, 1988.

231. Kwant G, Oeseburg B, Zijlstra WG: Unpublished experiments, Groningen 1988.

232. Kwant G, Oeseburg B, Zijlstra WG: Reliability of the determination of whole-blood oxygen affinity by means of blood-gas analyzers and multiwavelength oximeters. Clin Chem **35**, 773–777, 1989.

233. Lamberts R, Visser KR, Zijlstra WG: Impedance cardiography. Monograph, Van Gorcum & Comp, Assen, The Netherlands, 1984.

234. Landsman MLJ: Fiberoptic reflection photometry. Intracardiac and intravascular determination of oxygen saturation and dye concentration. Ph D thesis, University of Groningen, Groningen, The Netherlands, 1975.

235. Landsman MLJ, Knop N, Kwant G, Mook GA, Zijlstra WG: A fiberoptic reflection oximeter. Pflügers Arch **373**, 273–282, 1978.

236. Landsman MLJ, Knop N, Kwant G, Mook GA, Zijlstra WG: A fiberoptic reflection densitometer with cardiac output computer. Pflügers Arch **379**, 59–69, 1979.

237. Landsman MLJ, Kwant G, Mook GA, Zijlstra WG: Light-absorbing properties, stability, and spectral stabilization of indocyanine green. J Appl Physiol **40**, 575–583, 1976.

238. Langbroek AJM, Nijmeijer A, Rispens P, Zijlstra WG: Pitfalls in acid–base experiments in conscious dogs. Pflügers Arch **417**, 157–160, 1990.

239. Lee JCS, Cheung PW, Schoene RB, Kenny MA, Goldberg S: Inaccuracy of pulse oximeter readings due to carboxyhemoglobin. Ann Int Conf IEEE Eng Med Biol Soc **13**, 1608–1609, 1991.

240. Lewis SM, Burgess BJ: An evaluation of photoelectric haemoglobinometers. J Clin Pathol **23**, 805–810, 1970.

241. Lewis SM, Garvey B, Manning R, Sharp SA, Wardle J: Lauryl sulphate haemoglobin: a non-hazardous substitute for HiCN in haemoglobinometry. Clin Lab Haemat **13**, 279–290, 1991.

242. Lide DR (editor): Handbook of chemistry and physics. 76th Edition. Boca Raton 1995/1996 (CRC Press), p 14.14.

243. Liem KD: Neonatal cerebral oxygenation and hemodynamics. A study using near infrared spectrophotometry. Ph D thesis, University of Nijmegen, Nijmegen, The Netherlands, 1996.

244. Lister G, Walter TK, Versmold HT, Dallman PR, Rudolph AM: Oxygen delivery in lambs: cardiovascular and hematologic development. Am J Physiol **237**, H668–H675, 1979.

245. Livera LN, Spencer SA, Thorniley MS, Wickramasinghe YABD, Rolfe P: Effects of hypoxaemia and bradycardia on neonatal cerebral haemodynamics. Arch Dis Child **66**, 376–380, 1991.

246. Lynch PLM, Bruns E, Boyd JC, Savory J: Chiron 800 system CO-Oximeter module overestimates methemoglobin concentrations in neonatal samples containing fetal hemoglobin. Clin Chem **44**, 1569–1570, 1998.

247. Maas AHJ, Hamelink ML, de Leeuw RJM: An evaluation of the spectrophotometric determination of HbO_2, HbCO and Hb in blood with the CO-Oximeter IL182. Clin Chim Acta **29**, 303–309, 1970.

248. Maas BHA, Buursma A, Ernst RAJ, Maas AHJ, Zijlstra WG: Lyophilized bovine hemoglobin as a possible reference material for the determination of hemoglobin derivatives in human blood. Clin Chem **44**, 2331–2339, 1998.

249. Maas BHA, Buursma A: Unpublished experiments, 1998.

250. Malin MJ, Sclafani LD, Wyatt JL: Evaluation of 24-second cyanide-containing and cyanide-free methods for whole blood hemoglobin on the Technicon H*1™ analyzer with normal and abnormal blood samples. Am J Clin Pathol **92**, 286–294, 1989.

251. Malooly DA, Donald DE, Marshall HW, Wood EH: Assessment of an indicator-dilution technic for quantitating aortic regurgitation by electromagnetic flowmeter. Circ Res **12**, 487–507, 1963.

252. Matsubara T, Minura T: Reaction mechanism of SLS–Hb method. Sysmex J **13**, 379–384, 1990.

253. Matsubara T, Okuzono H, Senba U: A modification of van Kampen-Zijlstra's reagent for the hemiglobincyanide method. Clin Chim Acta **93**, 163–164, 1979.

254. Matsubara T, Okuzono H, Tamagawa S: Proposal for an improved reagent in the hemiglobincyanide method. In: Modern concepts in hematology, edited by G Izak and SM Lewis, New York 1972 (Academic Press), pp 29–43.

255. Matthes K: Untersuchungen über die Sauerstoffsättigung des menschlichen Arterienblutes. Arch exp Pathol Pharm **179**, 698–711, 1935.

256. Matthes K, Groß F: Zur Methode der fortlaufenden Registrierung der Lichtabsorption des Blutes in zwei verschiedenen Spektralbezirken. Arch exp Pathol Pharm **191**, 381–390, 1939.

257. Matthes K, Groß F: Zur Methode der fortlaufenden Registrierung der Farbe des menschlichen Blutes. Arch exp Pathol Pharm **191**, 523–528, 1939.

258. McConn R: The oxyhemoglobin dissociation curve in acute disease. Surg Clin North Amer **55**, 627–658, 1975.

259. McMillan DE, Utterback NG, Lee MM: Developing red cell flow orientation shown by changes in blood reflectivity. Biorheology **25**, 675–684, 1988.

260. Mendelson Y: Pulse oximetry: theory and applications for noninvasive monitoring. Clin Chem **38**, 1601–1607, 1992.

261. Merlet-Bénichou C, Sinet M, Blayo MC, Gaudebout C: Oxygen-combining capacity in dog. *In vitro* and *in vivo* determination. Resp Physiol **21**, 87–99, 1974.

262. Merriman JE, Wyant GM, Bray G, McGeachy W: Serial cardiac output determinations in man. Can Anaes Soc J **5**, 375–383, 1958.

263. Meyer-Wilmes J, Remmer H: Die Standardisierung des roten Blutfarbstoffes durch Hämiglobincyanid. I. Mitteilung: Bestimmung der spezifischen Extinction von Hämiglobincyanid. Naunyn-Schmiedebergs Arch exp Path u Pharm **229**, 441–449, 1956.

264. Miller ED, Gleason L, McIntosh HD: A comparison of the cardiac output determination by the direct Fick method and the dye-dilution method using indocyanine green dye and a cuvette densitometer. J Lab Clin Med **59**, 345ff., 1962.

265. Millikan GA: The oximeter, an instrument for measuring continously the oxygen saturation of the arterial blood in man. Rev Sc Instr **13**, 434–444, 1942.

266. Minkowski A, Swierczewski E: The oxygen capacity of human foetal blood. In: Oxygen Supply to the Human Foetus, edited by J Walker and AC Turnbull, Oxford 1959 (Blackwell), pp 237–253.

267. Montgomery GE, Geraci JE, Wood EH: Calibration of the Millikan compensated oximeter as used among white and colored persons. Fed Proc **7**, 81, 1948.

268. Mook GA: Directe oxymetrie tijdens hartcatheterisatie. Ph D thesis, University of Groningen, Groningen, The Netherlands, 1959.

269. Mook GA: Klinische Anwendung direkter Oxymetrie bei der Katheterisierung des Herzens. In: Oxymetrie. Theorie und klinische Anwendung, edited by K Kramer, Stuttgart 1960 (Georg Thieme Verlag), pp 187–192.

270. Mook GA, van Assendelft OW, Kwant G, Landsman MLJ, Zijlstra WG: Two-colour reflection oximetry. Proc Kon Ned Ak Wet **C79**, 472–491, 1976.

271. Mook GA, van Assendelft OW, Zijlstra WG: Wavelength dependency of the spectrophotometric determination of blood oxygen saturation. Clin Chim Acta **26**, 170–173, 1969.

272. Mook GA, Buursma A, Gerding A, Kwant G, Zijlstra WG: Spectrophotometric determination of oxygen saturation of blood independent of the presence of indocyanine green. Cardiovasc Res **13**, 233–237, 1979.

273. Mook GA, Knop N, Landsman MLJ: Further developments in fiber optic oximetry and densitometry. Pflügers Arch **328**, 253, 1971.

274. Mook GA, Osypka P, Sturm RE, Wood EH: Light reflection measurements on blood by fibre-optic catheter. Acta Physiol Pharmacol Neerl **14**, 67–69, 1966.

275. Mook GA, Osypka P, Sturm RE, Wood EH: Fibre optic reflection photometry on blood. Cardiovasc Res **2**, 199–209, 1968.

276. Mook GA, Zijlstra WG: Direct measurement of the oxygen saturation of human blood during cardiac catheterization. Proc Kon Ned Ak Wet **C60**, 158–166, 1957.

277. Mook GA, Zijlstra WG: Simultaneous use of reflection and transmission oximetry. A comparison of two cuvette oximeters. Acta Med Scand **169**, 141–148, 1961.

278. Mook GA, Zijlstra WG: On the principles underlying the single scale procedure of Wood's transmission cuvette oximeter. Acta Med Scand **169**, 149–153, 1961.

279. Mook GA, Zijlstra WG: Quantitative evaluation of intracardiac shunts from arterial dye dilution curves. Demonstration of very small shunts. Acta Med Scand **170**, 703–715, 1961.

280. Moore SJ, Norris JC, Walsh DA, Hume AS: Antidotal use of methemoglobin forming cyanide antagonists in concurrent carbon monoxide/cyanide intoxication. J Pharmacol Exp Ther **242**, 70–73, 1987.

281. Morikawa T, Tsujino Y, Hamaguchi Y: The application of the SLS–Hb method to animal blood. Sysmex J Intern **2**, 56–65, 1992.

282. Morningstar DA, Williams GZ, Suutarinen P: The millimolar extinction coefficient of cyanmethemoglobin from direct measurements of hemoglobin iron by X-ray emission spectrography. Am J Clin Pathol **46**, 603–607, 1966.

283. Muller CJ, Zijlstra WG: Weitere Untersuchungen zur spektrophotometrischen Methämoglobinbestimmung. Klin Wschr **35**, 356–358, 1957.

284. Nahas GG: Spectrophotometric determination of hemoglobin and oxyhemoglobin in whole hemolyzed blood. Science **113**, 723, 1951.

285. Nahas GG: A simplified lucite cuvette for the spectrophotometric measurement of hemoglobin and oxyhemoglobin. J Appl Physiol **13**, 147–152, 1958.

286. Nicolai L: Über Sichtbarmachung, Verlauf und chemische Kinetik der Oxyhaemoglobinreduktion im lebenden Gewebe, besonders in der menschlichen Haut. Arch Gesamt Physiol **229**, 372–384, 1932.

287. Nilsson NJ: Ein vereinfachtes Spectrophotometer zur Bestimmung des Hämoglobin- und Sauerstoffgehaltes im Blut. Pflügers Arch **262**, 595–615, 1956.

288. Nilsson NJ: Ein linear anzeigendes Oximeter. Pflügers Arch **263**, 374–400, 1956.

289. Nilsson NJ: Oximetry. Physiol Rev **40**, 1–26, 1960.

290. Nilsson NJ: A linear responding dichromatic cuvette densitometer for dye-dilution curves. Scand J Clin Lab Invest **15**, suppl 69, 181–192, 1963.

291. Nørgaard-Pedersen B, Siggaard-Andersen O, Rem J: Hemoglobin pigments. Mixing technique for preparation of known fractions of hemoglobin pigment. Clin Chim Acta **42**, 109–113, 1972.

292. Nijland R: Arterial oxygen saturation in the fetus. An experimental animal study with pulse oximetry. Ph D thesis, University of Nijmegen, Nijmegen, The Netherlands, 1995.

293. Nijland R, Jongsma HW, Nijhuis JG, Oeseburg B: The accuracy of a fiberoptic oximeter over a wide range of arterial oxygen saturation values in piglets. Acta Anesthesiol Scand **39**, suppl 107, 71–76, 1995.

294. Nijland R, Jongsma HW, Nijhuis JG, Oeseburg B, Zijlstra WG: Notes on the apparent discordance of pulse oximetry and multi-wavelength haemoglobin photometry. Acta Anaesthiol Scand **39**, suppl 107, 49–52, 1995.

295. Nijland R, Nierlich S, Jongsma HW, Nijhuis JG, Oeseburg B, Springer K, Mannheimer P: Validation of reflectance pulse oximetry: an evaluation of a new sensor in piglets. J Clin Monit **13**, 43–49, 1997.

296. Oeseburg B: Determination of O_2 affinity of hemoglobin by independent continuous measurement of S_{O_2}, P_{O_2}, and pH, with control of pH, P_{CO_2}, and temperature. Crit Care Med **7**, 396–398, 1979.

297. Oeseburg B, Kwant G, Schut JK, Zijlstra WG: Measuring oxygen tension in biological systems by means of Clark-type polarographic electrodes. Proc Kon Ned Ak Wet **C82**, 83–90, 1979.

298. Oeseburg B, Landsman MJL, Mook GA, Zijlstra WG: Direct recording of oxyhaemoglobin dissociation curve *in vivo*. Nature **237**, 149–150, 1972.

299. Oshiro I, Fuji M, Hatanaka T, Takenaka T, Maeda J: Evaluation of sodium lauryl sulfate (SLS) for hemoglobin determination. Studies of the SLS-Hb method using the automated hematology analyser "Sysmex K-1000". Sysmex J **13**, 220–226, 1990.

300. Oshiro I, Takenaka T, Maeda J: New method for hemoglobin determination by using sodium lauryl sulfate (SLS). Clin Biochem **15**, 83–88, 1982.

301. Ohta K, Gotoh F, Tomita M, Tanahashi N, Kobari M, Shinohara T, Terayama Y, Mihara B, Takeda H: Animal species differences in erythrocyte aggregability. Am J Physiol **262**, H1009–H1012, 1992.

302. van Oudheusden APM, van de Heuvel JM, van Stekelenburg GJ, Siertsema LH, Wadman SK: De ijking van de hemoglobinebepaling op basis van ijzer. Ned Tijdschr Geneesk **108**, 265–267, 1964.

303. Park CM, Nagel RL: Sulfhemoglobinemia. Clinical and molecular aspects. N Engl J Med **310**, 1579–1585, 1984.

304. Paul W: Oximeter for continuous absolute estimation of oxygen saturation. J Sc Instr **30**, 165–168, 1953.

305. van der Plas J, de Vries-van Rossen A, Bleeker WK, Bakker JC: Effect of coupling of 2-nor-2-formylpyridoxal 5′-phosphate to stroma-free haemoglobin on oxygen affinity and tissue oxygenation. Studies in isolated perfused rat liver under conditions of normoxia and stagnant hypoxia. J Lab Clin Med **108**, 253–260, 1986.

306. van der Plas J, de Vries-van Rossen A, Damm JLB, Bakker JC: Preparation and physical characteristics of a hemoglobin solution modified by coupling to 2-nor-2-formylpyridoxal 5′-phosphate. Transfusion **27**, 425–430, 1987.

307. van der Plas J, de Vries-van Rossen A, Koorevaar JJ, Buursma A, Zijlstra WG, Bakker JC: Purification and physical characterics of a hemoglobin solution modified by coupling to 2-nor-2-formylpyridoxal 5′-phosphate (NFPLP). Transfusion **28**, 525–530, 1988.

308. Polanyi ML, Hehir RM: New reflection oximeter. Rev Sc Instr **31**, 401–403, 1960.

309. Polanyi ML, Hehir RM: *In vivo* oximeter with fast dynamic response. Rev Sc Instr **33**, 1050–1054, 1962.

310. Power GG: Solubility of O_2 and CO in blood and pulmonary and placental tissue, J Appl Physiol **24**, 468–474, 1968.

311. Rahn H, Reeves RB, Howell BJ: Hydrogen ion regulation, temperature and evolution. Am Rev Resp Dis **112**, 165–172, 1975.

312. Ralston AC, Webb RK, Runciman WB: Potential errors in pulse oximetry. I. Pulse oximetry evaluation. Anaesthesia **46**, 202–206, 1991.

313. Ralston AC, Webb RK, Runciman WB: Potential errors in pulse oximetry. III. Effects of interference, dyes, dyshaemoglobins and other pigments. Anaesthesia **46**, 291–295, 1991.

314. Rapoport S, Lübering J: The formation of 2,3-diphosphoglycerate in rabbit erythrocytes: the existence of a diphosphoglycerate mutase. J Biol Chem **183**, 507–516, 1950.

315. Rebuck AS, Chapman KR, D'Urzo A: The accuracy and response of a simplified ear oximeter. Chest **83**, 860–864, 1983.

316. Reeves RB: A rapid micromethod for obtaining oxygen equilibrium curves on whole blood. Resp Physiol **42**, 299–315, 1980.

317. Refsum HE: Influence of hemolysis and temperature on the spectrophotometric determination of hemoglobin oxygen saturation in hemolyzed whole blood. Scand J Clin Lab Invest **9**, 85–88, 1957.

318. Refsum HE: Spectrophotometric determination of hemoglobin oxygen saturation in hemolyzed whole blood by means of various wavelength combinations. Scand J Clin Lab Invest **9**, 190–193, 1957.

319. Rem J, Siggaard-Andersen O, Nørgaard-Petersen B, Sørensen S: Hemoglobin pigments. Photometer for oxygen saturation, carboxyhemoglobin, and methemoglobin in capillary blood. Clin Chim Acta **42**, 101–108, 1972.

320. Remmer A: Die Standardisierung des roten Blutfarbstoffes durch Hämiglobincyanid. II. Mitteilung: Eisengehalt und O_2-Bindungsvermögen von menschlichem Blut. Naunyn-Schmiedebergs Arch exp Path u Pharm **229**, 450–462, 1956.

321. Rimington C: Haemoglobinometry. Brit Med J, 177–178, 1942(I).

322. Rispens P, van Assendelft OW, Oord J: Horizontal rotating tonometers for the equilibration of blood or plasma with gasmixtures at constant temperatures. Pflügers Arch **304**, 118–120, 1968.

323. Rispens P, Brunsting JR, Zock JP, Zijlstra WG: A modified Singer–Hastings nomogram. J Appl Physiol **34**, 377–382, 1973.

324. Rispens P, Hessels J, Zwart A, Zijlstra WG: Inhibition of carbonic anhydrase in dog plasma. Pflügers Arch **403**, 344–347, 1985.

325. Robert M: Affinité de l'hémoglobine pour l'oxygène. Bull Physiopath Resp **11**, 79–170, 1975.

326. Rodkey FL, Hill TA, Pitts LL, Robertson RF: Spectrophotometric measurement of carboxyhemoglobin and methemoglobin. Clin Chem **25**, 1388–1393, 1979.

327. Rodrigo FA: The determination of the oxygenation of blood *in vitro* by using reflected light. Amer Heart J **45**, 809–822, 1953.

328. Rodriguez LR, Kotin N, Lowenthal D, Kattan M: A study of pediatric house staff's knowledge of pulse oximetry. Pediatrics **93**, 810–813, 1994.

329. Rossi-Bernardi L, Perella M, Luzzana M, Samaja M, Raffaele I: Simultaneous determination of hemoglobin derivatives, oxygen content, oxygen capacity, and oxygen saturation in 10 μL of whole blood. Clin Chem **23**, 1215–1225, 1977.

330. Rossi-Bernardi L, Roughton FJW: The specific influence of carbon dioxide and carbamate compounds on the buffer power and Bohr effects in human haemoglobin solutions. J Physiol **189**, 1–29, 1967.

331. Roughton FJW: Some recent work on the interactions of oxygen, carbon dioxide and haemoglobin. The seventh Hopkins memorial lecture. Biochem J **117**, 801–812, 1970.

332. Roughton FJW, Severinghaus JW: Accurate determination of O_2 dissociation curve of human blood above 98.7% saturation with data on O_2 solubility in unmodified human blood from 0° to 37 °C. J Appl Physiol **35**, 861–869, 1973.

333. Sahli H: Bestimmung des Hämoglobingehaltes des Blutes. In: Lehrbuch der klinischen Untersuchungsmethoden, Leipzig 1894 (Deuticke).

334. Saito H, Shimizu Y, Shimazu C, Yasuda K: Investigation of the reagent SULFO-LYSER™ for hemoglobin analysis by the SLS–Hb method. Sysmex J **13**, 227–235, 1990.

335. Salvati AM, Tentori L, Vivaldi G: The extinction coefficient of human hemiglobincyanide. Clin Chim Acta **11**, 477–479, 1965.

336. Salvati AM, Tentori L, Vivaldi G: Il coefficiente di estinzione millimolare della cianometemoglobina di Rana esculenta e di Axolotl messicano. Ann Ist Super Sanità **3**, 165–174, 1967.

337. Sanford AH, Sheard C: The determination of haemoglobin with the photoelectrometer. J Lab Clin Med **15**, 483–489, 1930.

338. Saunders NA, Powless SCP, Rebuck AS: Ear oximetry: accuracy and practicability in the assessment of arterial oxygenation. Am Rev Resp Dis **113**, 745–749, 1976.

339. von Schenck H, Falkensson M, Lundberg B: Evaluation of "Hemocue", a new device for determining haemoglobin. Clin Chem **32**, 526–529, 1986.

340. Schmid-Schönbein H, Gallasch G, von Gosen J, Volger E, Klose HJ: Red cell aggregation in blood flow. I. New methods of quantification. Klin Wschr **54**, 149–157, 1976.

341. Schmid-Schönbein H, Volger E, Klose HJ: Microrheology and light transmission of blood. II. The photometric quantification of red cell aggregate formation and dispersion in flow. Pflügers Arch **333**, 140–155, 1972.

342. Schmitt JM: Simple photon diffusion analysis of the effects of multiple scattering on pulse oximetry. IEEE Trans Biomed Eng **38**, 1194–1203, 1991.

343. Schubart G, Bauereisen E, Berzon R, Conrad J: Spektralphotometrie von hämolysiertem Blut im nahen Infrarot. Pflügers Arch **265**, 1–10, 1957.

344. Sekelj P, Oriol A, Anderson NM, Morch J, McGregor M: Measurement of indocyanine green dye with a cuvette oximeter. J Appl Physiol **23**, 114–120, 1967.

345. Sehgal LR, Rosen AL, Noud G, Sehgal HL, Gould SA, DeWoskin R, Rice CL, Moss GS: Large volume preparation of pyridoxalated haemoglobin with high p_{50}. J Surg Res **30**, 14–20, 1981.

346. Sendroy jr J, Dillon RT, Van Slyke DD: Studies on gas and electrolyte equilibria in blood. XIX. The solubility and physical state of uncombined oxygen in blood. J Biol Chem **105**, 597–632, 1934.

347. Severinghaus JW: Blood gas calculator. J Appl Physiol **21**, 1108–1116, 1966.

348. Severinghaus JW, Astrup PB: History of blood gas analysis. Boston (Little, Brown and Company) 1987, pp 168–169.

349. Severinghaus JW, Honda Y: History of blood gas analysis. VII. Pulse oximetry. J Clin Monit **3**, 135–138, 1987.

350. Severinghaus JW, Kelleher JF: Recent developments in pulse oximetry. Anesthesiol **76**, 1018–1038, 1992.

351. Shepherd AP, Steinke JM: CO-Oximetry interference by perflubron emulsion: comparison of hemolyzing and non-hemolyzing instruments. Clin Chem **44**, 2183–2190, 1998.

352. Shimizu S, Enoki Y, Kohzuki H, Ohga Y, Sakata S: Determination of Hüfner's factor and inactive hemoglobins in human, canine and murine blood. Jap J Physiol **36**, 1047–1051, 1986.

353. Shinebourne E, Fleming J, Hamer J: Calibration of indicator dilution curves in man by the dynamic method. Brit Heart J **29**, 920–925, 1967.

354. Siggaard-Andersen O: The acid–base status of the blood (4th edition). Copenhagen 1974 (Munksgaard).

355. Siggaard-Andersen O: Experience with a new direct reading oxygen saturation photometer using ultrasound for hemolyzing the blood. Scand J Clin Lab Invest **37**, suppl 146, 3–8, 1977.

356. Siggaard-Andersen O, Garby L: The Bohr effect and the Haldane effect. Scand J Clin Lab Invest **31**, 1–8, 1973.

357. Siggaard-Andersen O, Nørgaard-Pedersen B, Rem J: Hemoglobin pigments. Spectrophotometric determination of oxy-, carboxy-, met- and sulfhemoglobin in capillary blood. Clin Chim Acta **42**, 85–100, 1972.

358. Siggaard-Andersen O, Siggaard-Andersen M: The oxygen status algorithm: a computer program for calculating and displaying pH and blood gas data. Scand J Clin Lab Invest **50**, suppl 50, 29–45, 1990.

359. Siggaard-Andersen O, Wimberley PD, Göthgen I, Siggaard-Andersen M: A mathematical model of the hemoglobin–oxygen dissociation curve of human blood and of the oxygen partial pressure as a function of temperature. Clin Chem **30**, 1646–1651, 1984.

360. Singer P, Hansen H: Suppression of fetal hemoglobin and bilirubin on oximetry measurement. Blood Gas News **8**, 12–17, 1999.

361. Sleater W, Elam JO, Killian DJ, Elam WN: Comparison of transmission spectra of blood in the human ear flushed by histamine with those obtained after heat flushing. Fed Proc **8**, 147, 1949.

362. van der Sluijs M, Colier WNJM, Houston RJF, Oeseburg B: A new and highly sensitive continuous wave near infrared spectrophotometer with multiple detectors. Spie **3194**, 63–72, 1998.

363. Small KA, Radford EP, Frazier JM, Rodkey FL, Collison HA: A rapid method for simultaneous determination of carboxy- and methemoglobin in blood. J Appl Physiol **31**, 154–160, 1971.

364. Smits TM, Aarnoudse JG, Zijlstra WG: Red blood cell flow in the fetal scalp during hypoxemia in the chronic sheep experiment: A laser Doppler flow study. Pediatr Res **20**, 407–410, 1986.

365. Smits TM, Aarnoudse JG, Zijlstra WG: Scalp blood flow, measured by laser Doppler flowmetry, and transcutaneous P_{O_2} and P_{CO_2} in the lamb. Pediatr Res **27**, 442–444, 1990.

366. Soret JL: Recherches sur l'absorption des rayons ultra-violet par diverses substances. Arch Sci Phys Natur **61**, 322–359, 1878.

367. Soret JL: Recherches sur l'absorption des rayons ultra-violet par diverses substances. Arch Sci Phys Natur **66**, 429–494, 1883.

368. Sparling CM: Recording and quantitative interpretation of dye dilution curves, obtained by reflectometry in red and infrared light. Ph D thesis, University of Groningen, Groningen, The Netherlands, 1961.

369. Speakman ED, Boyd JC, Bruns DE: Measurement of methemoglobin in neonatal samples containg fetal haemoglobin. Clin Chem **41**, 458–461, 1995.

370. Squire JR: An instrument for measuring the quantity of blood and its degree of oxygenation in the web of the hand. Clin Sc **4**, 331–339, 1940.

371. Stadie WC: A method for the determination of methemoglobin in blood. J Biol Chem **41**, 237–241, 1920.

372. Steinke JM, Shepherd AP: Effects of temperature on optical absorbance spectra of oxy-, carboxy- and deoxyhemoglobin. Clin Chem **38**, 1360–1364, 1992.

373. Steijermark A: Quantitative organic microanalysis, 2nd edition, New York 1961 (Academic Press).

374. Stewart GN: The output of the heart in dogs. Am J Physiol **57**, 27ff., 1921.

375. Stigbrand T: Molar absorbancy of cyanmethaemoglobin. Scand J Clin Lab Invest **20**, 252–254, 1967.

376. Stoddard JL, Adair GS: The refractometric determination of hemoglobin. J Biol Chem **57**, 437–454, 1923.

377. Stokes GG: On the reduction and oxidation of the colouring matter of the blood. Proc Royal Soc London **13**, 355, 1864.

378. Sweet CS, Emmert SE, Seymour AA, Stabilito II, Oppenheimer L: Measurement of cardiac output in anesthetized rats by dye dilution using a fiber optic catheter. J Pharmacol Meth **17**, 189–203, 1987.

379. Tallqvist TW: Ein einfaches Verfahren zur directen Schätzung der Farbestärke des Blutes. Z Klin Med **40**, 137–141, 1900.

380. Tammeling GJ, Zijlstra WG, Mook GA: Über die Grundlagen und Grenzen der "Zyklop"-Reflektionsmethode zur fortlaufenden Messung der arteriellen Sauerstoffsättigung. Thoraxchirurgie **5**, 118–138, 1957.

381. Taylor JD, Miller JDM: A source of error in the cyanmethemoglobin method of determination of hemoglobin concentration in blood containing carbon monoxide. Am J Clin Path **43**, 265–271, 1965.

382. Taylor MB, Whitham JG: The accuracy of pulse oximeters. A comparative clinical evaluation of five pulse oximeters. Anaesthesia **43**, 229–232, 1988.

383. Teisseire B, Loisance D, Soulard C, Hérigault R, Teisseire L, Laurent D: *In vitro* and *in vivo* studies of a stroma free hemoglobin solution as a potential blood substitute. Bull Europ Physiopath Resp **13**, 261–279, 1977.

384. Tentori L, Vivaldi G, Salvati AM: The extinction coefficient of human hemiglobin-cyanide as determined by nitrogen analysis. Clin Chim Acta **14**, 267–277, 1966.

385. Theye RA: Calculation of blood O_2 content from optically determined Hb and HbO_2. Anesthesiology **33**, 653–657, 1970.

386. Theye RA, Rehder K, Quesada RS, Fowler WS: Measurement of cardiac output by an indicator-dilution method. Anaesthesiol **25**, 71–74, 1964.

387. Thorsén G, Hint G: Aggregation, sedimentation and intravascular sludging of erythrocytes. Acta Chir Scand, suppl 154, 1950.

388. Toorop GP, Hardjowijono R, Dalinghaus M, Gerding AM, Koers JH, Zijlstra WG, Kuipers JRG: Effects of nitroprusside on myocardial blood flow and oxygen consumption in conscious lambs with an aortopulmonary left-to right shunt. Circulation **81**, 319–324, 1990.

389. Toorop GP, Hardjowijono R, Dalinghaus M, Wildevuur RH, Koers JH, Zijlstra WG, Kuipers JRG: Comparative circulatory effects of isoproterenol, dopamine, and dobutamine in conscious lambs with and without left-to-right shunts. Circulation **75**, 1222–1228, 1987.

390. Tyuma I: The Bohr effect and the Haldane effect in human blood. Jap J Physiol **34**, 205–216, 1984.

391. Usami S, Chien S: Optical reflectometry of red cell aggregation under shear flow. Bibl Anat **11**, 91–97, 1973.

392. Van Slyke DD: Gasometric determination of the oxygen and haemoglobin in blood. J Biol Chem **33**, 127–132, 1918.

393. Van Slyke DD, Hiller A, Weisiger RJ, Cruz WO: Determination of carbon monoxide in blood and of total and active hemoglobin by carbon monoxide capacity. Inactive haemoglobin and methaemoglobin contents of normal human blood. J Biol Chem **166**, 121–148, 1946.

394. Van Slyke DD, Neill JM: The determination of gases in blood and other solutions by vacuum extraction and manometric measurement. J Biol Chem **61**, 523–573, 1924.

395. Van Slyke DD, Plazin J: Micromanometric analyses. Baltimore 1961 (Williams & Wilkins), pp 1–62.

396. Vanzetti G: An azide–methemoglobin method for hemoglobin determination in blood. J Lab Clin Med **67**, 116–126, 1966.

397. Vanzetti G, Franzini C: Improved reagents and concentrated reference solutions in hemoglobinometry. In: Modern concepts in hematology, edited by G Izak and SM Lewis, New York 1972 (Academic Press), pp 44–53.

398. Vegfors M, Lennmarken C: Carboxyhaemoglobinaemia and pulse oximetry. Brit J Anaesthesia **66**, 625–626, 1991.

399. Versmold HT, Fürst K, Betke K, Riegel KP: Oxygen affinity of hemoglobins F and A partially oxidized to methemoglobin: influence of 2,3-diphosphoglycerate. Pediat Res **12**, 133–138, 1978.

400. Versteeg P: Circulation in liver disease. Bedside determination of cardiac output. Ph D thesis, University of Groningen, Groningen, The Netherlands, 1964.

401. Versteeg P, Palsma DMH, Schalm L: Le débit cardiaque dans les affections hépatique. Rev Int d'Hépatol **15**, 691–697, 1965.

402. Vierordt K: Die Anwendung des Spektralapparates zur Photometrie der Absorptionsspektren und zur quantitativen chemischen Analyse. H Lauppsche Buchhandlung, Tübingen 1873.

403. Vierordt K: Die quantitative Spektralanalyse in ihrer Anwendung auf Physiologie, Physik, Chemie und Technologie. H Lauppsche Buchhandlung, Tübingen 1876.

404. Visser KR, Lamberts R, Zijlstra WG: Investigation of the parallel conductor model of impedance cardiography by means of exchange transfusion with stroma-free haemoglobin solution in the dog. Cardiovasc Res **21**, 637–645, 1987.

405. Visser KR, Lamberts R, Zijlstra WG: Investigation of the origin of the impedance cardiogram by means of exchange transfusion with stroma-free haemoglobin solution in the dog. Cardiovasc Res **24**, 24–32, 1990.

406. Vliers ACAP, Oeseburg B, Visser KR, Zijlstra WG: Choice of detection site for the determination of cardiac output by thermal dilution: the injection-thermistor-catheter. Cardiovasc Res **7**, 133–138, 1973.

407. Vliers ACAP, Visser KR, Zijlstra WG: Analysis of indicator distribution in the determination of cardiac output by thermal dilution. Cardiovasc Res **7**, 125–132, 1973.

408. Ware PF, Polanyi ML, Hehir RM, Stapleton JF, Sanders JI, Kocot SL: A new reflection oximeter. J Thor Cardiovasc Surg **42**, 580–588, 1961.

409. Weatherburn MW, Logan JE: The effect of freezing on the potassium ferricyanide–potassium cyanide reagent used in the cyanmethemoglobin procedure for hemoglobin determination. Clin Chim Acta **9**, 581–584, 1964.

410. Webb RK, Ralston AC, Runciman WB: Potential errors in pulse oximetry. II. Effects of changes in saturation and signal quality. Anaesthesia **46**, 207–212, 1991.

411. Weidner VR, Mavrodineanu R, Mielenz KD, Velapoldi RA, Eckerle KL, Adams B: Spectral transmittance characteristics of holmium oxide in perchloric acid solutions. J Res Nat Bur Stand **90**, 115–125, 1985.

412. Welch JP, DeCesare R, Hess D: Pulse oximetry: Instrumentation and clinical applications. Resp Care **35**, 584–597, 1990.

413. Wever R: Untersuchungen zur Extinktion von strömendem Blut. Pflügers Arch **259**, 97–109, 1954.

414. Wimberley PD, Burnett RW, Covington AK, Fogh-Andersen N, Maas AHJ, Müller-Plathe O, Siggaard-Andersen O, Zijlstra WG: Guidelines for routine measurement of blood hemoglobin oxygen affinity. Scand J Clin Lab Invest **50**, suppl 203, 227–234, 1990.

415. Wimberley PD, Fogh-Andersen N, Siggaard-Andersen O, Lundsgaard FC, Zijlstra WG: Effect of pH on the absorption spectrum of oxyhemoglobin: a potential source of error in measuring the oxygen saturation of hemoglobin. Clin Chem **34**, 750–754, 1988.

416. Wimberley PD, Siggaard-Andersen O, Fogh-Andersen N: Accurate measurements of hemoglobin oxygen saturation, and fractions of carboxyhemoglobin and methemoglobin in fetal blood using Radiometer OSM3: corrections for fetal hemoglobin and pH. Scand J Clin Lab Invest **50**, Suppl 203, 235–239, 1990.

417. Wimberley PD, Siggaard-Andersen O, Fogh-Andersen N, Zijlstra WG, Severinghaus JW: Haemoglobin oxygen saturation and related quantities: definitions, symbols and clinical use. Scand J Clin Lab Invest **50**, 455–459, 1990.

418. Winslow RM: Vasoconstriction and the efficacy of hemoglobin-based blood substitutes. Lab Hematol **1**, 165–169, 1995.

419. Wood EH: A single scale absolute reading ear oximeter. Am J Physiol **159**, 597, 1949.

420. Wood EH: A single scale absolute reading ear oximeter. Proc Staff Meet Mayo Clin **25**, 384–391, 1950.

421. Wood EH: Oximetry. In: Medical Physics, Vol 2, edited by O Glasser, Chicago 1950 (Yearbook Publishers), pp 664–679.

422. Wood EH (ed): Symposium on use of indicator dilution technics in the study of the circulation. Circ Res **10**, 377–581, 1962.

423. Wood EH, Geraci JE: Photoelectric determination of the arterial oxygen saturation in man. J Lab Clin Med **34**, 387–401, 1949.

424. Wood EH, Knutson JRB, Taylor BE: A comparison of oximetric measurements on histaminized and heat flushed ears. Am J Physiol **159**, 597–598, 1949.

425. Wood EH, Sutterer WF, Cronin L: Oximetry. In: Medical Physics, Vol 3, edited by O Glasser, Chicago 1960 (Yearbook Publishers), pp 416–445.

426. Wootton IDP, Blevin WR: The extinction coefficient of cyanmethemoglobin. Lancet **I**, 434–436, 1964.

427. World Health Organisation (WHO): International biological standards and international biological reference preparations 1968. WHO Techn Rep Series **384**, 85, 1968.

428. Woudstra BR: Variability of arterial blood gases in the ovine fetus. Ph D thesis, University of Groningen, Groningen, The Netherlands, 1994.

429. Woudstra BR, Aarnoudse JG, de Wolf BTHM, Zijlstra WG: Nuchal muscle activity at different levels of hypoxemia in fetal sheep. Am J Obst Gynecol **162**, 559–564, 1990.

430. Woudstra BR, de Wolf BTHM, Smits TM, Nathanielsz PW, Zijlstra WG, Aarnoudse JG: Variability of continously measured arterial pH and blood gas values in the near-term fetal lamb. Pediatr Res **38**, 528–532, 1995.

431. Wu H: Studies on Hemoglobin I. The advantage of alkaline solutions for colorimetric determination of hemoglobin. J Biochem Japan **2**, 173, 1922.

432. Wukitsch MW, Petterson MT, Tobler DR, Pologe JA: Pulse oximetry: analysis of theory, technology, and practice. J Clin Monit **4**, 290–301, 1988.

433. Wyatt JS, Cope M, Delpy DT, Richardson CE, Edwards AD, Wray S, Reynolds EOR: Quantitation of cerebral blood volume in human infants by near-infrared spectroscopy. J Appl Physiol **68**, 1086–1091, 1990.

434. Wyman J: Linked functions and reciprocal effects on hemoglobin: a second look. Adv Prot Chem **19**, 223–226, 1964.

435. Yelderman M: Pulse oximetry. In: Monitoring in anesthesia and critical care medicine, edited by CD Blitt, New York 1990 (Churchill Livingstone), pp 417–427.

436. Yelderman M, New W: Evaluation of pulse oximetry. Anesthesiology **59**, 349–352, 1983.

437. Yoshiya I, Shimada Y, Tanada K: Spectrophotometric monitoring of arterial oxygen saturation in the fingertip. Med Biol Eng Comput **18**, 27–32, 1980.

438. Yoxall CW, Weindling AM, Dawani NH, Peart I: Measurement of cerebral venous oxyhemoglobin saturation in children by near-infrared spectroscopy and partial jugular venous occlusion. Pediatr Res **38**, 319–323, 1995.

439. Zander R, Lang W, Wolf HU: Alkaline haematin D-575, a new tool for the determination of haemoglobin as an alternative to the cyanhaemiglobin method. I. Description of the method. Clin Chim Acta **136**, 83–93, 1984.

440. Zander R, Lang W, Wolf HU: Alkaline haematin D-575, a new tool for the determination of haemoglobin as an alternative to the cyanhaemiglobin method. II. Standardisation of the method using pure chlorohaemin. Clin Chim Acta **136**, 95–104, 1984.

441. Zander R, Lang W, Wolf HU: The determination of haemoglobin as cyanhaemiglobin or as alkaline haematin D-575. Comparison of method-related errors. J Clin Chem Clin Biochem **27**, 185–189, 1989.

442. Zwart A: Spectral and functional properties of haemoglobin in human whole blood. Ph D thesis, University of Groningen, Groningen, The Netherlands, 1983.

443. Zwart A, Buursma A, van Kampen EJ, Oeseburg B, van der Ploeg PHW, Zijlstra WG: A multi-wavelength spectrophotometric method for the simultaneous determination of five haemoglobin derivatives. J Clin Chem Clin Biochem **19**, 457–463, 1981.

444. Zwart A, Buursma A, van Kampen EJ, Zijlstra WG: Multicomponent analysis of hemoglobin derivatives with a reversed-optics spectrophotometer. Clin Chem **30**, 373–379, 1984.

445. Zwart A, Buursma A, Kwant G, Oeseburg B, Zijlstra WG: Determination of total haemoglobin in whole blood: Further tests of the "Hemocue" method. Clin Chem **33**, 2307–2308, 1987.

446. Zwart A, Buursma A, Oeseburg B, Zijlstra WG: Determination of hemoglobin derivatives with the IL282 CO-oximeter as compared with a manual spectrophotometric five-wavelength method. Clin Chem **27**, 1903–1907, 1981.

447. Zwart A, van Kampen EJ, Zijlstra WG: Results of routine determination of clinically significant hemoglobin derivatives by multicomponent analysis. Clin Chem **32**, 972–978, 1986.

448. Zwart A, Kwant G, Oeseburg B, Zijlstra WG: Oxygen dissociation curves for whole blood, recorded with an instrument that continuously measures p_{O_2} and S_{O_2} independently at constant t, p_{CO_2} and pH. Clin Chem **28**, 1287–1292, 1982.

449. Zwart A, Kwant G, Oeseburg B, Zijlstra WG: Human whole-blood O_2 affinity: effect of carbon monoxide. J Appl Physiol **57**, 14–20, 1984.

450. Zwart A, Kwant G, Oeseburg B, Zijlstra WG: Human whole blood oxygen affinity: effect of temperature. J Appl Physiol **57**, 429–434, 1984.

451. Zwart A, van der Ploeg PHW, Oeseburg B, Zijlstra WG: Testing a practical method for the determination of total hemoglobin. Clin Chem **32**, 1152, 1986.

452. Zweens J, Frankena F, van Kampen EJ, Rispens P, Zijlstra WG: Ionic composition of arterial and mixed venous plasma in the unanaesthetized dog. Am J Physiol **233**, 412–415, 1977.

453. Zweens J, Frankena H, Zijlstra WG: Decomposition on freezing of reagents used in the ICSH-recommended method for determination of total haemoglobin in blood; its nature, cause and prevention. Clin Chim Acta **91**, 337–352, 1979.

454. Zijlstra WG: Fundamentals and applications of clinical oximetry. Ph D thesis, University of Groningen/Monograph, Van Gorcum & Comp, Assen, The Netherlands, 1951; 2nd edition 1953.

455. Zijlstra WG: The photometric determination of the oxygen saturation of human blood by transmission and by reflection techniques. Proc Kon Ned Ak Wet **C56**, 598–608, 1953.

456. Zijlstra WG: Determination of the oxygen saturation of human blood samples by transmission and reflection photometric techniques — the haemoreflector method. Arch Chir Neerl **5**, 300–310, 1953.

457. Zijlstra WG: Photoelektrische Bestimmung und fortlaufende Beobachtung der Blutsauerstoffsättigung unter benutzung reflektierten Lichtes. Anaesthetist **3**, 254–258, 1954.

458. Zijlstra WG: Die quantitative Bestimmung von Hämoglobin, Oxyhämoglobin, Kohlenoxydhämoglobin und Methämoglobin in kleinen Blutproben mittels Spektrophotometrie. Klin Wschr **34**, 384–389, 1956.

459. Zijlstra WG: A manual of reflection oximetry. Monograph, Van Gorcum & Comp, Assen, The Netherlands, 1958.

460. Zijlstra WG: Syllectometry, a new method for studying rouleaux formation of red blood cells. Acta Physiol Pharmacol Neerl **7**, 153, 1958.

461. Zijlstra WG: Further developements in syllectometry. Acta Physiol Pharmacol Neerl **8**, 427–428, 1959.

462. Zijlstra WG: Reflexionsoxymetrie. In: Oxymetrie. Theorie und klinische Anwendung, edited by K Kramer, Stuttgart 1960 (Georg Thieme Verlag), pp 98–116.

463. Zijlstra WG: The quantitative evaluation of intracardiac shunts from arterial dilution curves following injection of dye into the right heart. Memorias del IV Congreso Mundial de Cardiologia **I-A**, 419–426, 1962.

464. Zijlstra WG: Détermination de la saturation en oxygène du sang *in vitro* et *in vivo* a l'aide des méthodes photoélectriques. In: L'exploration fonctionelle pulmonaire, edited by H Denolin *et al.*, Paris 1964 (Flammarion), pp 607–625.

465. Zijlstra WG: Determination of interdependent ligand effects on human red cell oxygen affinity. Scand J Clin Lab Invest **42**, 339–345, 1982.

466. Zijlstra WG: Oximetry: a historical introduction. In: Fetal and neonatal physiological measurements, edited by HN Lafeber *et al.*, Amsterdam 1991 (Excerpta Medica), pp 97–102.

467. Zijlstra WG: Hüfner memorial lecture: the oxygen-binding capacity of haemoglobin. Lab Haematol **1**, 145–153, 1995.

468. Zijlstra WG: Standardisation of haemoglobinometry: history and new challenges. Comp Haematol Int **7**, 125–132, 1997.

469. Zijlstra WG, van Assendelft OW, Buursma A, van Kampen EJ: The use of an ion-selective electrode for checking the CN^- content of reagent solutions used in the HiCN method. In: Modern concepts in hematology, edited by G Izak and SM Lewis, New York 1972 (Academic Press), pp 54–57.

470. Zijlstra WG, van Assendelft OW, Houwen B: Unpublished results, 1992.

471. Zijlstra WG, van Assendelft OW, Rijskamp A: Oxygen capacity of normal human blood. Acta Physiol Pharmacol Neerl **13**, 229–230, 1965.

472. Zijlstra WG, Buursma A: Spectrophotometry of hemoglobin: a comparison of dog and man. Comp Biochem Physiol **88B**, 251–255, 1987.

473. Zijlstra WG, Buursma A: Multicomponent analysis of hemoglobin derivatives as an aid in the treatment of patients with cyanide poisoning. In: Methodology and clinical applications of blood gases, pH, electrolytes and sensor technology, Vol 12, edited by RF Moran *et al.*, Monterey 1990 (MVI Publishing), pp 119–126.

474. Zijlstra WG, Buursma A: Multicomponent analysis of hemoglobin derivatives including methemoglobin cyanide. Clin Chem **36**, 975, 1990.

475. Zijlstra WG, Buursma A: Rapid multicomponent analysis of hemoglobin derivatives for controlled antidotal use of methemoglobin-forming agents in cyanide poisoning. Clin Chem **39**, 1685–1689, 1993.

476. Zijlstra WG, Buursma A: Spectrophotometry of hemoglobin: absorption spectra of bovine oxyhemoglobin, deoxyhemoglobin, carboxyhemoglobin, and methemoglobin. Comp Biochem Physiol **118B**, 743–749, 1997.

477. Zijlstra WG, Buursma A, Falke HE, Catsburg JF: Spectrophotometry of hemoglobin: absorption spectra of rat oxyhemoglobin, deoxyhemoglobin, carboxyhemoglobin, and methemoglobin. Comp Biochem Physiol **107B**, 161–166, 1994.

478. Zijlstra WG, Buursma A, Koek JN, Zwart A: Problems in the spectrophotometric determination of HbO_2 and HbCO in fetal blood. In: Physiology and methodology of blood gases and pH, Vol **4**, edited by AHJ Maas *et al.*, Copenhagen 1984 (Private Press), pp 45–55.

479. Zijlstra WG, Buursma A, Kwant G, Oeseburg B, Zwart A: Carboxyhemoglobin: determination and significance in oxygen transport. In: Oxygen transport to tissues VII, edited by F Kreuzer *et al.*, New York 1985 (Plenum Press), pp 533–542.

480. Zijlstra WG, Buursma A, Meeuwsen-van der Roest WP: Absorption spectra of human fetal and adult oxyhemoglobin, de-oxyhemoglobin, carboxyhemoglobin, and methemoglobin. Clin Chem **37**, 1633–1638, 1991.

481. Zijlstra WG, Buursma A, Zwart A: Molar absorptivities of human hemoglobin in the visible range. J Appl Physiol **54**, 1287–1291, 1983.

482. Zijlstra WG, Buursma A, Zwart A: Performance of an automated six-wavelength photometer (Radiometer OSM3) for routine measurement of hemoglobin derivatives. Clin Chem **34**, 149–152, 1988.

483. Zijlstra WG, Heeres SG: The influence of plasma substitutes on the suspension stability of human blood. Proc Kon Ned Ak Wet **C68**, 412–423, 1965.

484. Zijlstra WG, van Kampen EJ: Standardization of haemoglobinometry. I. The extinction coefficient of haemiglobincyanide at $\lambda = 540$ mμ: $\varepsilon^{540}_{\mathrm{HiCN}}$. Clin Chim Acta **5**, 719–726, 1960.

485. Zijlstra WG, van Kampen EJ: Standardization of haemoglobinometry. III. Preparation and use of a stable haemiglobincyanide standard. Clin Chim Acta **7**, 96–99, 1962.

486. Zijlstra WG, van Kampen EJ, van Assendelft OW: Standardisation of haemoglobinometry: Establishing the reference point. Proc Kon Ned Ak Wet **C72**, 231–237, 1969.

487. Zijlstra WG, van Kampen EJ, Handelé MJ: Observations on the light reflection of red blood cells. Proc Kon Ned Ak Wet **C60**, 401–409, 1957.

488. Zijlstra WG, Kwant G, Oeseburg B, Zwart A: Oxygen affinity of human whole blood investigated by means of a new analytic set-up. In: Hemoglobin, edited by AG Schnek and C Paul, Bruxelles 1984 (Editions de l'Université de Bruxelles), pp 63–81, 1984.

489. Zijlstra WG, Maas AHJ, Moran RF: Definitions, significance and measurement of quantities pertaining to the oxygen carrying properties of human blood. Scand J Clin Lab Invest **56**, Suppl 224, 27–45, 1996.

490. Zijlstra WG, Mook GA: Medical Reflection Photometry. Assen, The Netherlands, 1962 (Van Gorcum).

491. Zijlstra WG, Mook GA, Elders RAR, Onderstal E: Simultaneous use of reflection and transmission oximetry. A comparison of the two techniques. Acta Med Scand **160**, 7–14, 1958.

492. Zijlstra WG, Mook GA, ten Hoor F, Kruizinga K: Reflection oximetry. Biomed Eng **1**, 52–57, 1966.

493. Zijlstra WG, Mook GA, Sparling CM, ten Hoor F: Oxymetrie und Herstellung von Farbstoffverdünnungskurven mittels Reflexometrie. Z Kreislaufforschung **53**, 1254–1275, 1964.

494. Zijlstra WG, Muller CJ: Spectrophotometry of solutions containing three components, with special reference to the simultaneous determination of carboxyhaemoglobin and methaemoglobin in human blood. Clin Chim Acta **2**, 237–245, 1957.

495. Zijlstra WG, Oeseburg B: Definition and notation of hemoglobin oxygen saturation. IEEE Trans Biomed Eng **36**, 872, 1989.

496. Zijlstra WG, Oeseburg B, Kwant G, Zwart A: Determination of interdependent ligand effects on human red cell oxygen affinity. Scand J Clin Lab Invest **42**, 339–345, 1982.

ABBREVIATIONS AND SYMBOLS

a:	arterial
American Optical:	American Optical Comp, Southbridge MA, USA
ASTM:	American Society for Testing Materials
Atlas:	Atlas Werke, Bremen, Germany
AVL:	AVL Medical Instruments, Schaffhausen, Switzerland
A_X:	absorbance of substance X
α_{O_2}:	concentrational solubility coefficient of oxygen
Beckman:	Beckman Instruments Inc., Fullerton CA, USA
Beckman DU:	Beckman DU, single beam, UV/VIS, quartz prism spectrophotometer
B_{O_2}:	oxygen capacity
β_{O_2}:	oxygen binding capacity in mL(STPD) per g of haemoglobin
Brinkhorst:	Brinkhorst High-Tech, Ruurlo, the Netherlands
Chiron:	Chiron Diagnostics (formerly Ciba-Corning), Walpole MA, USA
COHb:	carboxyhaemoglobin
Coleman:	Coleman Instruments, Maywood IL, USA
COSHb:	carboxysulfhaemoglobin
CRM:	Certified Reference Material
c_{O_2}(Hb):	concentration of haemoglobin-bound oxygen
c_{O_2}(free):	concentration of oxygen dissolved in blood according to Henry's law
ct_{Hb}:	total haemoglobin concentration
ct_{O_2}:	concentration of total oxygen
c_X:	concentration of substance X
Diagnostic Reagents:	Diagnostics Reagents, Thame, UK

Digital:	Digital Equipment Corp, Maynard MA, USA
4-DMAP:	4-dimethylaminophenol
DPF:	differential pathlength factor
DPG:	2,3-diphosphoglycerate
dysHb:	dyshaemoglobin
EDTA:	ethylenediaminetetraacetic acid
Electrofact:	Electrofact, Amersfoort, the Netherlands
Eschweiler:	Eschweiler, Kiel, Germany
Euro-Trol:	Euro-Trol bv, Wageningen, the Netherlands
$\varepsilon_X(\lambda)$:	millimolar absorptivity of substance X at wavelength λ
F_X:	fraction or percentage of substance X
GDS Diagnostics:	GDS Diagnostics, Elkhart IN, USA
Hamamatsu:	Hamamatsu Photonics KK, Hamamatsu, Japan
Hartmann-Leddon:	Hartmann-Leddon Comp, Philadelphia PA, USA
Hb:	haemoglobin
Hb_4:	haemoglobin tetramer
HbA:	adult haemoglobin
HbF:	fetal haemoglobin
Hb-NFPLP:	human haemoglobin cross-linked between the β-chains with NFPLP
HbS:	sickle cell haemoglobin
HHb:	deoxyhaemoglobin
Hi:	haemiglobin (methaemoglobin)
HiCN:	haemiglobincyanide
HiF:	haemiglobinfluoride
HiLS:	haemiglobinlaurylsulphate
HiN_3:	haemiglobinazide
$HiNO_2$:	haemiglobinnitrite
HP:	Hewlett-Packard Co., Palo Alto CA, USA
HP8450A:	Hewlett-Packard 8450A, single beam, reversed optics, diode array spectrophotometer
I_0:	incident light intensity
I:	transmitted light intensity
IBP:	isosbestic point
ICSH:	International Committee (Council) for Standardisation in Haematology

IL:	Instrumentation Laboratories Inc., Lexington MA, USA
Ingold:	Ingold, Urdorf, Switzerland
IR:	infrared (part of the electromagnetic radiation spectrum)
Kipp:	Kipp & Sons, Delft, the Netherlands
l:	lightpath length
LBH:	lyophilised bovine haemoglobin
LED:	light emitting diode
Leo Diagnostics:	Leo Diagnostics, Hälsingborg, Sweden
MCA:	multicomponent analysis
MFC:	mass flow controller
Millex-GV:	Millex-GV polyvinylidene difluoride filter membrane (Millipore)
Millipore:	Millipore Corp, Bedford MA, USA
MWM:	multiwavelength method
N:	number of specimens
n:	number of samples measured (Note that n is also used for the Hill coefficient; Chapter 20)
NCCLS:	National Committee for Clinical Laboratory Standards (USA)
Nellcor:	Nellcor, Pleasanton CA, USA
NFPLP:	2-nor-2-formylpyridoxal 5′-phosphate
NIRS:	near infrared spectroscopy
NIST:	National Institute for Standards and Technology (USA)
n-MCA:	n-component analysis
Nonidet P-40:	polyethylene glycol-P-ethyl/phenylglycol ether (Shell)
ODC:	oxygen dissociation curve
O_2Hb:	oxyhaemoglobin
$O_2(Hb)$:	haemoglobin-bound oxygen
Optica CF4:	Optica CF4, single beam, grating spectrophotometer
Optica CF4DR:	recording version of Optica CF4
OSM3M:	modified Radiometer OSM3 multiwavelength haemoglobin photometer
Oximetrics:	Oximetrics, Mountain View, CA, USA
p_{50}:	oxygen tension at $S_{O_2} = 50\%$

p_n: oxygen tension at $S_{O_2} = n\%$
p_{O_2}: oxygen tension
Puradisc 25 PP: Puradisc 25 PP polypropylene filter membrane (Whatman)
Q: flow rate
Radiometer: Radiometer Medical A/S, DK-2700 Brønshøj, Denmark
RIV(M): Rijksinstituut voor de Volksgezondheid (en Milieuhygiëne) [Dutch Institute of Public Health (and Environmental Protection)], Bilthoven
Schwarzer: Fritz Schwarzer GmbH, München, Germany
SD: standard deviation
SEM: standard error of the mean
SFH: stroma free haemoglobin
SHb: sulfhaemoglobin
SHi: sulfhaemiglobin
SHiCN: sulfhaemiglobincyanide
Shell: Shell International Chemical Comp, the Hague, the Netherlands
Sigma: Sigma Chemical Co., St Louis MO, USA
SLS: Sodium laurylsulphate
S_{O_2}: oxygen saturation
std: standard
Sterox SE: alkyl-phenol(thiol)polyethylene oxide (Hartmann-Leddon)
STPD: standard temperature and pressure, dry
Sysmex: Sysmex, Kobe, Japan
T: temperature (Note that T is also used for transmittance; Chapter 2)
TC: temperature coefficient
Tris: tris(hydroxymethyl)aminomethane
Triton X-100: octyl-phenyl-polyethylene glycol ether (Rohm and Haas)
USP: United States' Pharmacopoeia
UV: ultraviolet (part of the electromagnetic radiation spectrum)
Φ_C: carbamate Bohr factor
Φ_H: proton Bohr factor
Φ_{HC}: total Bohr factor

V:	volume
V_{O_2}:	oxygen consumption (volume of O_2 consumed per unit of time)
v:	venous
VIS:	visible (part of the electromagnetic radiation spectrum)
Waters:	Waters Corp, Rochester MN, USA
Whatman:	Whatman Inc, Fairfield NJ, USA
Wösthoff:	Wösthoff, Bochum, Germany

INDEX